TREATMENT, RECOVERY, AND DISPOSAL PROCESSES FOR RADIOACTIVE WASTES

Treatment, Recovery, and Disposal Processes for Radioactive Wastes

Recent Advances

Edited by J.I. Duffy

NOYES DATA CORPORATION

Park Ridge, New Jersey, U.S.A.

1983

Copyright © 1983 by Noyes Data Corporation
No part of this book may be reproduced in any form
without permission in writing from the Publisher.
Library of Congress Catalog Card Number: 82-22260
ISBN: 0-8155-0922-7
ISSN: 0090-516X; 0198-6880
Printed in the United States

Published in the United States of America by
Noyes Data Corporation
Mill Road, Park Ridge, New Jersey 07656

10 9 8 7 6 5 4 3 2 1

Library of Congress Cataloging in Publication Data
Main entry under title:

Treatment, recovery, and disposal processes for
radioactive wastes.

(Pollution technology review ; no. 95)
(Chemical technology review ; no. 216)
Includes index.
1. Radioactive waste disposal. I. Duffy,
J. I. (Joan Irene), 1950-   . II. Series.
III. Series: Chemical technology review ; no. 216.
TD898.T73 1983   621.48'38   82-22260
ISBN 0-8155-0922-7

FOREWORD

The detailed, descriptive information in this book is based on U.S. patents issued from January, 1975 through May, 1982 that relate to radioactive waste treatment, recovery, and disposal processes.

This book is a data-based publication, providing information retrieved and made available from the U.S. patent literature. It thus serves a double purpose in that it supplies detailed technical information and can be used as a guide to the patent literature in this field. By indicating all the information that is significant, and eliminating legal jargon and juristic phraseology, this book presents an advanced commercially oriented review of recent developments in the field of radioactive waste treatment and handling.

The U.S. patent literature is the largest and most comprehensive collection of technical information in the world. There is more practical, commercial, timely process information assembled here than is available from any other source. The technical information obtained from a patent is extremely reliable and comprehensive; sufficient information must be included to avoid rejection for "insufficient disclosure." These patents include practically all of those issued on the subject in the United States during the period under review; there has been no bias in the selection of patents for inclusion.

The patent literature covers a substantial amount of information not available in the journal literature. The patent literature is a prime source of basic commercially useful information. This information is overlooked by those who rely primarily on the periodical journal literature. It is realized that there is a lag between a patent application on a new process development and the granting of a patent, but it is felt that this may roughly parallel or even anticipate the lag in putting that development into commercial practice.

Many of these patents are being utilized commercially. Whether used or not, they offer opportunities for technological transfer. Also, a major purpose of this book is to describe the number of technical possibilities available, which may open up profitable areas of research and development. The information contained in this book will allow you to establish a sound background before launching into research in this field.

Advanced composition and production methods developed by Noyes Data are employed to bring these durably bound books to you in a minimum of time. Special techniques are used to close the gap between "manuscript" and "completed book." Industrial technology is progressing so rapidly that time-honored, conventional typesetting, binding and shipping methods are no longer suitable. We have bypassed the delays in the conventional book publishing cycle and provide the user with an effective and convenient means of reviewing up-to-date information in depth.

The table of contents is organized in such a way as to serve as a subject index. Other indexes by company, inventor and patent number help in providing easy access to the information contained in this book.

16 Reasons Why the U.S. Patent Office Literature Is Important to You

1. The U.S. patent literature is the largest and most comprehensive collection of technical information in the world. There is more practical commercial process information assembled here than is available from any other source. Most important technological advances are described in the patent literature.

2. The technical information obtained from the patent literature is extremely comprehensive; sufficient information must be included to avoid rejection for "insufficient disclosure."

3. The patent literature is a prime source of basic commercially utilizable information. This information is overlooked by those who rely primarily on the periodical journal literature.

4. An important feature of the patent literature is that it can serve to avoid duplication of research and development.

5. Patents, unlike periodical literature, are bound by definition to contain new information, data and ideas.

6. It can serve as a source of new ideas in a different but related field, and may be outside the patent protection offered the original invention.

7. Since claims are narrowly defined, much valuable information is included that may be outside the legal protection afforded by the claims.

8. Patents discuss the difficulties associated with previous research, development or production techniques, and offer a specific method of overcoming problems. This gives clues to current process information that has not been published in periodicals or books.

9. Can aid in process design by providing a selection of alternate techniques. A powerful research and engineering tool.

10. Obtain licenses—many U.S. chemical patents have not been developed commercially.

11. Patents provide an excellent starting point for the next investigator.

12. Frequently, innovations derived from research are first disclosed in the patent literature, prior to coverage in the periodical literature.

13. Patents offer a most valuable method of keeping abreast of latest technologies, serving an individual's own "current awareness" program.

14. Identifying potential new competitors.

15. It is a creative source of ideas for those with imagination.

16. Scrutiny of the patent literature has important profit-making potential.

CONTENTS AND SUBJECT INDEX

INTRODUCTION

Few matters are of greater concern to environmentalists than the handling and disposition of radioactive waste. The half life of many radioactive substances is in the range of thousands of years, posing dangers of contamination for generations to come. The consequences of the mishandling or careless storage of radioactive waste can be both grave and long-lasting, affecting the quality of life now and in the distant future.

Radioactive waste, or radwaste, is a complex mixture. It may contain actinide fuel elements such as uranium, thorium, and plutonium, as well as fission products including rubidium and palladium, along with various other contaminants. Various procedures have been proposed to store these high level radioactive wastes so as to isolate them from the environment. The means of isolation must withstand attack essentially indefinitely not only from the radioactive material, but also from the chemical and physical stresses of the natural environment. Most of these proposals have contemplated first reducing the volume of the effluents, and thereafter incorporating the concentrates into a surrounding matrix. The matrix may be of glass, cement, bitumen, or polymer resins. When placed in suitable containers, then, the leaking of the contaminating material into the soil and surrounding atmosphere would be prevented.

Not only the waste itself, but all apparatus and equipment used in the handling and processing of such waste must be kept as free of contamination as possible. There must be a minimal risk of exposure to the radioactivity of the personnel involved. The waste must be deposited in its final repository with no leakage of radioactivity into the environment at any step of the process.

Another approach to the problem of radwaste handling and processing is that of recovering and recycling. It is more economical and safer to reprocess nuclear fuel than to discard it. Spent nuclear fuel rods can be treated in a variety of ways to recover the particles of uranium and plutonium and separate them from fission products and other contaminants. This recovered fuel can then be reused in the reactor system.

1

This book contains over 200 new processes relating to the handling and disposal of radioactive waste. It will be an invaluable source of information for any professional concerned with nuclear facilities or the control of any radioactive substance.

IMMOBILIZATION TECHNIQUES

GLASS AND CERAMICS

Conversion of Ferrocyanide Compounds to Glasses

W.W. Schulz and A.L. Dressen; U.S. Patent 4,020,004; April 26, 1977; assigned to U.S. Energy Research and Development Administration describe a process whereby a radioactive ferrocyanide of the formula $^{134-137}Cs_aM_b[Fe(CN)_6]_c \cdot xH_2O$, where M represents Ni, Zn, Cu, Fe, Co, Cd, UO_2 and Mn; a, b, and c are integers, and x is zero or a small number, is converted to a dense immobile glass of low leachability by melting it with sodium carbonate (Na_2CO_3) and a mixture of (a) basalt and boron trioxide (B_2O_3) or (b) silica (SiO_2) and lime (CaO).

The process involves the use of finely ground constituents. They are mixed together in the dry state, melted, and allowed to solidify. The melting may be carried out in the canister or other receptacle in which the product will be stored, or it can be carried out in a separate melter and the molten product poured into the storage canister.

Example 1: In carrying out the "basalt" process, the basalt is finely ground and mixed with the complex ferrocyanide, sodium carbonate, and boron trioxide. The latter two constituents lower the melting point of the mixture and, in addition, the boron has been found to lessen the volatilization of the cesium. However, too much boron has been found to increase the leachability of the glass. The B_2O_3 may constitute from 5 to 15% by weight of the charge. The Na_2CO_3 may range from 15 to 25% by weight.

While the sodium carbonate and boron trioxide lower the melting point of the basalt to about 1000°C, it is desirable, in order to secure good incorporation of the cesium and other elements to heat the mixture to about 1200°C.

The molten glass can be poured into stainless steel canisters and allowed to harden. The canisters can then be stored with adequate circulation of air or water provided to remove the heat generated. See for example U.S. Atomic Energy Report ARH-2888 Rev. July 1974, " Retrievable Surface Storage Facility Alternative Concepts – Engineering Study."

3

While the stored precipitate may be more complex, it is reasonably represented by the compound $Cs_2Ni[Fe(CN)_6]$. For purposes of this experiment that compound was prepared by adding appropriate amounts of $K_4Fe(CN)_6$ and $Ni(NO_3)_2$ reagents to a nonradioactive 0.01 M $CsNO_3$ solution which was 5.5M in $NaNO_3$ and had a pH of 10. The resulting precipitate was washed with water and dried overnight at 100°C.

Basalt having the composition by weight 52% SiO_2, 14% FeO, 13% Al_2O_3, 8% CaO, 4% MgO, 3% Na_2O, 2.5% TiO_2 and 1.5% K_2O and melting at about 1200°C was crushed and screened. The portion finer than 30 mesh (595 microns) was used. The crushed basalt was mixed with B_2O_3, Na_2CO_3 and $Cs_2Ni[Fe(CN)_6]$ to form 100 g charges. Each charge contained, by weight, 10% B_2O_3 and 20% Na_2CO_3. The proportions of the other constituents are shown in the table below.

| | | | . . . . Product . . . . | | |
Charge Composition $Cs_2Ni[Fe(CN)_6]$	Wt % Basalt	Percent Cesium Volatilized	Density g/cm^3	Ap- pear- ance	Leach Rate in Water g/cm^2-day
10	60	0.20	2.67	Glass	8.62×10^{-6}
20	50	0.20	2.69	Glass	1.86×10^{-5}
30	40	0.23	2.84	Glass	3.71×10^{-5}

The charges were placed in a graphite-clay crucible which in turn was placed in a furnace maintained at 1200°C and heated for an hour. An inverted quartz funnel covered the crucible and was connected by a condenser and traps to a vacuum pump. Any cesium volatilized was condensed and its weight determined.

The glass product was crushed and screened. The 14 to 20 mesh (U.S. Standard Sieve Series) fraction was taken for leach tests.

Leach tests were performed with 15 to 25 grams of the dried 14 to 20 mesh material which, for calculation of surface area, was assumed to consist of 0.11 cm diameter spheres. (The value 0.11 cm is the average of the width of the openings, 0.14 and 0.084 cm, respectively, of 14 and 20 mesh screens). Total surface area of the weighed leach samples was estimated from the weight and surface area of a counted number of the (assumed) 0.11 cm diameter pieces.

The test material was supported on stainless steel screen and airlift circulators were used to circulate 200 ml of distilled and deionized water over the sample pieces. Test samples were leached initially for 24 hours at 25°C and then, after changing of the leach liquor, for 96 hours more at 25°C. Cesium was determined by atomic absorption methods.

The leach rate was determined by the formula:

$$\text{Leach rate (g/cm}^2\text{-day) based on Cs} = \frac{\text{g of Cs leached}}{(\text{g of Cs/g of sample})(\text{sample area, cm}^2)(\text{time, days})}.$$

The final product in all cases was a dense, emerald green colored glass very resistant to leaching by water. Leach rates of radioactive glass generally decrease by one or two orders of magnitude as leaching continues. Hence, leach rates listed in the table may be taken as maximum values. The volume of glass obtained with a charge containing 20% by weight $Cs_2Ni[Fe(CN)_6]$ is about 1.3 times the volume of dry $Cs_2Ni[Fe(CN)_6]$ but only about half that of the wet precipitate.

The small amounts of cesium volatilized (see above table) can be recovered by washing the equipment with water and reprecipitating the complex ferrocyanide, which may be recycled to the process.

While the metal indicated by "M" in the general formula is nickel in the above example, the ferrocyanides in which the metal is zinc, copper, iron, cobalt, cadmium or manganese, or in which the radical UO_2^{2+} is substituted, can be used instead.

As is shown by the above table, the leachability of the product increases with increasing proportions of the ferrocyanide in the mixture. For this reason, and also to obtain a good quality glass, the upper limit of the ferrocyanide in the mixture is set at 30%. There is no lower operative limit. The lower the proportion of ferrocyanide, the greater the bulk for a given cesium content, however, and the desirable lower limit is about 10%. The preferred proportions by weight are about 20% ferrocyanide, 10% B_2O_3, 20% Na_2CO_3, 50% basalt.

Example 2: In the "soda-lime" process fine sand or crushed quartz, calcium oxide and sodium carbonate are employed. The following ranges or proportions of charge constituents can be employed in percent by weight:

Ferrocyanide precipitate	10-30
SiO_2	40-60
Na_2CO_3	15-25
CaO	5-10

A preferred charge composition, in percent by weight is:

$Cs_2Ni[Fe(CN)_6]$	20
SiO_2	50
Na_2CO_3	21
CaO	9

The manner of preparing and handling the soda-lime glass is the same as for the basalt glass, except that slightly higher temperatures (1250° to 1350°C) are employed.

Sintering of Mordenite Containing Cs-137

M. Tanno; U.S. Patent 4,087,375; May 2, 1978; assigned to Shin Tohoku Chemical Industry Co., Ltd., Japan describes a process for adsorbing and capturing ^{137}Cs contained in radioactive wastewater by mordenite, followed by elution to concentrate ^{137}Cs in the eluant. The mordenite, in which the radionuclides contained in the radioactive wastewater are adsorbed is sintered to a ceramic form at high temperatures, thereby fixing and encasing the radionuclides in the sintered product.

According to the process, the reactor cooling water containing corrosive radionuclides of heavy metals such as Fe, Co, Mn, Cr and Ni, and liquid wastewater discharged from the fuel reprocessing operations containing low, intermediate and high level fission products such as ^{137}Cs, [89-90]Sr, [85]Kr, [132]Xe, ^{90}Y, [141-144]Ce, [147]Pm, [95]Zr, ^{131}I, [103-106]Ru, Ba and Am, are passed through a mordenite column at a velocity of 2-50 cm/min, thereby allowing the corrosive radionuclides and fission-products to be adsorbed and captured by the mordenite, followed by, if necessary, elution of the thus adsorbed and captured radioactive substances by means of an eluant such as alkali metal salts or polyvalent metal salts like NaCl, KCl and NH_4Cl, thus obtaining the radioactive substances concentrated in the eluant. In addition to these chlorides, solutions of salts such as soluble sulfates and nitrates can be used.

Particularly advantageous eluants are, for example, solutions of chlorides of sodium, potassium and ammonium salts. Such solutions of salts are employed after they have been stabilized to be acid, neutral or alkaline. But solutions of bivalent to polyvalent metal salts should not be used under alkaline conditions because they form insoluble precipitates.

Further, Cs-137 can be selectively and efficiently recovered from acidic to alkalinic (preferably within the range of pH 2-11) radioactive wastewater containing a variety of ions by use of mordenite having ion exchange groups mainly consisting of monovalent metal ions through the steps of ion exchange adsorption, separation and concentration without damage to the structure of the mordenite.

Cs-137 and other nuclides concentrated in the eluant in the manner described above may again be adsorbed on the mordenite to saturation. Alternatively, the mordenite which has captured the Cs-137 and other nuclides may be sintered to a ceramic form at 1000°-1400°C to destroy its three dimensional structure, thereby closing an infinite number of micro or macro pores present in the mordenite to seal and fix the Cs-137 and other nuclides. Since the zeolite that has adsorbed a great amount of Cs-137 is solidified to a ceramic body as mentioned above, there is no possibility of leakage of the radioactive element. Such ceramic body can be utilized efficiently and safely as a source of heat, power and radiation. Other nuclides can also be recycled to useful applications by similar treatments. Therefore, the process is more efficient and economical than the conventional radioactive waste disposal methods, such as encasement in concrete which permits considerable leakage of radioactivity, and encasement in glass that requires a troublesome melting pot. Another great advantage of this process is that it allows for the treatment of Cs-137 and other nuclides in such a manner that they can be recycled to useful applications.

Addition of Reducing Agent to Melt

F. Kaufmann, S. Weisenburger, H. Koschorke, H. Seiffert, D. Sienel, and K.-H. Weiss; U.S. Patent 4,202,792; May 13, 1980; assigned to Gesellschaft fur Kernforschung mbH, Germany describe a method for the solidification, in a manner which protects the environment against contamination, of waste materials obtained during reprocessing of irradiated nuclear fuel and/or breeder materials in a matrix of borosilicate glass, in which highly radioactive solutions or slurries containing the waste materials in dissolved or suspended form are evaporated in a vessel in the presence of glass former substances until they are dry, the dry residue is calcinated and the calcinate is melted together with the glass formers while waste gases are discharged to the environment, comprising: introducing, in a controlled and continuous manner, a waste liquid which has been obtained from a reprocessing system without pretreatment and which has been mixed with glass formers and a reduction agent, into the center of a borosilicate glass melt disposed in a melting crucible at a temperature in the region of 1000° to 1400°C to form an island-like drying and calcinating zone (island zone) on the surface of the melt while avoiding contact of the waste liquid with the walls of the crucible, to form a reducing atmosphere, and to avoid the presence of components in the waste gases which would radiologically and/or chemically contaminate the environment.

In a preferred embodiment, a concentration maximum of the reduction agent is formed continuously in the gaseous phase in the region of the island zone, with a concentration gradient in the reducing atmosphere which decreases with increasing radial distance from this maximum. In a further embodiment, a positive heat input is provided which rapidly penetrates the melt radially from the outside toward the center because of the temperature radiation of the heated walls. Advantageously the reduction agent used is formic acid.

Example: In a recipient vessel of about 2 m^3, finely ground borosilicate glass frit (<200 μ) was added as the glass former substance to a simulated highly radioactive fission product solution. The resulting suspension was continuously mixed by means of a pulsating column operating at a pulse repetition frequency of about 16 to 18 pulses per minute, thus preventing the deposit of solids. The pulsating column had a diameter of 200 mm, a height of 870 mm and was filled with 350 ℓ which were pulsed at an amplitude of 13 mm. The mixed suspension was transported and introduced into a glass melting trough via an airlift conveying device. The suspension was mixed with formic acid immediately before being introduced into the melt at a quantity ratio of suspension to formic acid which corresponded to a mol ratio of nitrate ions to HCOOH of 1:1.2 to 2.5. The throughput of suspension plus HCOOH added to the melt was 20 ℓ per hour, with an accuracy of ±5%.

The addition of the suspension to the melt occurred continuously into the center of the melting bath, either through an atomizer nozzle or through an inlet pipe. The solution, which had already been predried by evaporation during its introduction into the melting crucible, formed an island-like drying or calcinate coating on the melt. The calcinate coating was continuously melted into the melt at about 1150°C. Through a preheated discharge disposed at the side of the bottom of the melting trough, about 50 kg of glass melt were filled into an ingot mold disposed on a mount once every 8 hours via an electrically heated plug. Then, the ingot mold was subjected to controlled cooling of about 5° to 10°C per hour in a normalizing system. The continuous introduction of the suspension to the melt was maintained for more than 1000 hours. This was the first time in this field that a solidification system was operated in a long-term experiment for more than 1000 hours.

The ratio of the dried solids of the liquid waste to the glass formers is 1:4 in weight. The glass formers are added in the form of fine powdered premelted glass frit. A typical composition of the glass frit is as follows: 51.8 wt % SiO_2, 21.5 wt % Na_2O, 1.3 wt % Al_2O_3, 8.8 wt % TiO_2, 2.6 wt % CaO, 14.0 wt % B_2O_3.

Addition of Diatomaceous Earth

S. Drobnik, H. Koschorke, F. Kaufmann and J. Saidl; U.S. Patent 4,119,561; October 10, 1978; assigned to Gesellschaft fur Kernforschung mbH, Germany describe a method for preventing malfunctions in the solidification of radioactive wastes contained in an aqueous waste solution in a glass, glass-ceramic or glass-ceramic-like matrix wherein the waste solution is spray-dried and calcinated, comprising, adding to the aqueous waste solution, before the spray drying, diatomaceous earth or diatomaceous earth-like substances in a solid form in quantities of 45 to 70 g/ℓ and in grain sizes of which more than 85% by weight of the grain size distribution are grain sizes from 6 to 75 μ.

Preferably, the diatomaceous earth which is added has a composition of about 90% by weight SiO_2; 4% by weight Al_2O_3; 3.3% by weight $Na_2O + K_2O$; 1.3% by weight Fe_2O_3; the remainder the sum of $MgO + CaO + TiO + P_2O_5$; and a grain size distribution in which more than 70% by weight of the grains are between 10 and 40 μ.

It has been found that the most favorable range for the addition of diatomaceous earth or of diatomaceous earth-like substances to the aqueous waste solution is an addition in quantities of 50 to 60 g/ℓ.

Example: Two aliquots were taken from a simulated waste solution. At room temperature 52.5 g/ℓ Aerosil were added to the first aliquot under stirring or pulsating. Similarly, 58.7 g/ℓ diatomaceous earth were added to the second aliquot at room temperature under stirring or pulsating. The mixing process was continued for a few minutes after completion of the addition process. Thereafter, the two aliquots were separately transported to a spray nozzle and were atomized, dried and calcinated with the aid of atomizing vapor at about 450°C.

The first aliquot to which Aerosil had been added produced larger, adhering deposits at the spray nozzle and the waste gas filter candles than the second aliquot which contained diatomaceous earth in accordance with this process. A free flowing calcinate was obtained from the second aliquot containing diatomaceous earth and it fell into the melting crucible disposed below the spray nozzle. When working with the second aliquot, a small portion of the calcinate was retained at the filter candles and this could easily and completely be removed from the filter candles by a blowback.

Glass-Ceramic in Metal Matrix

W. Heimerl, E. Schiewer and A.K. Dé; U.S. Patent 4,209,421; June 24, 1980; assigned to Gelsenberg AG, Germany describe a method of making bodies which contain radioactive substances and which comprise a glass-ceramic embedded in a metal matrix. In accordance with the process, the glass, which is made in particle form in a known manner (e.g., German Patent 24 53 404), is transformed by heat treatment in a metal bath to a glass-ceramic.

For this purpose, the glass particles, whose composition is adjusted to the desired glass-ceramic, are placed in a molten metal which is contained in a suitable vessel. In particular, the vessel for this purpose can be one in which glass-ceramic and metal conglomerate will ultimately be stored, since in this case there will be no need to transfer an intermediate product or the end product to another container.

It is also, of course, possible to place the glass particles in the heat treatment vessel and then fill the interstices with molten metal. In this case the metal can also be put into the heat treatment vessel in solid form, e.g., in the form of scraps or rods, and can then be melted. In any case, a packing of virtually maximum density of the glass particles is achieved, in which the interstitial volume is completely filled with the molten metal or metal alloy. Suitable metals are lead and its alloys or aluminum and its alloys.

The glass particles embedded in the molten metal are then subjected to a suitable heat treatment program. Since the incorporation of the particles into the molten metal is performed as a rule at the lowest possible temperature, the temperature is at first raised and sustained at a relatively high level.

If the composition of the glass particles is suitable, this heat treatment initiates a controlled devitrification, in which a glass-ceramic product is formed from the glass. After the ceramization is completed, the molten metal and the glass-ceramic contained in it are cooled. The embedding of the particles in metal and their ceramization are thus accomplished in a single step; if the ultimate container is used for the heat treatment as mentioned above, the end product is obtained without further manipulation.

The method has the advantage over the heat treatment of a monolithic glass block that, as a result of the smaller dimensions of the glass particles, their wall temperatures and internal temperatures are nearer one another, so that the heat treatment is easier to accomplish. If the glass particles should be brought to devitrification before incorporation into the melt, additional difficulties would be created on the one hand by the formation of high temperature gradients in the mass of particles on account of their poor heat conductivity, and on the other hand by the cohesion of particles which would make it difficult or impossible to transfer them to the end product vessel. In this method, however, cohesion of the particles due to softening of the glass does not occur.

Example: 100 grams of lenticular borosilicate particles (composition 35 wt % SiO_2, 16% Al_2O_3, 8% B_2O_3, 2% Na_2O, 3% Li_2O, 5% CaO, 1.5% MgO, 18.5% BaO, 1% ZrO_2, 5% TiO_2, 4.5% ZnO, 0.5% As_2O_3, plus 20% of fission product oxides) having a diameter of 4 to 5 mm were introduced into 25 ml of molten pure lead of a temperature of approximately 400°C. Then the temperature was raised to 800°C and maintained at this level for 12 hours. Then the furnace was shut off and allowed to cool. The end product was an aggregate of borosilicate glass ceramic and lead.

Glass-Ceramics of Two Crystal Phases

G.H. Beall and H. L. Rittler; U.S. Patent 4,314,909; February 9, 1982; assigned to Corning Glass Works describes a process whereby glass-ceramics suitable for the incorporation of radioactive wastes are prepared from precursor glasses which consist essentially, expressed in weight percent on the oxide basis, of 5-40% Cs_2O, 15-50% Al_2O_3, 0-30% La_2O_3 + CeO_2, 0-20% P_2O_5, 0-30% ZrO_2, 12-65% La_2O_3 + CeO_2 + P_2O_5 + ZrO_2, and 15-50% SiO_2. Monazite with or without zirconia crystals, and preferably both phases, ought to be present in the glass-ceramic to provide for the incorporation of actinides and rare earths. Accordingly, the preferred compositions will contain at least 5% La_2O_3 + CeO_2 and sufficient P_2O_5 up to 20% to react therewith to form monazite. The precursor glasses are crystallized in situ to glass-ceramic bodies by heat treating at temperatures between about 1250°-1550°C. Customarily, the parent glass bodies are crystallized via a two-step heat treatment to insure a uniformly fine-grained, final product. This two-step treatment involves nucleation at about 900°-1100°C followed by crystallization at 1250°-1550°C.

Several embodiments of the basic method are possible: First, radwaste can be mixed with the glass batch materials, the composition of the batch being governed by the analysis of the waste, the mixture melted under controlled conditions to prevent contamination of the environment outside of the melting unit, the melt cooled to a glass and thereafter crystallized to a phase assemblage tailored to incorporate the radioactive species and provide a body demonstrating high thermal stability and excellent corrosion resistance to hostile environments. The controlled conditions referred to customarily involves melting in a closed system.

Second, the precursor glass-forming batch, exclusive of radwaste, can be melted under conventional conditions and the glass resulting therefrom comminuted to a fine powder [commonly finer than a No. 200 U.S. Standard Sieve (74 microns)]. This powder is then combined with the radwaste and the mixture melted under controlled conditions to uniformly dissolve the radwaste in the melt. Thereafter, the melt is cooled to a glass body which is heat treated to develop the desired crystal assemblage therein. Inasmuch as melting of the powdered glass can be accomplished at temperatures below those demanded for melting the batch ingredients, this method embodiment permits the use of melting temperatures of about 1600°-1700°C rather than at least about 1800°C.

This practice has the additional advantage of providing flexibility in the treatment of radwaste. For example, the precursor glass-forming batch can be melted, powdered, and the powdered glass stored until needed to be mixed with the radwaste. Or, the powdered precursor glass powder can be produced at one site and then shipped to the source of the radwaste and the second melting plus crystallization steps conducted there.

Third, the precursor glass-forming batch, exclusive of radwaste, can be melted and powdered in like manner to the practice described immediately above. Thereafter, the powder is thoroughly blended with the radwaste and the mixture essentially simultaneously crystallized and sintered under controlled conditions into a solid body. Standard ceramic forming techniques such as slip casting, dry pressing, extrusion, and hot pressing can be utilized to shape the body. Because glass compositions can be formulated to flow somewhat and sinter prior to crystallization, this practice allows sintering to be accomplished at temperatures as low as about 1400°C. Hence, this embodiment provides the advantage of requiring lower firing temperatures but, as with any sintering process, a small amount of residual porosity may be present in the final product. Neverless, judicious formulation of the precursor glass composition can reduce this porosity to a minimum.

Silica Coated with Alkali Metal Monoxide

R.D. Wolson and C.C. McPheeters; U.S. Patent 4,234,449; November 18, 1980; assigned to U.S. Department of Energy describe a method of treating radioactive alkali metals or radioactive solid salts thereof, the method comprising mixing particulate silica substrate material having a particle size such that the majority of the substrate material passes through a 200 mesh sieve and the radioactive material in a rotary drum calciner and converting the radioactive material to alkali metal monoxide by reaction with oxygen present in a diluent at a temperature sufficient to initiate the reaction thereby forming particulate substrate particles coated with alkali metal monoxide, the reaction temperature being controlled by the amount of oxygen present in the diluent to ensure the reaction product remains flowable for easy handling.

The preferred weight ratio of silica to alkali metal is 7 to 1 in order to produce a feed material for the final glass product having a silica to alkali metal monoxide ratio of about 5 to 1.

Preconditioning of Radioactive Waste

W. Hild, H. Krause, K. Scheffler, H. Gusten, E. Gilbert and R. Köster; U.S. Patent 3,971,717; July 27, 1976; assigned to Gesellschaft fur Kernforschung mbH, Germany describe a method for conditioning highly radioactive, solidified waste products which are incorporated in shaped bodies of glass, ceramic or basaltic masses before they are transported to nonpolluting ultimate storage locations.

Highly radioactive solidified waste, incorporated in molded bodies of glass, ceramic or a basaltic mass, is introduced into wastewater and/or settling or precipitated sludge from community or industrial systems [in order to sterilize, improve the filterability of and condition (a) the wastewater before it is introduced into a receiving stream and (b) the sludge before it is composted] and is left to remain there until the heat energy produced by the radioactive radiation has been reduced in its order of magnitude to one tenth of the amount available in the molded bodies upon their completion. The molded bodies, which may be sheathed or coated, are alternatively introduced into raw water used for preparing drinking water in order to sterilize it and are left to remain therein until heat energy produced by the radioactive radiation has been reduced in its order of magnitude to one tenth of the amount available in the freshly prepared molded bodies.

An object of the process is to find a better way to utilize available ultimate storage locations in salt stocks and to reduce the cost of noncontaminating removal or final storage of highly radioactive waste.

In an advantageous embodiment of the process the molded body is introduced into wastewater which contains at least one contaminant which is a monochloro-, di-chloro-, trichloro-, tetrachloro- or pentachlorophenol and the wastewater is treated in a continuous or discontinuous operation, the molded body remaining in the wastewater until a dose of the maximum order of magnitude of 1.5×10^6 rad (the unit dose of absorbed radiation) has been reached. With continuous operation, a stream of air is introduced upstream of the molded bodies with respect to the direction of flow of the wastewater.

Example 1: The storage of glass blocks having a diameter of 200 mm and a length of 800 mm in a salt stock requires spacing bores accommodating the glass blocks at about 10 m intervals when the permissible quantity of fission product mixture (originating from a light water reactor fuel element which has been burnt to about 33,000 MWd/ton) incorporated therein has a heat output of 40 W/dm^3 of glass and the storage takes place soon after manufacture of the glass blocks without conditioning (introducing and maintaining the glass-blocks in wastewater or sludge). With conditioning for more than 10 years in a wastewater or sludge container with continuous or discontinuous wastewater or sludge changes, the heat output drops to about 4 W/dm^3 of glass. The spacing for blocks conditioned in such a manner with the same originally incorporated quantity of fission products is only less than one half the previously required minimum spacing. This means that in a salt surface previously accommodating 7 bores, 10 times the number of bore holes can be made for conditioned blocks.

Example 2: In a glass flask with a frit for the intake of air, aqueous solutions (1 ℓ of each having an initial concentration of 10^{-3} mol/ℓ of substance) were irradiated [at atmospheric pressure and at room temperature by means of a radiation source having an output of $1.88 \times 10^{+5}$ roentgens per hour (r/h)] as model solutions of wastewater with biologically resistant organic compounds. Glass-blocks containing a mixture of gamma-emitting fission nuclides were used. In the respective solutions the substance is:

  (1) 4-monochlorophenol;

  (2) 2,4-dichlorophenol;

  (3) 2,4,6-trichlorophenol;

  (4) 2,3,5,6-tetrachlorophenol;

  (5) pentachlorophenol.

The sequence of the radiochemical decomposition of these substances was observed under oxidizing conditions with the aid of the following analytical procedures:

gas chromatography: the decrease in chlorophenols;

spectral photometry: proof of phenolic functions;

ion-specific chloride electrode: proof of chloride;

gravimetry: proof of the decomposition end product, oxalic
acid.

The test results confirmed that, in all cases, gamma radiation of the chloro-substi-
tuted phenols in aqueous solution under oxidative conditions leads to complete dechlorination. With a dose of 1.5×10^6 rads complete dechlorination is achieved, for
example, with monochlorophenol. The chlorinated phenols are radiochemically
more efficiently decomposed with increasing chlorine content. Thus with an increasing number of chlorine atoms in the molecule the dose of irradiation required
to effect complete dechlorination is continuously decreased. For tetrachlorophenol
and pentachlorophenol a dose of 0.4×10^6 rads is sufficient. Moreover, the biologic
oxygen requirement (BOR) which, before irradiation, was 0 increases with increased
duration of irradiation until it reaches its maximum at complete dechlorination. The
phenolic functions decrease in parallel with increased dose and advancing dechlorination.

The G values (the G value indicates the number of formed or decomposed molecules
per 100 eV of absorbed radiation energy) for the decrease of the corresponding
chlorophenol and for the formation of chloride ions or hydrogen peroxide, respectively, in solution are:

	Decrease in Phenol	G Values for Chloride Formation	Formation of H_2O_2
4-chlorophenol	2.4	1.4	1.7
2,4-dichlorophenol	3.2	3.8	1.8
2,4,6-trichlorophenol	4.4	6.1	1.0
2,3,5,6-tetrachlorophenol	4.8	6.7	–
pentachlorophenol	4.8	9.9	–

With all chlorinated phenols the main decomposition product is oxalic acid, the
yield of oxalic acid increasing for pentachlorophenol to more than 90% with reference to the TOC (total organic carbon, i.e. the total content of organically bound
carbon) value after irradiation. Further decomposition products are CO_2 and substances containing keto groups.

In contradistinction to photochemical primary processes, the energy of ionizing
radiation is not selectively absorbed. Ionizing radiation of aqueous solutions primarily decomposes the water. Only with the onset of subsequent reactions of sol-
vated electrons and primary radicals resulting from water radiolysis, e.g., OH radicals, with dissolved substances are the latter attacked and decomposed.

In a further exemplary experiment technical wastewater from an industrial plant
containing, as its main component of organic substances, 2,4-dichlorophenol and 4-
chloro-p-cresol (together they constitute 90% by weight of the organic components)
was treated under the conditions outlined above. The contaminants therein were
also completely dechlorinated.

An addition of salts to this technical wastewater had no influence on the dechlorination. The following salts and NaOH were examined in the concentrations listed:

NaCl – 10^{-3} mol/ℓ;

NaCl – 10^{-1} mol/ℓ;

KNO_3, K_2SO_4, NaOH, each at 10^{-1} mol/ℓ.

The G values for the dechlorination were not influenced by such a load of salt in the wastewater. It is thus possible to dilute organic wastewater with salt-containing wastewater or with alkali or acid containing wastewater, respectively, in industrial operation.

Glass Cables

H.R. Leuchtag; U.S. Patent 4,320,028; March 16, 1982 describes a process whereby nuclear waste is incorporated into a glass by any convenient or conventional method, and the molten mixture is drawn into fibers. Preferably, a pool of water is provided for relatively short-term storage of the fibers, during which the intensity of radiation will decrease rapidly, the disposal of the heat generated in the process being more readily effected at this stage of the processing than at a more advanced stage.

The fibers are then made into a bundle or cable, the diameter of the fibers and the number of fibers in the cable being such that the cable is flexible and can be wound on a support. The cable is fed through an underground duct to a well-head and through a well leading deep into the earth to a storage chamber in which winding apparatus winds the flexible cable onto a support for long-term storage. A buffer device for the temporary storage of a portion of the cable is provided between the cable-fabricating plant and the duct to accommodate momentary inequalities between the rates of fabrication and transport through the duct; another such buffer device is installed, for a similar reason, at the well-head. The manufacturing and transport process are remotely controlled throughout. During both transport and storage, critical information about the state of the cable is received from an in-cable monitoring system.

The cable is so devised that it can be withdrawn from the support to the surface of the earth should the monitoring system indicate that the integrity of the storage chamber is endangered or breached. The cable may also be retrieved for harvesting isotopes or for incorporating additional nuclear waste after the activity of the cable has decreased. As long as no cause for concern or economic motive for retrieval exists, the cable may be left in place indefinitely. Feeding the cable to the waste-receiving facility and then to the storage chamber is facilitated by provision of leaders, that is, nonradioactive segments of the cable, at the forward and rearward ends of the cable. The flexibility of the cable is enhanced by making it in flattened form, that is, in the form of a belt. The cable may be color-coded for identification and may have a coating therearound for retention of glass fragments. Longitudinal variations in the radiation spectra may be used to provide further information for characterizing and identifying locations along the cable.

Transfer of the nuclear waste from the fabrication plant and its buffer device to the well-head is facilitated by provision of an underground conduit through which the cable may be fed with the aid of the leaders. As is evident, a cable is much more readily transported automatically from the melting tank to the well-head than would be glass billets. With the nuclear waste in the form of a cable transportable through a conduit, the problem of transporting nuclear waste above ground is completely eliminated. It is similarly evident that the problem of accounting for a single cable is much less severe than that of keeping track of the thousands of billets it replaces.

The integrity of the cable may be monitored by means of light pulses transmitted through optical fibers associated with the cable.

The underground conduit is positioned far enough below the surface of the earth so that the earth serves to screen out all but a minimal portion of the radiation from the nuclear waste. The optimal depth for the duct is best determined either by measurements on simulated configurations or the adaptation of existing computer codes to a linear source model. The principles of shielding are discussed in L. Wang Lau, *Elements of Nuclear Reactor Engineering,* Gordon & Breach, NY, 1974, where the formal solution for such a model is given. However, rough calculations based on rules of thumb are also given by Lau suffice to give an estimate on the depth required. These indicate that a depth equivalent to 2.5 meters of concrete will attenuate the gamma radiation by a factor of 10^8 and the neutron flux by 10^{25}. Beta and alpha particles and heavy ions are stopped much more readily than these. The actual depth required will depend on the nature of the soil covering the duct, and an additional safety factor to account for such eventualities as soil erosion and digging by animals and human intruders will be necessary, but a depth of 4 to 5 meters can be entirely satisfactory.

CEMENT

Addition of Lime and Cement to Borate-Containing Liquids

The cementing of radioactive waste liquids is known as a suitable and approved process for conversion into a solid, transportable and final storable form. However, it has been found problematical for solutions containing more than 5% of solid substances and more than 5% boric acid or borate because cracked and crumbly compositions form and the liquid frequently is not completely set.

N. Iffland, H.-J. Isensee, G. Wagner and H. Witte; U.S. Patent 4,122,028; Oct. 24, 1978; assigned to Nukem Nuklear-Chemie und Metallurgie GmbH, Germany have found that this problem can be solved by adding to 100 parts by weight of the radioactive, boron-containing aqueous solution first 5-30 parts by weight of slaked lime and then 30-80 parts by weight of cement, i.e., Portland cement. Thereby there can be replaced up to 30% of the cement portion by addition of 5-30 parts by weight of silica and/or kieselguhr per 100 parts by weight of solution. An increased strength, accelerated setting and a better resistance to leaching of the waste substance is obtained if there is added 3-30 parts by weight of water glass and especially 1-15 parts by weight of phosphoric acid or hydrogen phosphate, e.g., sodium hydrogen phosphate or potassium hydrogen phosphate, in each case based on 100 parts by weight of the liquid to be eliminated.

The process is advantageously suited for aqueous solutions which contain 5-25% of borate, especially 15%, and 5-30% of solids, especially 20%. The borate is usually present as sodium borate, but may be present as potassium borate or boric acid. Unless otherwise indicated all parts and percentages are by weight.

The addition of lime apparently leads to the formation of low soluble calcium borates, while the cement is indispensable as hydraulic binder for development of mechanical strength. The addition of water glass causes a better and quicker cross-linking and is also advantageous through the formation of high polymer mixed borate-silicate-anions, however, by itself in combination with cement only in none of the investigated cases it is sufficient to cause solidification. The known use of silica or kieselguhr as water binding fillers likewise only in combination with other additives leads to a better compacted block. By addition of phosphoric acid or hydrogen phosphates there is produced a definite acceleration of the solidification process and a reduction of the total amount of additive. The reasons for this are the ability of the phosphate to form polymeric anions and its buffering properties.

Example: There were stirred into a 100 ℓ drum 50 ℓ, corresponding to 63 kg of radioactive borate-containing (sodium borate) evaporator concentrate with a borate content of 5% and a solids portion of 30%, 10 kg of calcium hydroxide (slaked lime). After complete homogenization, there were mixed in succession 30 kg of iron-Portland cement, 5 kg of kieselguhr and 10 kg of sodium water glass (about 38°Bé). The volume then amounted to about 77 ℓ. The mixture solidified in the drum at room temperature within about 20 hours and reached the complete mechanical final strength after 2 to 4 weeks.

Polymer-Impregnated Cement

P. Colombo, R.M. Neilson, Jr., and W.W. Becker; U.S. Patent 4,174,293; Nov. 13, 1979; assigned to U.S. Department of Energy describe a process for disposing of aqueous solutions containing radioactive wastes whereby the waste solution is dispersed in situ throughout a mass of powdered Portland cement, the cement being placed in a leak-proof container and is densified therein to a bulk density ranging from about 1.3 to about 1.8 g/cc and a practice size ranging from about 120 mesh to about 400 mesh; the amount of water dispersed in the powdered cement being in a weight ratio thereto of from about 15 to 30 wt % based upon the weight of the powdered cement so used. The process further comprises sealing the container containing the cement with the aqueous solution dispersed therein to prevent evaporation of the aqueous solution contained therein during cement curing, thereafter impregnating the cured cement in the container with a mixture of a monomer and polymerization catalyst and polymerizing the monomer impregnated in the mixture in situ within the cured cement body, thereafter storing the container containing the polymer-impregnated cement in a storage facility suitable for storing such containers. Throughout the process the temperature of the cement body containing the aqueous solution containing the radioactive waste material is maintained at a temperature below 99°C. In the preferred embodiment, the cement-aqueous solution mix is maintained at a temperature below 90°C throughout the processing.

While cooling mechanisms can be employed if it is desired to make very large bodies of cement in accordance with this process, none are needed, when the body is formed within conventional 55 gallon industrial drums and type II Portland cement is used. No mixing apparatus is required to work the mixture of cement and aqueous solution but rather the water is dispersed in situ within a quiescent body of the densified powdered cement in the container through porous tubes strategically placed throughout the powdered cement. These tubes may be left in the cement after processing.

The impregnation of the concrete with monomer and catalyst is accomplished simply by covering the entire surface of the cured cement within the container with a mixture of monomer and polymerization catalyst and allowing the monomer-catalyst solution to seep down through and completely impregnate the concrete body in the container.

Some of the monomer systems which have been used successfully to demonstrate this process are:

Monomer*	Initiator**	Promoter
Styrene	benzoyl peroxide	heat
Methyl methacrylate	benzoyl peroxide	heat
90% Styrene-10% TMPTMA***	benzoyl peroxide	dimethyl aniline
60% Styrene-40% acrylonitrile	benzoyl peroxide	heat
Vinyl ester	methyl ethyl	cobalt
Dow 470	ketone peroxide	naphthenate

    *Comonomers can be used in varying proportions although those indicated
        above gave excellent results.
    **Any initiator capable of initiating free radical polymerization can be used.
    ***Trimethylolpropanetrimethylacrylate.

The following example is given to illustrate a practice of a preferred embodiment of the process. A cylindrical glass battery jar with a diameter of 11½ inches and a height of about 24 inches was filled to a height of 20 inches with dry Portland type III cement having a fineness (as measured by air-permeability testing) of 2,800 cm^2/g with less than 3 weight percent greater than 170 mesh. This required 43.6 kg of cement.

This cement was vibrated in its container with a Syntron Vibra-Flow Model V51 vibrator at a Controller setting of 5 for a period of 5 minutes to densify. The height of the concrete in the container after vibration was 17 inches, reflecting an increase in density of from 1.28 g/cm^3 to 1.51 g/cm^3.

A glass injector tube was inserted along the centerline of the cement cylindrical axis to a depth of 10 inches. The injector was a ½ inch diameter x 12 inch long glass tube containing twelve injector ports. The ports were located in groups of four at three levels along the injector length. In each group, ports were located 90° from one another about the axis of the injector. Taking the bottom of the injector as x equals 0, the port groups were located at x equals 0.5 inch, 5 inches, and 9.5 inches. Injector hole size was increased towards the top of the injector such that the hydraulic head was equal for all ports. A Chromel-Alumel thermocouple was inserted into the cement near the injector for thermal measurements of cement during curing and polymerization exotherms.

Tritiated aqueous waste was added through the injector to produce a water to cement ratio of about 0.25 by weight. This required 10.9 kg of water and gave a loading of 10.7 ℓ of waste per ft^3 of concrete. Waste was added at a rate of 15 ℓ/hr.

The cement casting was allowed to cure at 70°F for 24 hr. The peak temperature of the concrete casting was 70°C.

Styrene monomer to which one-half weight percent 2,2-azobis-(2-methylpropionitrile) was added as a polymerization catalyst was added to the top of the cement casting and allowed to soak in. Monomer was also added to fill the injector tube. The monomer was allowed to soak for approximately 4 hr, after which time, it was observed that no additional monomer was soaking in. The excess monomer was siphoned off the top of the casting.

The casting was heated to 50°C for 16 hr in order to polymerize the monomer. The peak casting temperature as indicated by the thermocouple was 75°C. The resultant increase in casting weight due to polymer was 7.2 kg, corresponding to a polymer loading of 13 weight percent in the cement casting. There was a thin film of polymer on the top of the casting.

A second experiment was carried out wherein high alumina cement was substituted for the Portland cement in the above experiment. A thermocouple placed in the center of the sample indicated a peak temperature of 186°C, approximately 8 hr after the water was dispersed in the cement. Substantial water was observed to be boiling off from the sample at that time.

$Ba(IO_3)_2$ in Concrete

W.E. Clark and C.T. Thompson; U.S. Patent 4,017,417; April 12, 1977; assigned to U.S. Energy Research and Development Administration describe a method for immobilizing fission product radioactive iodine recovered from irradiated nuclear reactor fuel, the method comprising combining material comprising water, Portland cement and about 3-20 wt % iodine as $Ba(IO_3)_2$ to provide a fluid mixture and allowing the fluid mixture to harden, the $Ba(IO_3)_2$ comprising the radioactive iodine. For purposes of this process wt % iodine is with respect to the total weight of hardened concrete. $Ba(IO_3)_2$ is meant to include hydrates as well as anhydrous salts. This method provides a highly water leach resistant concrete article which is suitable for encasement and underground storage. Preferably, a waterproofing agent such as butyl stearate is added to the fluid concrete mixture to retard leach from the resulting article.

The preferred method of carrying out the process is to contact an aqueous radioactive waste solution containing HIO_3 or HI_3O_8 (such as is readily provided from radioiodine gas scrubbing processes such as Iodox) with a source of barium ions such as a soluble barium salt to cause the precipitation of $Ba(IO_3)_2$. Preferably the liquid mixture is then contacted with sufficient Portland cement to form a concrete article using the entire waste stream for hydration. The precipitated $Ba(IO_3)_2$ functions as the aggregate of the concrete.

Example: One thousand liters of solution from the Iodox waste decay tank consisting of 12.4 M HNO_3 and containing 93 g/ℓ iodine (75% ^{129}I, 25% ^{127}I) as iodate is distilled to remove water and nitric acid and to yield 123.9 kg of solid metaiodic acid HI_3O_8. This solid is dissolved in 66 ℓ of water. A separate slurry is made up by mixing together 123.5 kg of $Ba(OH)_2 \cdot 8H_2O$ and 140 ℓ of water at about 80°C. To this slurry the solution of HI_3O_8 is added with constant stirring to neutralize the acid and to precipitate barium iodate as a solid phase. The stirring is continued to prevent caking of the slurry and to help dissipate the heat of neutralization.

During this period 7.8 kg of solid butyl stearate is added. When the temperature has dropped to 35° to 40°C this slurry is added to 320 kg of type I Portland cement. The mixture is thoroughly mixed by conventional techniques and poured into steel or plastic drums while the latter are vibrated to remove the entrained air bubbles. The cement is then allowed to harden and then is cured for 28 days in an atmosphere saturated with water vapor. The top of the concrete plugs are then rinsed with a small volume of water to remove any surface HIO_3. This water is drained off and used in making up the next batch. The tops of the plugs are then treated with a water-repelling compound (e.g., a polybutene). The drum is then sealed in the conventional manner. The article containing about 11.9 wt % I is now suitable for long-term underground storage.

The cement-to-water ratio for the concrete is not critical so long as sufficient mechanical strength is assured. Sufficient mechanical strength for underground disposal can be attained by cement-to-total water ratios of about 0.6 to 1.3 with about 0.7-0.9 preferred.

Heavy-Metal-Containing Sludge Wastes

K.S. Chen and H.W. Majewski; U.S. Patent 4,113,504; September 12, 1978; assigned to Stauffer Chemical Company describe a method of treating heavy-metal-containing sludge waste to produce a cementitious solid product having reduced heavy metal content leachability. The heavy metal content of the waste is effectively fixed to restrict release into the environment.. The solid product is produced by mixing sludge waste with fixing ingredients comprising vermiculite and cement. Upon setting, the product may be disposed of as landfill.

In particular, it is desirable to remove heavy metals with undesirable toxicologic properties such as arsenic, selenium, cadmium, mercury, bismuth, thorium, uranium, plutonium, or mixtures thereof. The heavy metal contained in the sludge may be any compound, complex, or elemental form of the heavy metal.

The essential ingredients necessary to fix the heavy metal content of the sludge are vermiculite and cement. The ratio of vermiculite to cement is not critical but it is typical practice to use a vermiculite to cement weight ratio in the range from 5:1 to 1:5. The vermiculite ingredient is particulate exfoliated magnesium silicate with a density as low as 0.09. Industrial grades of vermiculite having average sizes in the range of about 0.25 mm to about 10 mm diameters are suitable.

The cement ingredient is any particulate inorganic material containing calcareous, siliceous, and argillaceous components which forms a plastic mass upon mixing with the proper proportion of water. The plastic mass sets (solidifies) by a complex mechanism of chemical combination, gelation, and crystallization over a period of hours or days. Examples of suitable cements are natural cement, alumina cement, and Portland cement. In particular, it is desirable to use Portland cement such as ASTM Types I, II, III, IV, or V.

The essential ingredients may also be combined with optional ingredients such as sand, gravel, gypsum, lime, clays, or plastics to provide desired properties. Generally, such optional ingredients should be present in a ratio of less than one part by weight per one part by weight of cement fixing ingredient.

The leachability of heavy metals in the sludge waste is reduced by fixing both the liquid and solid phases of the sludge in the solid cementitious matrix of vermiculite and cement. Since acidic substances are generally deleterious to the stability of cement, it is recommended that basic substances be added to neutralize acidic sludges to give a mixture of sludge and fixing ingredients which has a pH of at least 7 and preferably above pH 8. Examples of useful basic materials are oxides, hydroxides, and carbonates of magnesium and calcium (e.g., limestone).

Sludge waste usually contains sufficient free water in its liquid phase to combine with the cement ingredient and form a neat cement paste capable of setting to a solid cementitious product. The effective proportions of sludge and fixing ingredients may be adequately determined by the visual appearance of the mixture resulting from mechanically blending sludge waste, cement, vermiculite, and any desired optional ingredients. The desired appearance of the mixture is a neat cement paste of uniform consistency without a separate liquid phase. The quantity of sludge which may effectively be combined with the cement and vermiculite is conveniently estimated from the ratio of liquid sludge phase to the quantity of cement ingredient.

Generally, sludge waste is mixed with cement ingredient in proportion such that the weight ratio of aqueous liquid phase of the sludge to the cement ingredient is in the range from 10:1 to 1:3.

Sludge waste of low (e.g., less than 30 weight percent) aqueous liquid phase content may also be treated by the process by either initially adding water to the sludge or adding water to the fixing ingredients.

Alkali Silicate Additive

D.H. Curtiss and H.W. Heacock; U.S. Patent 3,988,258; October 26, 1976; assigned to United Nuclear Industries, Inc. have found that the addition of alkali or alkaline earth silicate to a radwaste-cementing material mixture produces a number of unexpected benefits and important advantages over the known radwaste disposal process. These include:

(1) direct solidification of all common nuclear power industry radioactive wastes, including boric acid solutions;

(2) rapid hardening to a gel in less than 2 minutes, eliminating requirements for continuous mixing to insure homogeneity;

(3) solidification of maximum hardness in less than 7 days, compared to 28 days for cement alone without the alkali silicate additive;

(4) increased water retention over nonsilicated processes due to the high capacity of silicates for water fixation by hydration;

(5) production of more fluid mixes causing ready adaptability to batch or continuous processing of radioactive wastes, and

(6) minimum operator training and control required to obtain solidified wastes in a form suitable for safe handling and shipment.

Cementing materials that can be used in the process include Portland cement (all types), natural cement (all types), masonry cement (all types), gypsum, gypsum cement or plaster, plaster of Paris, lime (slaked or unslaked) and pozzolans, all materials which harden by a combination of hydrolysis and hydration reactions upon the addition of water. The preferred cementing material is Type II Portland cement, as it is inexpensive and easily obtainable.

While in general any alkali or alkaline earth silicate can be employed as the additive, sodium silicate is the preferred additive because of its low cost and ready availability.

The proportions of the radwaste-cement-additive mixture can be varied over a rather wide range. In general, for solidifying 100 parts (by weight) of radwaste, it is preferred to use 20-100 parts (by weight) of the cementing material, and 5-50 parts (by weight) of the silicate additive. The silicate additive will preferably constitute

3-15% by weight of the final mixture. Since in the radwaste disposal process the emphasis is on increased liquid fixation in the solidified product allowing a maximum of waste to be incorporated into a minimum final volume, it is preferred to use relatively high proportions of the silicate additive, constituting at least 20% by weight of the combined cement-additive weight. What is optimum for certain wastes may not be optimum for other wastes, but simple experimentation with different waste samples will enable those skilled in this art to arrive at optimum proportions with little effort.

Treatment of Cesium with Silicate and Shale

J.F. Hayes; U.S. Patent 4,173,546; November 6, 1979 has found that when liquid waste material which is contaminated with radioactive cesium isotopes is solidified by adding thereto a mixture of aqueous alkali metal silicate, an alkali metal silicate hardening agent and a plurality of shale particles the radioactive cesium isotopes in the resultant solidified mass are rendered essentially immobile. That is, they essentially cannot be leached out of the so-produced mass by bringing it into contact with an aqueous environment.

As is seen from a study of the table below, the results realized by the practice of the process are spectacular. This table clearly shows that synergistic results are realized when radioactive cesium isotopes are rendered immobile by solidifying (treating) such cesium-containing liquid waste material by adding thereto a mixture of an aqueous solution of alkali metal silicate, an alkali metal silicate hardening agent and a plurality of particles of shale which have the ability to ion exchange with the cesium.

Sample	Ingredients	Counts	Net Counts per Minute
1	Cement 20 g Silicate*0 Shale 0	(20 min counts) 43,590 44,996 45,953	2239
2	Cement 5 g Silicate*2 ml Shale 0	(20 min counts) 68,771 72,217 75,011	3597
3	Cement 5 g Silicate*2 ml Shale 1 g	(20 min counts) 10,982 11,520 12,078	523

*Specific gravity of 1.4.

In practice, the waste material is placed into a suitable container, such as a steel barrel or the like. To this waste material is then added an aqueous solution of alkali metal silicate, an alkali metal silicate hardening agent and a plurality of shale particles. The exact sequence in which these ingredients are added to the waste material is not critical. However, when the alkali metal silicate hardening agent is cement, it is preferred to add the cement before the alkali metal silicate. The shale particles can be added at any time before solidification occurs. If desired, it is possible to simply mix the waste with the various components as they are fed into the desired container.

In the process, any alkali metal silicate can be utilized. All that is required is that it be soluble in water. For example, potassium silicate and lithium silicate are suitable, but they are generally too expensive to be practical and are often difficult to obtain. Sodium silicate is ideal because it is relatively inexpensive and is generally available throughout the United States in either liquid or solid form. The liquid silicate is commercially available in a variety of ratios of Na_2O to SiO_2.

The sodium silicate will ordinarily be used in liquid form, but if for any reason it is desired to use solid silicate, water may be added to the mixture in the form of a solution of hardening agent or simply as water.

Various hardening agents can be used. In general, acids or acidic materials act promptly to cause gelation, or hardening of the silicate. If the hardening agent is to be added to the mixture, it should be a polyvalent metal compound; that is, a composition containing polyvalent metal ions. It has been found that hardening agents which are only slightly soluble or compositions containing only small amounts of soluble hardening agents are most desirable for commercial use with this process. Typical hardening agents are Portland cement, lime, gypsum and calcium carbonate, which are the least expensive and most available, although aluminum, iron, magnesium, nickel, chromium, manganese or copper compounds could be used, but they are more expensive and difficult to obtain. Portland cement, lime and gypsum have a quick gel forming reaction, which is highly desirable, and then continue with a hardening reaction over a period of time. The properties of Portland cement as a setting agent are excellent. In addition, it is economical and readily available in large quantities throughout the United States. Also, its reaction rate with the silicate is easily controllable.

As is well known, shale is a material which has a definite geological form. Basically, shale is a fine-grained sedimentary rock whose original constituents were clays or muds. It is characterized by thin laminae breaking with an irregular curving fracture, often splintery, and parallel to the often indistinguishable bedding planes. In the process, shale having a particle size ranging from about 8 mm to through 200 mesh have been used successfully. The exact particle size of the shale is not critical. All that is required is that the shale have a relatively high cation exchange capacity for cesium and that enough shale be used to immobilize essentially all of the radioactive cesium isotopes which may be present. That is, an effective amount of suitable sized shale particles is added to the waste material together with the alkali metal silicate and silicate hardening agent. Obviously, the optimum amount and size of shale in any given situation can be determined empirically.

BITUMEN

Low Temperature Processing of Plastic Wastes

A. Puthawala, L. Hoffmann, H. Emmert, W. Hild, H.-E. John and W. Kluger; U.S. Patent 4,204,974; May 27, 1980; assigned to Kraftwerk Union AG, Germany describe a method for removing radioactive plastic wastes, particularly radioactively contaminated ion exchange filter material admixed with water, by drying and encapsulating in bituminous solidification substances and pouring into a container, which includes subjecting the radioactive waste separate and apart from the bituminous substance to drying to produce a substantially dry radioactive waste, introducing regulated amounts of the dry radioactive waste into the bituminous substance, subjecting the mixture of dry radioactive waste and bituminous substance to kneading to effect encapsulation of the ratioactive waste by the mechanical action of the kneading, maintaining the mixture at a temperature below 120°C during the kneading, and discharging the kneaded, encapsulated radioactive plastic waste into a container.

The apparatus for removing radioactive plastic wastes includes a vertical collecting and drying vessel having an inlet for the introduction of radioactive waste, a source of hot drying gas, a hot gas inlet in the drying vessel for the introduction of the gas for passage in contact with radioactive waste, a gas outlet in the drying vessel for discharge of the gas after contact with the radioactive waste, a dry radioactive waste outlet at the bottom of the drying vessel for the discharge of dry radioactive waste, a dosage device connected to the radioactive waste outlet for regulating measured amounts of dry radioactive waste, a substantially vertical conduit connected at its upper end to the dosage device for receiving the measured amounts of dry radioactive waste and downward passage by gravity through the conduit, a horizontal kneader at a level below the drying vessel connected at its inlet end with the lower end of the conduit to receive the dry radioactive waste, a source of bituminous substance, a bitumen inlet at the inlet end of the kneader for the introduction of the bituminous material in admixture with the dry radioactive waste, power means for mechanically kneading the mixture to encapsulate the radioactive waste, heating means for heating the mixture during kneading, an outlet at the other end of the kneader for the discharge of encapsulated radioactive waste, and a container filling station adjacent the kneader discharge.

In one advantageous embodiment of the process, flowable dried spherical resins, i.e., resins of spherical shape with diameters of 0.1 to 1 mm, are mixed with the bitumen in the ratio 1:1. This mixture is discharged from the kneader at 110°C. It is sent directly into barrels serving for the final storage, e.g., the standardized 200 ℓ barrels, which also serve for shipping. While the ratio of resin to bitumen may vary, good results are usually obtained in the range of one part resin to 0.7 to 2 parts bitumen by volume.

Waste Pellets Impregnated with Asphalt

M. Hirano and S. Horiuchi; U.S. Patent 4,299,721; November 10, 1981; assigned to Hitachi, Ltd., Japan describe a method of producing a radioactive waste package comprising the steps of filling a container with a predetermined amount of pellets of the radioactive waste, filling the container with a thermoplastic composition in the molten state while heating the container, heating the container over a predetermined period, cooling the container, filling again the container with the molten thermoplastic composition, and then sealingly fitting a lid member to the container.

In the preferred embodiment of the process, an asphalt material is used as the thermoplastic composition.

At the same time, in the preferred embodiment, the temperature to which the pellets of the radioactive waste and the container are heated is about 160°C, and the heating time is about 1 hour. A test showed that the block of the pellets is satisfactorily impregnated with the asphalt material, when the drum filled with the pellets of the radioactive waste and then with the molten asphalt is heated to and held at an elevated temperature of 160°C.

The radioactive waste package produced by this method, sufficiently impregnated with the thermoplastic material, exhibits a large enough strength to withstand any external force which may be exerted on the package when it is subjected to an offshore or other natural environmental disposal.

Figure 1.1, shown on the following page, is an illustration of steps of a process for producing a radioactive waste package from a radioactive waste.

Figure 1.1: Method of Producing Radioactive Waste Package

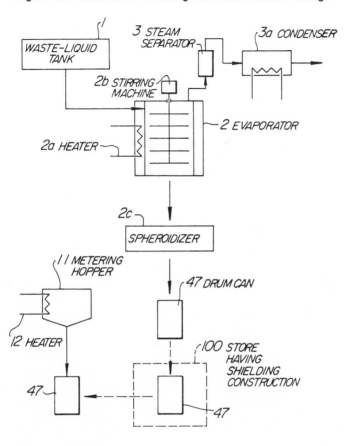

Source: U.S. Patent 4,299,721

Azeotropic Distillation

K.T. Norell; U.S. Patent 4,123,380; October 31, 1978; assigned to AB Bofors, Sweden describes a method of transferring the solid constituents from aqueous waste containing low and/or medium active radioactive solid constituents into solid units intended for longtime depositing wherein the solid constituents contain ion-exchange masses which comprises:

   (A) conveying the aqueous waste to a container provided with a stirrer;

   (B) adding an organic compound which forms an azeotrope with water to the aqueous waste;

   (C) distilling off the azeotrope and continuing the distilling until most or all of the water has been removed;

   (D) during the distillation returning the organic compound to the aqueous waste after the water portion of the azeotrope has been removed;

(E) adding while stirring a substance which subsequently either through hardening or reaction with another substance is transformable into a solid state;

(F) distilling off most or all of the organic compound;

(G) pouring the waste from the container into a vessel for transformation into solid units for subsequent deposition.

The aqueous waste in question in the example of the procedure which is described consists of:

(I) An evaporation concentrate with approximately 20% by weight dry substance

(II) A suspension of granular ion exchange masses with approximately 17% by weight dry substance

(III) A suspension of powdered ion exchange masses with approximately 14% by weight dry substance.

According to the procedure illustrated in Figure 1.2, appropriate proportions of the abovementioned aqueous waste are fed to a buffer tank (1) which appropriately has a capacity of approximately 2.5 m^3 and is provided with a propeller stirrer, via the pipes (2), (3) and (4). The quantity fed of (I) is approximately 200 kg, of (II) approximately 700 kg and (III) approximately 1,000 kg, whereby the total quantity of dry substance will amount to approximately 300 kg.

Figure 1.2: Waste Disposal Apparatus Employing Azeotropic Mixture

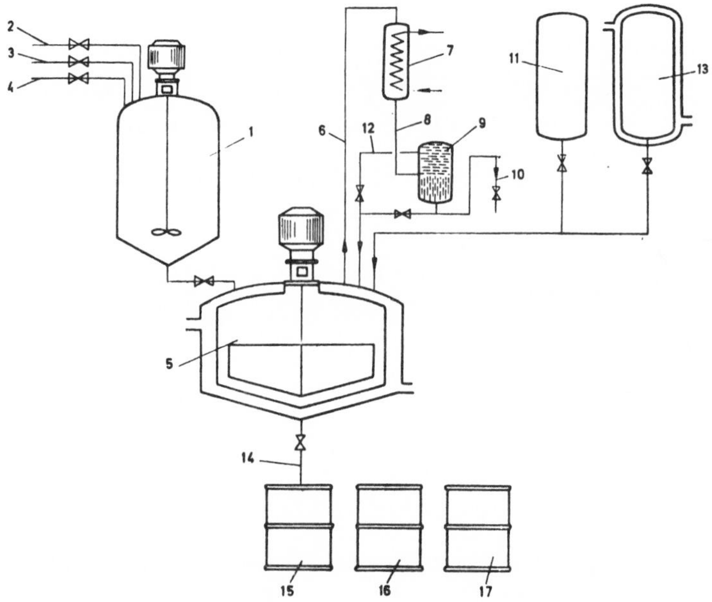

Source: U.S. Patent 4,123,380

The process is, of course, also applicable to other proportions of (I), (II) and (III) and can also be used for one only, or for two of these types of waste.

From the buffer tank (1), the mixture of waste is fed successively to the container (5), which can be heated, and which is provided with an effective stirrer. In conjunction therewith, the container (5) is heated so that water evaporates and the pace for the feeding from the buffer tank (1) is adapted so that the liquid volume in the container (5) is kept more or less constant. In this way, among other things, favorable heat transfer conditions are obtained. The water vapor which is removed at the evaporation is conveyed via the pipe (6) to the cooler (7), where the water vapor is condensed, and via the pipe (8), the vessel (9) and the pipe (10) the water is thereafter returned to the closed system from which the waste has come.

When the entire quantity of aqueous waste which has been put into the buffer tank (1) has been fed to the container (5), the evaporation continues until the dry substance content of the liquid present in the container (5) amounts to approximately 40% by weight. Thereafter, approximately 300 kg toluene is fed from the storage container (11), and through continued heating of the container (5), an azeotrope consisting of water and toluene will be conveyed off through the pipe (6), condensed in the cooler (7), and separated in the vessel (9). The heavier water will be collected in the lower part of the vessel (9), and is drained off, as described above, through the pipe (10). The lighter toluene is collected in the upper part of the vessel (9), and is returned to the container (5) via the pipe (12). The azeotropic distillation is continued until the major portion of the water has been removed from the container (5) and the water content in this container should be reduced to less than 0.5% by weight and preferably less than 0.2% by weight before the azeotropic distillation is discontinued. This can be carried out here without the temperature of the liquid in the container (5) exceeding 150°C.

When the major portion of the water has been removed from the liquid present in the container (5), in the abovementioned way, 300 kg of melted bitumen with a temperature of approximately 110°C is added successively from the storage vessel (13), which can be heated. At the addition of the bitumen, through the continued heating, the toluene will be distilled off through the pipe (6), condensed in the cooler (7), and via the pipe (8) will be collected in the vessel (9), from which it can be added to the next batch.

The distillation of the toluene continues until the major portion of this has been removed from the liquid present in the container (5), and the content has been reduced to less than 10% by weight. Thereafter the melted bitumen (which now contains the waste) is poured via the pipe (14) into three drums (15, 16, 17) each with a capacity of 200 ℓ, and is thereafter allowed to stiffen there. In this way, the aqueous waste has been transformed into solid units intended for long-time depositing.

Apparatus Providing Shortened Mixing Time

A. Puthawala, O. Meichsner and E. Marr; U.S. Patent 4,280,922; July 28, 1981; assigned to Kraftwerk Union AG, Germany describe a method and device for removing radiant, pulverulent synthetic wastes wherein danger of decomposition and gas formation of the synthetic wastes is reduced without requiring the temperature of the bitumen to become undesirably low. Simultaneously, a dangerous concentration of partly explosive gas in the degassing dome is prevented. In this regard, the dried filter residues are to be metered to the bitumen so that no blockages or obstructions occur.

The radiant, pulverulent synthetic wastes are mixed dry with a thermoplastic mass in a kneader and then delivered from a discharge opening of the kneader into a container capable of providing a final storage therefor, while gases and/or vapors are withdrawn from degassing domes of the kneader, while delivering fluidic dried wastes by mechanical movement through a metering tube into a degassing dome in the kneader disposed next to the discharge opening, and admitting scavenging gas into the metering tube at least temporarily in direction toward the kneader.

Through this process, the mixing time is considerably shortened and, accordingly, the thermal loading or stressing of the material being embedded is reduced because the transit time of the synthetic wastes is not provided any more by the movement of the wastes through the entire kneader but rather only by the movement thereof through the short distance between the "last" degassing dome and the discharge opening of the kneader. In this manner, the mixing time can be shortened to one-third or even less. Thus, the decomposition, which indeed, does not occur abruptly, is accordingly reduced. In this regard, due to the mechanical movement in the metering tube, the accuracy of the metering operation is improved so that it is actually possible to manage with such brief mixing times. Furthermore, due to the scavenging gas, the metering tube is kept free of contaminating vapors so that the removal of gases or vapors limited in essence to this degassing dome can effect no obstruction or blockage.

In accordance with another feature of the process, the method comprises applying vibratory motion to the metering tube as the mechanical movement, and the vibratory motion is applied transversely to the longitudinal axis of the metering tube.

In accordance with an additional feature of the process, the method comprises continuously admitting the scavenging gas into the metering tube in order to avoid the formation of clumps by the wastes. If necessary, by varying the amount of scavenging gas, obstructions in the metering tube can be removed. Besides, ordinary air as well as other gases can be introduced which, if need be, due to the content of inert components therein, can avoid danger of explosion even when disruptions occur in the feed of the scavenging air.

The method includes periodically heating the metering tube which preferably projects vertically into the degassing dome, with steam. Thereby, all at once, a given heating of the tube is achieved which facilitates expulsion of condensation products and possible bitumen splashes or spray along the short path to the discharge opening. The steam is passed through the metering tube, and is admitted from the metering tube into the degassing dome for scavenging synthetic powders adhering therein. For this purpose, a steam-conducting jacket disposed around the metering tube is provided with suitable discharge or exhaust openings.

The method comprises mixing the wastes in a ratio of about 60:40 with bitumen which is at a temperature of from 110° to 150°C and which is preferably distillation bitumen (B 15 or B 25).

In accordance with the device of the process, there is provided a kneader for embedding radiant synthetic wastes in bitumen comprising a plurality of degassing domes and formed with a discharge opening, one of the degassing domes being located adjacent the discharge opening and having a metering tube displaceable with respect to the one degassing dome.

Figure 1.3 is a diagrammatic side elevational view, partly in section, of a device in the form of a kneader for removing radiant, pulverulent synthetic wastes, in accordance with the process, and lines with the radiant synthetic wastes leading thereto.

Figure 1.3: Apparatus for Delivering Bitumen-Mixed Waste to Container

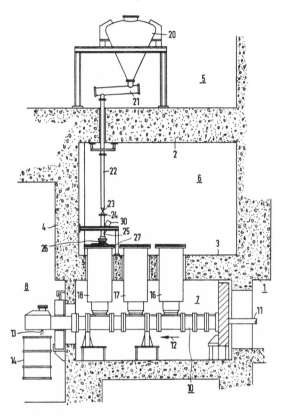

Source: U.S. Patent 4,280,922

Mixing Technique Employing Partial Vacuum

K. Knotik, P. Leichter and H. Jakusch; U.S. Patent 4,139,488; February 13 1979; assigned to Vereinigte Edelstahlwerke AG (VEW), and Oesterreichische Studiengesellschaft fuer Atomenergie GmbH, Austria describe a binder mixing technique wherein a highly homogeneous mixture of the binder with the waste is accomplished to assure uniform impregnation of the latter. Illustratively, the hardenable binding agent is added to a container filled with the waste while a partial vacuum is applied thereto. Such partial vacuum has been found to be especially effective in aiding the mixing and wetting of the waste by the softened binding agent, and thereby avoiding the presence of porous areas in the resulting matrix.

Preferably, the container is maintained at a temperature above the softening point of the binding agent during the addition of the agent to the waste. If it is desirable to dry the waste prior to the application of the binding agent, this can be accomplished by the simultaneous application of heat and a partial vacuum to the waste container before the agent is added.

For certain types of wastes, i.e., spent ion exchange resins which have embedded therein radioactive inorganic compounds through prior exposure of the resin to wastes from a nuclear processing installation or the like, it may be further preferable, prior to the addition of the binder to the waste, to thermally decompose the resin via carbonization in the presence of a medium that does not support combustion.

Further enhancement of the homogeneity of the waste-binder matrix may be assured by removing the partial vacuum after the addition of the binder to the waste and before the hardening of the binder.

In one feature of the process, the homogeneous waste-binder matrix remaining in the container after the treatment discussed above may be afforded additional protection against release of pollutants to the environment by placing such container within a larger container prior to storage, and filling the space between the two containers with a suitable binding agent such as bituminous material or concrete.

RESINS

Conditions for Resin Encapsulation

J.L. Arnold and R.W. Boyle; U.S. Patent 4,077,901; March 7, 1978 describe a method for encapsulating liquid or finely divided solid toxic waste substances into a form suitable for burial. In essence the method involves uniformly dispersing the waste in a liquid thermosettable polymer composition and thereafter curing the waste/polymer dispersion under thermal and catalytic conditions such that the exotherm developed during the cure never rises above the temperature at which the integrity of the encapsulating material is destroyed.

The method finds wide utility with the diverse wastes. It is particularly useful with the radioactive wastes resulting from nuclear powered plants. Thus, it is adaptable for encapsulating wastes of high or low level radioactivity, of high or low acidity, of a wide variety of solutes and dispersed substances and of large or small amounts of waste. The method uses readily available materials that are easily handled in a safe manner.

The thermosettable polymer compositions include a vinyl ester resin or an unsaturated polyester or blends and mixtures of those two materials.

Preferably the thermosettable resin phase comprises from 40 to 70 weight percent of the vinyl ester of polyester resin and from 60 to 30% of a copolymerizable monomer. Suitable monomers must be essentially water insoluble to maintain the monomer in the resin phase in the emulsion, although complete water insolubility is not required and a small amount of monomer dissolved in the emulsified water does no harm.

Suitable monomers include vinyl aromatic compounds such as styrene, vinyl toluene, divinyl benzene and the like; saturated alcohols such as methyl, ethyl, isopropyl, octyl, etc; esters of acrylic acid or methacrylic acid; vinyl acetate, diallyl maleate, dimethallyl fumarate; mixtures of the same and all other monomers which are capable of copolymerizing with the vinyl ester resin and are essentially water insoluble.

Another embodiment of this process utilizes a modified vinyl ester resin wherein about 0.1 to 0.6 mol of a dicarboxylic acid anhydride per equivalent of hydroxyl is reacted with the vinyl ester resin. The stability of the water-in-resin emulsion prepared from the modified vinyl ester resin is somewhat less, comparatively, than that found with the unmodified vinyl ester resins, yet the stability is significantly improved over the art. Both saturated and unsaturated acid anhydrides are useful in the modification.

Suitable dicarboxylic acid anhydrides containing ethylenic unsaturation include maleic anhydride, citraconic anhydride, itaconic anhydride and the like and mixtures thereof. Saturated dicarboxylic acid anhydrides include phthalic anhydride, anhydrides of aliphatic unsaturated dicarboxylic acid and the like. The modified vinyl ester resin is utilized in this process in the same manner for the unmodified vinyl ester resin.

A wide variety of unsaturated polyesters which are readily available or can be prepared by methods well known to the art may also be utilized in the method. Such polyesters result from the condensation of polybasic carboxylic acids and compounds having several hydroxyl groups. Generally, in the preparation of suitable polyesters, an ethylenically unsaturated dicarboxylic acid such as maleic acid, fumaric acid, itaconic acid or the like is interesterified with an alkylene glycol or polyalkylene glycol having a molecular weight of up to 2000 or thereabouts. Frequently, dicarboxylic acids free of ethylenic unsaturation such as phthalic acid, isophthalic acid, adipic acid, succinic acid and the like may be employed within a molar range of 0.25 to as much as 15 mols per mol of the unsaturated dicarboxylic acid. It will be understood that the appropriate acid anhydrides when they exist may be used and usually are preferred when available.

The glycol or polyhydric alcohol component of the polyester is usually stoichiometric or in slight excess with respect to the sum of the acids. The excess of polyhydric alcohol seldom will exceed 20-25% and usually is about 10-15%.

These unsaturated polyesters may be generally prepared by heating a mixture of the polyhydric alcohol with the dicarboxylic acid or anhydride in the proper molar proportions at elevated temperatures, usually at about 150° to 225°C for a period of time ranging from about 1 to 5 hours.

Polymerization inhibitors such as tert-butyl catechol may be advantageously added. It is also possible to prepare unsaturated polyesters directly from the appropriate oxide rather than the glycol, e.g., propylene oxide may be used in place of propylene glycol. Generally, the condensation (polymerization) reaction is continued until the acid content drops to about 2 to 12% (—COOH) and preferably from 4 to 8%.

Yet another embodiment of this process utilizes a vinyl ester/unsaturated polyester resin composition wherein the weight ratio of the polyester to the vinyl ester ranges up to 2:3. The composition may be prepared either by physically mixing the two resins in the desired weight proportions or by preparing the vinyl ester resin in the presence of the unsaturated polyester. These vinyl ester/unsaturated polyester resin compositions readily form waste-in-resin dispersions in the same manner as previously described for the vinyl ester resins even though the unsaturated polyesters, alone, at times do not form stable emulsions with liquid waste materials.

In the practice of the method, waste material-in-resin dispersions, may be prepared in a variety of ways. Generally a free-radical-yielding catalyst is blended with the phase and the waste then dispersed in that resin under conditions to form a uniform dispersion. When the waste is a solid, it should be finely divided of a size generally less than about ⅛ inch or less. When the waste is a liquid, it is preferred to form a liquid waste-in-resin emulsion. In that instance the liquid is added to the liquid un-cured resin under shearing conditions to form the emulsion. While the shear condi-tions may be widely varied, generally with liquid wastes sufficient shear should be applied to produce a relatively uniform emulsion of small droplet size.

The dispersions, whether of liquid or solid diperse phase, should have sufficient storage stability to last at least through the initial gelation of the resin. The dis-persions made with vinyl ester resins, particularly those within the previously de-scribed monomer proportions, generally exhibit adequate stability without added emulsifier. Emulsions made with unsaturated polyesters frequently will require added emulsifier. Such emulsifiers are known in the art and judicious selection can be made with simple routine experiments.

The proportions of liquid waste in the resin phase are also important by reason that these emulsified liquids serve as a heat sink and assist in control of exotherm and final temperature. Preferably the compositions (waste-in-resin emulsions) are prepared to contain from about 30 to 75% by weight of liquid waste with the bal-ance comprising the resin phase.

Catalysts that may be used for the curing or polymerization are preferably the per-oxide and hydroperoxide catalysts such as benzoyl peroxide, lauroyl peroxide, tert-butyl hydroperoxide, methyl ethyl ketone peroxide, tert-butyl perbenzoate, potas-sium persulfate and the like. The amount of the catalyst added will vary preferably from 0.1 to about 5% by weight of the resin phase.

Preferably, the cure of the emulsion can be initiated at room temperature by the addition of known accelerating agents or promoters, such as lead or cobalt naph-thenate, dimethyl aniline, N,N-dimethyl-p-toluidine and the like usually in con-centration ranging from 0.1 to 5.0 weight percent. The promoted emulsion can be readily cured in about 3 to 30 minutes, depending on the temperature, the catalyst level and the promotor level. Cure of the emulsion can also be initiated by heating to temperature of below 100°C. The common practice of post-curing thermoset articles at elevated temperatures for varying periods of time may be utilized with this process.

The conditions of selection of catalyst, catalyst concentration and promoter selec-tion and concentration must be such that the temperature of the exotherm does not exceed 100°C. If the exotherm exceeds 100°C, the water in the liquid waste will boil which may cause waste material to be released.

The solidification may be carried out in any suitable vessel such as a 55 gallon drum. Larger or smaller vessels may be used depending on the amount of waste to be dis-posed of, on the equipment available and on the limitations of handling and trans-portation stock.

Example: A simulated radioactive evaporator waste was prepared in water and 2.0 microcuries cobalt 60 and 0.92 microcurie cesium 137 were added as the chloride salts.

422.5 grams of the waste was solidified with the following ingredients: 338 g of a vinyl ester resin made by reacting 32.6 parts of the diglycidyl ether of bisphenol A extended with 8.7 parts of bisphenol A then reacted with 1.2 parts maleic anhydride and 7.5 parts methacrylic acid, the resin dissolved in 50 parts styrene; 8.45 g of 40% benzoyl peroxide emulsified in dibutyl phthalate; 1.125 g of N,N-dimethyl-p-toluidine.

The vinyl ester resin and benzoyl peroxide solution were measured into a large metal vessel and mixed thoroughly with an air stirrer. The radioactive waste was slowly added to the above blend with the air stirrer at high speed to assure good emulsification. The dimethyl toluidine was added to the emulsion and mixed thoroughly for 30 to 60 seconds. The stirrer was removed and the emulsion poured into plastic containers of 4.75 cm diameter and 7.3 cm length. The emulsion cured to a hard homogeneous solid.

Specimens of the cured solids were tested accordingly to the special tests for massive solids as listed in Section 173.398, Hazardous Materials Regulations of the Department of Transportation.

In the water leaching test the specimen is immersed for 1 week in water at pH 6-8 and 68°F and a maximum conductivity of 10 micromhos/centimeter, and by immersion in air at 86°F. To pass this test the product must not dissolve or convert into dispersible form to the extent of more than 0.005% by weight.

When so tested the specimen of this example did not dissolve or convert into dispersible form and in fact showed 0% weight loss.

When tested according to the International Atomic Energy Agency Safety Standards, Safety Series Number 6, Regulations for the Safe Transport of Radioactive Materials, 1973 Revised Edition the leaching water showed 0.085 microcurie of ^{60}Co and 0.042 microcurie ^{137}Cs. This is less than 10% of the radioactive material present in the original sample.

Monovinyl and Polyvinyl Compounds

W. Bähr, S. Drobnik, W. Hild, R. Kroebel, A. Meyer and G. Naumann; U.S. Patents 4,009,116; February 22, 1977; assigned to Bayer AG, and 4,131,563; December 26, 1978; assigned to Steag Kernenergie GmbH, both of Germany describe a process of preparing substantially organic waste liquids containing radioactive or toxic substances for safe, nonpollutive handling, transportation and permanent storage.

This object can be achieved by mixing the liquids with polymerizable mixtures consisting of one or more monomeric monovinyl compounds and one or more polyvinyl compounds and polymerization catalysts, and converting the resulting mixture into solid blocks by polymerization at temperatures in the range of from 15° to 150°C preferably of from 15° to 50°C.

The waste liquids are preferably used in quantities of from 20 to 75%, particularly of from 30 to 60% by weight, based on the total weight of waste liquid and polymerization mixture. The monovinyl compounds are preferably used in a quantity of from 70 to 99%, particularly of from 80 to 95% by weight, based on the weight of the monomers, the polyvinyl compounds are preferably used in a quantity of from 1 to 30%, particularly of from 5 to 20% by weight, based on the weight of the monomers and the polymerization catalysts are preferably used in a quantity of from 0.05 to 6% particularly of from 0.2 to 4% by weight based on the weight of the monomer. The process may be carried out at normal pressure but also at elevated pressure.

Particularly preferred monovinyl compounds are styrene, vinyl toluene and methyl acrylate.

Particularly preferred polyvinyl compounds are divinyl benzene and trivinyl benzene. The polymerization catalysts used in the process are the catalysts usually applied in the polymerization of monovinyl and polyvinyl compounds, e.g., acetyl peroxide, tert-butyl hydroperoxide, cumene peroxide, lauryl peroxide, methyl ethyl ketone peroxide, tetralin peroxide and persulfates. Preferred are those polymerization catalysts which react already at low temperatures, e.g., room temperature such as azo-bis-isobutyronitrile.

Example: 100 kg of radioactive tri-n-butyl phosphate are added with stirring to 41 kg of 53.7% divinyl benzene, 108 kg of styrene and 3 kg of azo-bis-isobutyronitrile accommodated in a 250 ℓ vessel. The mixture to be polymerized is left standing at 20°C. In this way, a homogeneous block is obtained after 21 days.

Urea-Formaldehyde Resin

K.A. Gablin and L.J. Hansen; U.S. Patents 4,010,108; March 1, 1977 and 4,167,491; September 11, 1979; both assigned to Nuclear Engineering Co., Inc., describe a process for making a disposable waste product material in which improvements are made over the use of Portland cement in order to increase the range of disposable materials, provide reduction of weight required for shipping, and generally provide a more reliable disposal from the standpoint of safety and the like.

These and other objects are achieved by solidifying the wet or water-carried waste product through the steps of adding a hydrophilic resinous material to the waste in an amount sufficient to set up and cure into a solid block, mixing the materials together to provide the desired distribution of waste materials therein, and curing the material to a solid mass.

The preferred hydrophilic resin to be used in accordance with this process is any of the urea-formaldehyde resins available from a plurality of commercial sources as standard articles of commerce. These resins are prepared by reacting urea and formaldehyde in mol ratios between about 1:1 and 1:4 respectively, and preferably between 1:1.5 and 1:2.5. For optimum results, the mol ratio is about 1 part urea to about 2 parts formaldehyde. Typically, solid urea and an aqueous solution of formaldehyde are reacted with one another to produce a resin syrup that is in the thermosetting state but capable of being converted to a thermoset state. These resins are available in syrup form, and sometimes available in a spray-dried form, which may be redispersed in water to a desired solids content. Since part of the water will come from the waste material, the urea-formaldehyde should be in a concentrated form with the final ratio of resin solids and water being present in the final dispersion in a ratio of about 2½ to about 5 parts water per part resin by weight and preferably from about 3 to about 4 parts water per part resin solids.

A typical catalytic material used to convert the urea resin to a thermoset state at ambient temperature is an acidic material having a dissociation constant between about 10^0 to 10^{-5}. The amount of catalytic material used will depend upon the strength of the acidic material used and upon the nature of the composition in which it is used. For example, materials like boric acid tend to inhibit the polymerization, and therefore increased catalyst is required to achieve the same cure time. However, generally the amount of acidic catalytic material will be between about 0.3 and 20% by weight of the resin solids in the mixture. In general, any acid capable of providing a pH below 5 in the dispersion may be utilized, as is well known in the art, and it is preferred to utilize sodium bisulfate, since it is available as a solid and provides an excellent strength acid.

Certain materials such as filter aids, ion exchange resins and materials that act as one of these or both are usually added in order to improve processing and provide the most economical and practical way to eliminate waste. However, any of these materials which are compatible with the urea-formaldehyde are suitable, and considerable latitude is permissible in this area.

Example: Water from a reactor cooling loop containing radioactive isotopes of the iron family is passed through a conduit packed with 1200 ml of resin beads, which beads are composed of a cation exchange resin available commercially (specifically a sulfonated styrene-divinylbenzene polymer). In this way, radioactive cationic materials are removed from the water and collected by the resin beads. The water is allowed to drain from the beads and the wet beads are then placed in a five gallon container. A 2000 ml solution or dispersion of urea-formaldehyde resin is then prepared by adding 1200 ml water to 800 ml of a dispersion containing about 63-66% solids. This diluted dispersion is then added to the wet beads in the container, and the mixture stirred by an electric stirrer at a speed sufficient to keep the resin beads substantially evenly distributed in the mixture. 50 ml of a saturated solution of sodium bisulfate is then added gradually with the stirring being continued. After the sodium bisulfate is added and the mixture gels sufficiently to hold the resin beads from sinking by gravity, the stirring is discontinued and the stirring blades disconnected and left in the mixture. The gel is then allowed to set until the cure is complete, whereupon the unit is ready for disposal.

System Employing Urea-Formaldehyde Resin

In a related process, *K.A. Gablin and L.J. Hansen; U.S. Patents 4,056,362; Nov. 1, 1977; 4,168,243; Sept. 18, 1979; and 4,196,169; April 1, 1980; all assigned to Nuclear Engineering Company, Inc.* describe a system for disposing of radioactive waste material from nuclear reactors by solidifying the liquid components to produce an encapsulated mass adapted for disposal by burial. The method contemplates mixing of radioactive waste materials, with or without contained solids, with a setting agent capable of solidifying the waste liquids into a free standing hardened mass, placing the resulting liquid mixture in a container with a proportionate amount of a curing agent to effect solidification under controlled conditions, and thereafter burying the container and contained solidified mixture.

The setting agent is a water-extendable polymer consisting of a suspension of partially polymerized particles of urea-formaldehyde in water, and the curing agent is sodium bisulfate. Methods are disclosed for dewatering slurry-like mixtures of liquid and particulate radioactive waste materials, such as spent ion exchange resin beads, and for effecting desired distribution of nonliquid radioactive materials in the central area of the container prior to solidification, so that the surrounding mass of lower specific radioactivity acts as a partial shield against higher radioactivity of the nonliquid radioactive materials. The methods also provide for addition of non-radioactive filler materials to dilute the mixture and lower the overall radioactivity of the hardened mixture to desired Lowest Specific Activity counts. An inhibiting agent is added to the liquid mixture to adjust the solidification time, and provision is made for adding additional amounts of setting agent and curing agent to take up any free water and further encapsulate the hardened material within the container.

The equipment utilized includes:

> a supply tank adapted to contain a polymerizable setting agent in liquid form,
>
> a catalyst tank adapted to contain a curing agent for the setting agent in liquid form,
>
> a receiving tank adapted to contain solidified radioactive waste material and formed with an inlet thereto,
>
> manifold means connected to the supply and catalyst and receiving tanks,
>
> slurry pump means interposed in the manifold means and adapted for connection to a source of liquid slurry containing radioactive waste material,
>
> catalyst pump means in the manifold connected to pump the curing agent from the catalyst tank,
>
> setting agent pump means in the manifold connected to pump setting agent from the supply tank,
>
> the pump means being formed to pump proportioned amounts of the setting agent and the curing agent and the radioactive slurry into the receiving tank for regulating the specific activity radiation characteristics per unit of volume of the materials in the tank,
>
> radioactive control means connected to the pump means and responsive to specific activity levels in the system to regulate pump speed thereby to maintain the specific radioactivity level of the material in the receiving tank at no more than a predetermined Low Specific Activity rating,
>
> a first mixing means connected to the pumping means and formed for intermixing the setting agent and the radioactive slurry before delivery to the receiving tank,
>
> and a second mixing means positioned at the inlet to the receiving tank and formed for intermixing the curing agent with the intermixed setting agent and slurry from the first mixing means as they enter the receiving tank.

Method for Determining Solidification Point

K.A. Gablin; U.S. Patent 3,986,977; October 19, 1976; assigned to Nuclear Engineering Company, Inc., describe a related method of disposing of at least partially radioactive waste material, comprising:

intermixing such waste material with a liquid containing a setting agent consisting of an aqueous suspension of partially polymerized urea-formaldehyde capable of forming with the waste material, upon mixing with a curing agent, a free standing hardened mass;

placing the resulting mixture in a container;

maintaining two spaced apart electrodes in contact with the mixture at or near the free surface thereof;

monitoring the resistance between the electrodes;

adding to the mixture a proportionate amount of a curing agent capable of hardening the mixture to a free standing hardened mass;

retaining the mixture in the container until the resistance reaches a maximum and slightly declines therefrom; and

burying the container and its solidified mixture for disposal.

Figure 1.4 illustrates the apparatus, which includes an ohmmeter attached to sensing contacts across the surface of the liquid waste, which is used in the process.

Figure 1.4: Device for Measuring Resistance Across Surface Liquid Waste

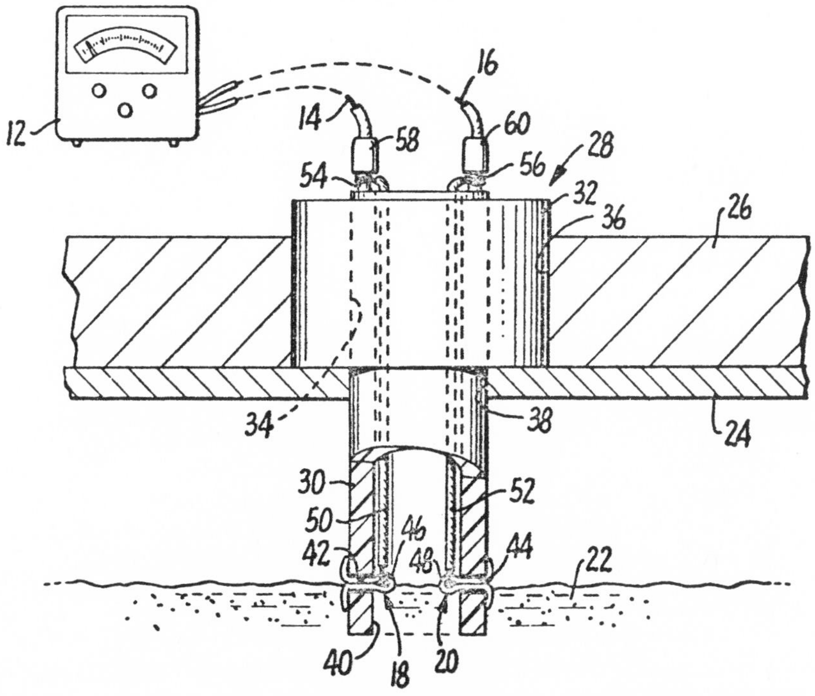

Source: U.S. Patent 3,986,977

Urea-Formaldehyde Polymer Matrix

T.S. Bustard and C.S. Pohl; U.S. Patent 4,230,597; October 28, 1980; assigned to Hittman Corporation describe a urea-formaldehyde based formulation which when solidified, forms a solid, substantially rigid matrix which effectively contains and entraps radioactive wastes. The solidified mass or matrix, which contains the waste, is physically and chemically stable over a wide range of conditions and can be stored or disposed of either above or below the ground.

The radioactive waste-binding composition comprises from between about 30 to 48%, by weight, urea-formaldehyde; from between about 25 to 45%, by weight methylated urea-formaldehyde; from between about 15 to 30%, by weight, urea and from between about 0.1 to 2.5%, by weight, plasticizer. A defoaming agent may also be incorporated in the composition to prevent foaming.

A preferred and particularly advantageous formulation comprises from between about 31.2 to 46.3% by weight, urea-formaldehyde; from between about 28.5 to 43.5% by weight, methylated urea-formaldehyde; from between about 19 to 28% by weight urea; from between about 0.4 to 2.4% by weight plasticizer and from between about 0 to 2% by weight defoamer.

Although the method steps employed for preparing the solidification agent are not necessarily limiting, both the urea-formaldehyde and methylated urea-formaldehyde should be first thoroughly mixed or blended prior to the addition of the solid urea. The apparatus or equipment used to form the desired homogeneous mixture of the various ingredients are those which are conventionally known and used in the chemical industry.

In accordance with known technology, the radioactive waste materials may be initially pretreated to adjust the solids content thereof, adjust its pH, or the like. Thereafter, the pretreated radioactive waste, in the form of a liquid or slurry, is mixed with the polymeric formulation.

The proportions of the waste to that of the polymeric formulation should be on the order of from between about 100 parts by weight waste to from about 40 to 120 preferably 60 to 100 parts by weight polymeric formulation. In this regard, the precise catalyst used may also affect the time required for solidification. In this regard, it has been found that particularly advantageous results are obtained if the acidic catalytic curing agent comprises a strong mineral acid catalyst selected from the group consisting of sulfuric acid and hydrochloric acid.

Example: As a control, 120 g of a simulated waste, i.e., water was introduced into a 1 ℓ disposable beaker. 100 ml of a polymeric formulation of the composition shown below was continuously introduced into the beaker with agitation until a substantially homogeneous mixture was formed. Thereafter 10 ml of 6 N H_2SO_4 was added to the beaker and mixed well, then the mixture was allowed to stand. In three minutes a solid dry, hard body having no free liquid was formed.

Polymeric Formulation	Weight Percent
Methylated urea-formaldehyde	36.586
Urea-formaldehyde	37.885
Urea	24.169
Defoamer	0.529
Plasticizer	0.831

The polymeric formulation was prepared by forming a mixture of the methylated urea-formaldehyde and urea-formaldehyde and thereafter adding urea which had been ground to a fine particle size. This mixture was stirred for 5 minutes. The plasticizer (Cyanamer P-35) and a defoamer (Bubble Breaker 748) were then added. The mixture was cured for several days with occasional stirring.

Alkali Salts of Gelatinized Starch-Polyacrylonitrile Graft Polymers

A process by *J.F. Pritchard and R.R. Komrow; U.S. Patent 4,235,737; Nov. 25, 1980* describes a process for treating radioactive liquids by adding to the liquids a highly absorbent starch-containing polymer composition to form a solid or semi-solid product or gel which renders any subsequent treatment, handling, transportation or disposal of the liquds safer and more efficient. The polymeric composition comprises water-insoluble alkali salts of aqueous alkali saponified gelatinized starch-polyacrylonitrile graft polymers. Only a small amount of the polymer is required to form the gel and the consistency of the gel can be modified by varying the amount and type of the polymer added to the liquid. The process is reversible and the polymer can be recovered and reused.

The polymer which is employed in the process is a starch-containing polymer composition which can absorb many times its own weight of water or an organic liquid. The polymer is water-insoluble and is prepared by saponifying a gelatinized starch-polyacrylonitrile graft polymer in an aqueous slurry with an alkali in amounts such that the molar ratio of alkali to the acrylonitrile repeating element of the polymer is from about 0.1:1 to 7:1 to form a solution or dispersion of water-soluble saponified graft polymer. The absorbent water-insoluble form of the polymer is prepared by simply drying the solution or dispersion obtained from the saponification step. The process for preparing such polymers is more specifically set forth in U.S. Patent 3,935,099.

The solution or dispersion may be dried by casting a film therefrom and drying by a known method such as oven drying. The resulting dry films are water-insoluble and absorb many times their weight in water forming clear, cohesive, self-supporting sheets. The dry films can, if desired, be ground or milled to flakes or powders which have greatly increased surface areas over the films, and consequently higher absorption rates. The water-insoluble character is retained by the flakes or powder.

The process comprises adding a small amount of the polymer to the radioactive liquids to form a solid, semi-solid or gel product which can easily and safely be handled and transported. The amount of polymer added to the liquid is not particularly critical, and only a small amount is required since the polymer can absorb many times its own weight in liquid. The preferred amount of polymer is typically 1 part by weight polymer to 2000 parts or less, preferably 300-400 parts, by weight of the radioactive liquids. Higher or lower amounts of polymer can be employed depending upon the results desired, with the higher amounts providing a more solid product.

An unexpected advantage of the process is the greatly reduced emission of radioactive nucleides, such as airborne, particulate or gaseous, from the gelled radioactive liquid.

The temperature is not particularly critical and it is preferred that the temperature be elevated to increase the rate of gel formation. The preferred temperature range is 0° to 95°C. It is not absolutely essential to heat the liquid upon the addition of the polymer, and the process can be practiced at room or ambient temperatures.

The pH of the liquid is not significant as long as it is not substantially acidic. With a strongly acidic solution, it is very difficult to obtain a gel product, and therefore an alkaline pH is preferred. Specifically, it is preferred that the pH be at least 5.5, although the process can be successfully operated at lower pHs.

It is also preferred, although not necessary, that agitation be provided in order to ensure that the polymer is uniformly dispersed in the liquid. The presence of agitation decreases the time required for forming the gel product.

The product resulting from the addition of the polymer to the liquid is a semi-solid or gel-type material, depending upon the amount of polymer added to the liquid. The product can be rendered almost solid by, i.e., evaporating the water or adding an inert filler such as cellulose, or like material, to the polymer prior to or after the addition of the polymer to the liquid. The amount or type of the filler is not critical, and the preferred amount is dependent on the selected filler. Any filler can be employed as long as it does not affect the property of the polymer to absorb the water contained in the liquid waste solution. Other exemplary fillers are sawdust, vermiculite, cement, etc.

The absorption rate of the polymer can also be varied by "cleaning" the polymer prior to use. The polymer may be cleaned by washing with alcohol. The absorption rate of the polymer can be increased by this procedure, and although not necessary, smaller amounts of the polymer can be used if its absorption rate is increased and hence the cleaning of the polymer may be preferred prior to use.

The nature of the polymer is such that no reaction occurs between the contents of the liquid waste solution and the polymer. The gel or semi-solid product, or solid product if desired, remains in that state for an indefinite period of time, and additional ingredients can be added such as agents which inhibit or prevent biological attack which might cause a reversal in the absorption reaction. The reaction is reversible, and the product can be converted back into the liquid state by any suitable procedure. A typical procedure is to acidify the product by the addition of any suitable acidic material, such as hydrochloric acid. The amount of the acidic material added is not critical, as long as the result is that the product is rendered acidic. It is preferred that the pH be lowered to below 3 in order to increase the rate of recovery. Once the product is acidified, and it is converted back into the liquid form, the water is removed and the polymer can be recovered for subsequent reuse by, for example, drying.

Since the polymer does not usually react with the solids contained in the radioactive liquids, and since the polymer is not adversely affected by the absorption operation, it can be recovered for subsequent reuse. The polymer used in the process is nontoxic and can therefore be used safely. If desired, as in U.S. Patent 3,935,099, if the liquid contains an organic liquid in substantial amount, an amine salt of the polymer can be formed by reacting the free acid form thereof with a suitable amine, which amine salt of the polymer will absorb the organic constituent in the liquid. The alkali salt can be used together with the amine salt if such a liquid contains both water and an organic liquid.

The apparatus used in the process can be any apparatus conventionally available in the chemical engineering art. Of course, special precautions may have to be taken concerning the apparatus, such as by lining the same with lead, glass or other suitable material for safety reasons.

Example: Five hundred milliliters of tap water and impurities (for example, derived from a typical soil sample) are mixed to form a simulated radioactive waste. The amount of soil added is about ¼ teaspoon. A small amount, about ½ teaspoon, of the polymer described above is added to the simulated radioactive waste with agitation in a blender. The liquid begins to set into a gel in about 30 seconds and is completely gelled in about 15 minutes. The particular polymer employed was H-Span hydrolized starch polyacrylonitrile.

As illustrated in the foregoing example, the time necessary for the gel to be formed is dependent to an extent upon the amount of polymer added. When the amount of polymer used is about 1 part by weight based on 300-400 parts by weight of the liquid radioactive wastes, it is expected that the time necessary to form a gel or like product is on the order of 2 minutes. The amount of time is not critical, and can be varied as desired by those of ordinary skill in the art.

Fixation of Ions in Ion Exchange Gels

B.W. Mercer, Jr. and W.L. Godfrey; U.S. Patent 4,156,658; May 29, 1979; assigned to U.S. Department of Energy describe a method for fixing radioactive ions in soils. Generally, the method comprises injecting a chemical grout into the soil which contains the ions, such as radioactive ions, whose mobility is desired to be limited and fixing the ions in place in the soil. The chemical grout contains water-soluble organic monomers which are polymerizable to gel structures with ion exchange sites. Water-soluble monomers are used, as they have proven very satisfactory, whereas water-insoluble monomers are not acceptable due to inconsistent setting characteristics and difficulty in polymerizing these materials in soil.

Particular types of ion exchange sites which have been found to be effective are carboxyl groups or carboxyl salts. An initiator and a catalyst for the polymerization of the monomer are injected into the soil to cause polymerization and the formation of the ion exchange gel in the soil. The soil and ions are physically fixed in place by the gel structure which surrounds the soil particles, thereby encapsulating the material and preventing mobility of the radioactive contamination through the open soil. In addition, the ion exchange properties of the gel chemically fix the ions onto the ion exchange sites, further preventing leaching and diffusion of the radioactive ions through the gel. Strontium and cesium ions are of particular concern both because the radioisotopes of these elements are very common in radioactive wastes and radioisotopes of these elements have extremely long half-lives which render them particularly hazardous. Strontium and cesium ions are readily exchanged onto and retained by the ion exchange gels.

Water-soluble organic monomers which have been found to be particularly useful in the process include acrylic acid, acrylates, methacrylic acid, and methacrylates. N,N'-methylene bisacrylamide has been found to be a particularly good agent for crosslinking with these monomers to form gel structures with ion exchange sites. N,N'-methylene bisacrylamide has proven particularly useful because it has a relatively high solubility in water. The high solubility of this material and the water-solubility of the other organic monomers is extremely important, as it has been found that water-soluble materials are particularly effective for in situ polymerization and fixation of soils. Dispersion of these water-soluble materials from the point of injection through the soil containing the contamination is far better then with water-insoluble materials; and it has also been found that the water-soluble organic monomers are far more readily polymerized in soil than the water-insoluble materials.

The effectiveness of the particular method for fixing ions and the effect of the ion exchange properties of the material in reducing the mobility of ions through the gel is illustrated by the following examples.

A chemical grout which forms an ion exchange gel was prepared by dissolving sufficient AM-9 (a product of American Cyanamid Co. containing acrylamide and N,N'-methylene bisacrylamide, and which, based on solubility tests, is believed to contain less than 10% N,N'-methylene bisacrylamide and mostly acrylamide), sodium acrylate and N,N'-methylene bisacrylamide to give a solution containing 9% AM-9, 1% acrylate as sodium acrylate, and 1.8% N,N'-methylene bisacrylamide. Radioactive strontium was added to give a concentration of 0.0007 microcurie of ^{85}Sr per liter of grout. The grout was then treated while mixing with 1.5% of catalyst, β-dimethylaminopropionitrile (DMAPN) followed by 1% of initiator, ammonium persulfate (AP), to begin the polymerization process. The acrylamide and sodium acrylate cross-links with the N,N'-methylene bisacrylamide to produce a stable gel. The sodium acrylate forms ion exchange sites (carboxyl group) on the gel structure. The solution was divided into 75 ml portions placed in polyethylene containers and allowed to gel.

A grout which produces a gel without ion exchange properties was prepared by repeating the above steps with the exception that 10% AM-9 was used and no sodium acrylate was added.

When gelation was complete, the gel surface was leached with a 50 ml aliquot of tap water to determine the ^{85}Sr leach rate as a function of time. After various time intervals, the supernatant liquid was decanted and replaced with a fresh 50 ml aliquot of tap water. The leach water was not agitated or stirred during contact with the gel except during addition and removal. Measured portions of the leach water were analyzed to determine the ^{85}Sr concentrations. The leach rate, which is a function of the diffusion rate, is computed from the following equation:

$$\text{leach rate, g/cm}^2\text{/day} = \frac{\text{activity leached/cm}^2\text{-day}}{\text{activity/g of gel mixture}}$$

The leach rate as shown above is an apparent dissolution rate of the gel which does not actually take place. Only soluble substances (i.e., ions) are removed from the gel. Appropriate corrections were included in the computation to account for the decay of ^{85}Sr which has a half-life of 65 days.

The ^{85}Sr leach rate for the ion exchange gel is compared with that of the gel without ion exchange properties in Figure 1.5a which is a graph of the leach rate in g/cm^2/day versus time in days. As is apparent from the graph, the strontium leach rate from the ion exchange gel was only 10% or less than that for the gel without ion exchange sites.

The cesium leach rate from soil-gel mixtures was also determined using soil based with cesium traced with ^{134}Cs. Chemical grout solutions were prepared as above

Figure 1.5: Graphs of Strontium and Cesium Leach Rates

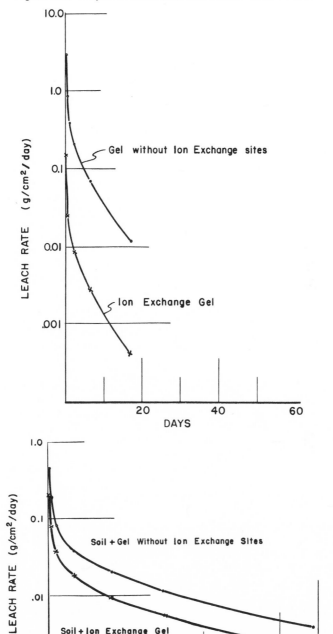

Source: U.S. Patent 4,156,658

except no ^{85}Sr tracer was added and the amount of AP was double to promote gelation in the presence of soil. 100 cm^3 of packed soil was mixed with 50 ml of catalyzed grout, separated into two portions, and placed in polyethylene containers to gel. The surfaces of the soil-gel mixtures were scraped to remove a very small excess of gel and were then leached with successive volumes (33 ml) of tap water over a two-month period. The results of these soil-gel leaching experiments are illustrated in Figure 1.5b, which is a graph of the leach rate in g/cm^2/day versus time in days.

It can be seen from this graph that the cesium leach rate from the ion exchange gel is only about half that from the gel without ion exchange properties. The ion exchange gels are expected to be more effective for reducing the leaching rate of divalent ions, such as strontium, than of univalent ions, such as cesium, because of the high attractive forces in the gel for the divalent ions where in contact with low ionic strength solutions, such as tap water.

Consequently, it can be seen, that, while present techniques exist for polymerizing material in soil to immobilize radioactive materials, the technique of using water-soluble monomers which are polymerizable to gel structures with ion exchange sites provides advantages of improved dispersion and polymerization in the soils and improved immobilizing ability by the chemical retention of radioactive ions over and above the physical fixation.

Addition of PVC to Phosphoric Acid Esters

G. Rudolph and W. Hild; U.S. Patent 4,148,745; April 10, 1979; assigned to Gesellschaft für Kernforschung mbH, Germany describe a method of preparing, for nonpulluting storage, phosphoric acid esters used in reprocessing spent nuclear fuels and/or blanket materials and to obtained products. In accordance with the process, phosphoric acid esters which have been separated from hydrocarbon and which contain radionuclides and decomposition products are contacted with a solidification matrix which consists of crushed polyvinyl chloride (PVC), in a weight ratio of phosphoric acid ester to PVC of 5 to 1 or less.

In a further aspect of the process, a method is provided which comprises adding spent ion exchanger contaminated with radionuclides to the liquid mixture of phosphoric acid ester and PVC by stirring to form a resulting liquid mixture having a minimum PVC content of about 14 weight percent and then permitting the resulting liquid mixture to cool until a solid mass is formed which consists of the phosphoric acid ester, the PVC, and the spent ion exchanger.

In the embodiment where spent ion exchange materials contaminated with radionuclides are added and form part of the final solid mass, the use of such ion exchange materials enables the amount of radioactive material in the solid mass to be increased up to 86%, comprised of the phosphoric acid ester and spent ion exchanger, and enables the amount of PVC used to be decreased down to 14%.

Example 1: 320 g of TBP (tributyl phosphate) are heated to 170°C and poured into 150 g of PVC chips which are at room temperature. After a short period of stirring, a homogeneous mixture is generated. This mixture solidifies, forming a jellylike mass, on cooling.

Example 2: 150 g of PVC chips and 700 g of TBP are heated to 110°C; 210 g of cation exchange resin, polystyrene resin containing sulfuric acid groups, (previously dried in a drying closet) are stirred into the resulting liquid mixture. After cooling, a soft mass in which the ion exchanger is embedded is obtained. This soft mass does not flow.

Wax and Polyolefin

M.W. Standart and W.E. Spargo; U.S. Patent 4,021,363; May 3, 1977; assigned to Aerojet-General Corporation describe the immobilization of toxic solids with a mixture of wax and a low molecular weight polyolefin. In a preferred embodiment, a mixture of between about 60-90% wax and 10-40% of either polyethylene or polypropylene is used to immobilize the low-level radioactive particulate discharged from water treatment in nuclear power plants and similar facilities. This mixture is strong, nontoxic, inexpensive, readily available, easy to handle, and resistant to breakdown, leaching, and combustion. It also mixes well with many different toxic materials. And it will hold up to about four times its weight of a toxic material having a density similar to that of the mixture and a greater weight of more dense materials.

Many different waxes and polyolefins can be used. It is preferred to use a wax having melting temperature between about 140°-170°F to immobilize the radioactive salts discharged by water treatment systems in nuclear facilities because those salts are often packaged in barrels and stored in the open at some remote location. Direct sunlight may produce temperatures around 120°F inside the barrels. A wax having a lower melting temperature might melt during storage. Those with higher melting temperatures are hard to melt and mix with the waste initially. The wax may be of either a natural such as a petroleum base, or vegetable base; or a synthetic base. Many examples of each of these waxes melt at appropriate temperatures, mix well with many different toxic materials, and are compatible with polyolefins. Petroleum base paraffin waxes are preferred because they are slightly better solvents for polyethylene and polypropylene than the other waxes.

As for the polyolefin, those with relatively low molecular weights and low melting temperatures around 250°F, are preferred because they are easiest to melt without charring. The polyethylenes and polypropylenes have these characteristics.

The wax-polyolefin mixture described can be used to immobilize many different radioactive and non-radioactive toxic materials. The toxic materials should be dry when added though, because water will steam or foam a liquefied mixture of wax and polyolefin, and thus prevent formation of a dense package containing substantial quantities of solid particles. The solids should also be preheated. A temperature between about 200°-350°F is preferred when they are to be added to a mixture of a wax having a melting temperature between 140°-170°F and a polyethylene or polypropylene having a melting temperature around 250°F. When the salts are cooler than this, they tend to solidify the surface of the immobilization material before any substantial quantities can be inserted. Particles at a higher temperature tend to char the polyolefin.

No special processing is required for use of this mixture to immobilize the low-level radioactive particulate discharged by many water treatment systems because those systems dry the salts and other particles and discharge them at temperatures between 200°-350°F. The material being immobilized must also be chemically compatible with the wax and polyolefin, but most solids that are either toxic or likely to become so, are compatible with these materials. The toxic solids can be of any size, but mechanical mixing may be needed for small dust-size particles having insufficient weight for gravity settling. If the material being immobilized is toxic by reason of radioactivity, it should not be added to a mixture of wax and either polyethylene or polypropylene for storage in a closed container if the radiation level is high, above 100 μCi/cc or so. When radiation exceeds this level, there are sufficient number and energy of radioactive decay particles for likelihood of release of hydrogen atoms from the hydrocarbon molecules. Hydrogen buildup could be an explosive hazard.

Conditioning of Solid Waste of Large Dimensions

On incorporating large solid radioactive waste into a resin polymerizable at ambient temperature a shrinkage phenomenon occurs during the polymerization of the resin which, in view of the large dimensions of the solid waste which it is desired to encase, causes the formation of cracks in the solid blocks obtained.

B. Morin and D. Thiéry; U.S. Patent 4,315,831; February 16, 1982; assigned to Commissariat a l'Energie Atomique, France describe the addition to the thermosetting resin of an inert filler which makes it possible to significantly reduce shrinkage occurring at the time of crosslinking and thus obviates any dislocation or cracking in the solid blocks obtained.

This inert filler can comprise a sand such as a silica sand or blast furnace residue called slag. Preferably silica sand with a continuous grain size between 0.1 and 1.2 mm is used.

According to an advantageous feature of the process, a plasticizer such as polystyrene or polyethylene is previously added to the thermosetting resin in addition to the inert filler. This plasticizer is added in proportions preferably ranging from 0.1 to 1 part of plasticizer for 1 part of resin. The prior addition of this plasticizer to the resin leads to a further improvement of this process due to its plastic deformation capacity. In fact it makes the resin more supple at the time of crosslinking and obviates subsequent cracking in the solid blocks obtained.

According to another advantageous feature of the process, an expanding agent, that is to say a mineral substance which has the property of expanding the resin during crosslinking is added to the thermosetting resin beforehand in addition to the inert filler. This expanding agent can comprise borax ($Na_2B_4O_7 \cdot 10H_2O$). If such an expanding agent is added in a sufficient quantity it eliminates any supplementary risk of shrinkage and therefore subsequent cracking. The borax can be added to the resin preferably in a proportion of 30 to 60% by weight relative to the resin.

It is also possible to add beforehand to the thermosetting resin a thixotropic agent which is able to prevent a possible settling of the inert filler in the resin prior to crosslinking. This thixotropic agent can comprise, for example, a silica gel, hydrogenated castor oil or a mixture of silica gel and asbestos fibers. The thixotropic agent is preferably used in a proportion of 0.5 to 4% by weight based on the resin. The thixotropic agent used thickens the solution which then has a viscosity such that there is sufficient time for crosslinking to take place without there being any settling.

The process is performed in the following manner. Firstly the encasing mixture is prepared which is constituted by the selected thermosetting resin to which has been added the inert filler and optionally the plasticizer or expanding agent and also the thixotropic agent as defined hereinbefore. The mixture is then placed in a mold containing the solid waste which it is desired to condition. Alternatively, the mixture is placed in a mold and the radioactive waste which it is desired to condition is then introduced into the same—the radioactive waste generally being in a basket which is introduced into the mold. This is followed by the crosslinking thereof.

The encasing operation can be performed either in air or under water. In the case where the process is performed under water, borax cannot be used as the expanding agent because it has the disadvantage of being water-soluble. In the case where the process is performed in air ventilation can be provided.

The mixing of the various components must be homogeneous in order to ensure good mechanical and physico-chemical qualities of the material. To this end it is possible to use a vertical turbine with four blades driven by a motor operating at 0 to 400 rpm.

There must be sufficient time to place the solid radioactive waste in the resin before it sets solid. Therefore the time at which the polymerization catalyst/accelerator mixture is introduced into the resin is forecasted and adjusted. The time necessary for the resin to set solid is also a function of the temperature at which the reaction is performed (for the same proportions of the reactive mixture, two or three minutes are required to obtain solid setting when the ambient temperature is 25°C, whereas more than 24 hours are required when the ambient temperature is below 16°C).

Example: About 60 ℓ of solid radioactive waste are to be dry-encased. The following are introduced into a cylindrical mold:

> 130 kg of a glycol-maleophthalate-based polyester resin dissolved in styrene, marketed as "NS 574" (CDF Chimie Co.);
>
> 150 kg of silica sand of grain size ranging between 0.1 and 1.2 mm;
>
> 70 kg of borax ($Na_2B_4O_7 \cdot 10H_2O$);
>
> 2.5 kg of silica gel;
>
> 1.95 kg or 1.5% by weight based on the resin of methyl-ethyl-ketone peroxide.

The complete mixture is mixed homogeneously for 20 minutes, followed by the addition of 120 g or 0.8% by weight based on the resin of cobalt naphthenate.

The basket containing 60 ℓ of solid radioactive waste is then introduced and crosslinking takes place.

A homogeneous block is obtained without cracks whose diameter is 60 cm, height 77 cm and volume 220 ℓ. Polymerization shrinkage is small and no cracks are encountered.

Prevention of Borate Precipitation

H.E. Filter and K. Roberson; U.S. Patent 4,253,985; March 3, 1981; assigned to The Dow Chemical Company describe a process for treating radioactive acidic, aqueous wastes containing high temperature-dependent, precipitatable borates (precipitation occurring at temperatures below about 120°-140°F) to render such borates soluble in the waste solution whenever the solution is cooled to around ambient temperatures, herein defined as meaning the range of from about 60°-90°F.

A primary aspect of the process relates to the addition of an alkali metal hydroxide or ammonium hydroxide (including nondetrimentally substituted ammonium hydroxides) to the circulation system by which such wastes are routed from the reactor to heat exchange means and thereafter to cleanup means such as an evaporator to be deionized and returned for reuse in cooling the reactor. An effective amount of such base or bases is added to render the temperature dependent, precipitatable borates of such waste solution soluble in the ambient temperature range.

Example: Preparatory to treatment on a larger scale, a laboratory analysis is performed on high borate content radioactive evaporator bottoms, using a 350 ml sample, of about 2.5 pH, and which upon cooling to ambient temperature forms a cloudy blue solution estimated to contain about 5-8 weight percent white precipitate. The solution is successfully treated with 2.5 ml of a 50 weight percent aqueous sodium hydroxide additive to completely solubilize the sample. 150 ml of the treated sample, having a pH of about 7.0, is subsequently solidified with the following ingredients: 100 ml of a fluid thermosettable vinyl ester resin which is made by reacting 32.6 parts of the diglycidyl ether of bisphenol A extended with 8.7 parts of bisphenol A then reacted with 1.2 parts maleic anhydride and 7.5 parts methacrylic acid, the resin dissolved in 50 parts styrene; 2.5 ml of 40% benzoyl peroxide emulsified in dibutyl phthalate; 0.2 ml of N,N-dimethyl-p-toluidine.

The vinyl ester resin and benzoyl peroxide are measured into a large paper cup and mixed thoroughly with an electric stirrer. The radioactive waste sample is slowly added to the blend with the stirrer at high speed to assure good emulsification. The dimethyl toluidine is added to the emulsion and mixed thoroughly for 30 to 60 seconds. The stirrer is removed and gelling is observed in 4.5 minutes. A homogeneous rock-hard solid is achieved in about 1 hour.

Microwave Processing

G. Sato; U.S. Patent 4,242,220; December 30, 1980 describes a method to process sludge into resin clad capsules so as to seal poisonous substances using a microwave technique wherein the capsules can be kept unchanged substantially eternally, and prevent poisonous substances from leaching or eluting when contacted with water, and allow the processed sludge to be bonded into a solid shape.

The following description refers to sludge as a typical waste.

Water Removal: Sludge is reduced by means of any conventional equipment to a water content of 65-70% by weight. Sludge squeezers, vacuum filters, centrifugal filters or the like may be used in this step.

Drying: The sludge is directly or indirectly heated to remove residual water in a dryer by burning fuel oil. During this heating, the sludge is stirred and vibrated so that organic material such as bark, straw, sawdust and other material liable to burn on heating may not be ignited or burnt. If these materials are burnt, the desired end product cannot be obtained. This drying will result in a residual water content of the sludge of about 12-13% by weight as far as burning is prevented. It is to be noted that a minimum water content of about 4% will eventually remain in the form of water of crystallization.

Microwave Drying: The sludge is then placed in a vessel equipped with 4-6 microwave oscillators having an output of 5 kW. The vessel is also equipped with a rotary feeder. Microwaves are applied to the sludge to heat the sludge through dielectric heating upon transmission to further reduce the residual water content thereof to 0.1-5% by weight. Stirring is also conducted throughout this heating. Reduction of the water content to such low levels weakens the coalescence between sludge particles and the sludge thus becomes powdery. This process is assisted by stirring. For microwave irradiation, a klystron may be used at a frequency of 915 or 2450 Mc/s, for example.

Cooling: The thus dried sludge powder is conveyed, for example, by a screw conveyor through which cold air is blown to cool the sludge to room temperature or about 30°C.

Mixing: The cooled sludge powder is introduced in a mixer to which a thermoplastic resin is also charged. Unless the sludge has been cooled at this stage, the thermoplastic resin will prematurely melt to agglomerate sludge powder into a mass before the subsequent microwave irradiation. In this case, encapsulation is only partly accomplished and heavy metal will leach out where encapsulation is imperfect.

Microwave Heating: The cooled mixture of sludge powder and thermoplastic resin is introduced into a vessel. With stirring, the mixture is irradiated with microwaves. The thermoplastic resin is melted to encapsulate the fine particles of sludge. Upon cooling, the resin sets to cover the sludge particles completely. Each sludge particle is enclosed or confined by a resin coating.

Molding: A suitable amount of the resin encapsulated sludge particles is molded into an article having a desired shape.

METALS

Metal Oxides in Metallic Matrix

T.C. Quinby; U.S. Patent 4,072,501; February 7, 1978; assigned to U.S. Department of Energy describes a powder mixture useful in a method for immobilizing radioactive waste metal values as oxides within a nonradioactive metallic or metal oxide matrix.

Finely divided powders are prepared according to this process by first reacting an aqueous solution containing dissolved metal values with excess urea. The urea reacts with the water present in the solution, including water of hydration of soluble species, according to the reaction

$$NH_2-CO-NH_2 + H_2O \rightleftharpoons 2NH_3 \uparrow + CO_2 \uparrow$$

The reaction of urea with water in the solution takes place at a temperature sufficient to melt the urea. Depending upon the metal salt in the solution, the urea/salt mixture melts at about 120°-132°C, whereupon urea reacts with the water forming volatile products, leaving the urea present in excess of the water content as a molten urea solution containing the metal values. The aqueous solution can be a solution of any soluble metal salt or mixture of metal salts. Preferred are metal salts such as halides, nitrites, or nitrates whose anions form volatile products upon subsequent heating with molten urea. The preferred method of carrying out the reaction is to add solid urea to a concentrated aqueous solution and heat the resulting mixture to a temperature sufficient to melt the urea. It is preferred that the initial aqueous solution be essentially saturated to minimize the amount of urea reacted.

After the reaction of water in the solution with urea is complete, as evidenced by a volume reduction representative of the original aqueous volume present (volume of water plus stoichiometric urea), the resulting molten urea solution is heated to cause metal values dissolved therein to precipitate. Metal values present in the molten urea solution precipitate homogeneously at about 180°C, forming a mixture of urea and the precipitated metal values. As the precipitation goes to completion, the temperature of the mixture rises, indicating an exothermic reaction. The exact chemical composition of the precipitate has not been determined. It is believed to be either a hydrated oxide or an amine compound. It is believed that a part of the urea in the molten urea/metal salt mixture is decomposed as it is heated above its melting point, leaving a liquid product. For purposes of this process, the term "molten urea" will be used to indicate the liquid phase present after the urea/water reaction and after the precipitation reaction, regardless of its actual chemical composition.

After the completion of the precipitation reaction the resulting mixture containing precipitated metal values is heated to evaporate volatile material from the mixture, leaving a dry powder containing the metal values. In actual practice, at the completion of the precipitation step, the product is virtually a solid mass of precipitate bound together with residual molten urea. The volatile products, e.g., NH_4NO_3, are evaporated by heating to about 400°C at atmospheric pressure.

The exact chemical composition of the dry powder has not been determined. In order to form metal from reducible metal precipitates, the dry powder is heated in the presence of a reducing gas such as hydrogen. In order to provide metal oxide particles, the dry powder can be calcined at above about 650°C. For purposes of this process, the term calcining is used in its general sense to mean heating the dry powder to a high temperature but below its fusing point to cause it to lose moisture or other volatile material and/or to cause it to be oxidized. The product of the calcination step is a fine oxide powder or mixed oxide powder. When halide salt solutions are used, it is believed that the calcination step should be performed in the presence of oxygen. When oxysalts are used such as nitrates, sulfates, nitrates, phosphates, etc., oxygen is probably not needed for the calcination step.

It is not necessary that the heating steps of this process be carried out in a stagewise manner. All that is required is that the water in aqueous solution react with urea prior to the precipitation step and that the precipitation be complete prior to evaporating to dryness. Accordingly, it is a simple matter to devise a continuous or varying heating rate to accomplish these objectives.

Any metal cation or mixture of metal cations can be converted to oxides according to this process. The particle size of the oxide product is a function of the amount of excess urea present (excess with respect to reaction with water). Since excess molten urea is the solvent for the metal values, the greater the amount of excess urea, the smaller will be the precipitated particles. It is a matter of routine experimentation to determine the oxide/urea ratio needed to provide a particular particle size. For example, oxide particles in the 0.2-0.5 micron range are prepared from molten urea solution containing the equivalent of 20 grams metal oxide per liter of molten urea. Oxide particles in the 10-12 micron range are prepared from molten urea solution containing the equivalent of 50 grams metal oxide per liter of molten urea. The very small particle sizes obtained by the subject process and the fact that all metal species precipitate simultaneously enable it to be used to provide homogeneous oxide powder mixture, having less than ±1 wt % variance in composition throughout.

Example: In this example a cermet article containing simulated fission product oxides in a nickel matrix is prepared. The fission products simulate the composition of fission products from Purex reprocessing of irradiated light water reactor fuel after a burnup of 30,000 MW days/ton. This cermet article is useful for the immobilization of nuclear reactor fission products for long term storage. The article contains 25 volume percent simulated fission product oxides and the remainder is nonradioactive nickel as a metallic matrix. The composition of the simulated fission products was as follows:

Fission Products	(g-mol)	Corrosion Products	(g-mol)
La	1.04×10^{-2}	Fe	3.15×10^{-3}
Ce	1.59×10^{-3}	Ni	1.2×10^{-3}
Cs	2.05×10^{-3}	Cr	7.4×10^{-4}
Sr	1.9×10^{-3}	Na	3.8×10^{-4}
Zr	3.58×10^{-3}		
Ru	2.03×10^{-3}		
U	4.1×10^{-3}		

Plus, Al_2O_3 (2.67×10^{-2} g/cc) and SiO_2 (7.02×10^{-2} g/cc) were added to form pollucite (cesium aluminum silicate) to stabilize cesium.

The simulated fission products and nickel were dissolved as nitrates in a minimum amount of 1 M HNO_3, about 180 ml. Sufficient solid urea was added to provide about 20 g of oxides per liter of urea. The solution was heated to 130°-140°C for 1 hour to remove the water of dissolution and hydration, after which the temperature was raised to about 180°C for 40-50 minutes to precipitate the metal values. All the metal values precipitated homogeneously. The molten urea mixture was heated to 400°C for sufficient time to evaporate the urea leaving a dry powder. This dried powder was heated to 800°C in air for calcining, forming a mixture of nickel oxide and simulated fission and corrosion product oxides. The dried powder was heated at 850°C for 1 hour in hydrogen atmosphere, whereupon the nickel oxide was reduced to metal powder. The metal values from simulated fission products were still present as oxides having typical particle size of 0.2 micron. The mixture was consolidated by hot pressing at 1100°C and 4000 psi for 1 hour to form a cermet waste immobilization article containing fission products within a continuous nickel matrix. Scanning electron microscopy revealed that each oxide particle was completely surrounded by the metal matrix. For waste immobilization articles, the matrix metal can be 50 to 70 volume % of the article.

If desired, a ceramic waste immobilization article can be prepared by eliminating the reduction step and hot pressing the oxide powder mixture. Cermet waste immobilization articles are substantially more ductile and thermally conductive than ceramic waste articles. The process is particularly advantageous for preparing radio-active powders and for other applications requiring remote handling, since the solids are separated from solution by evaporation rather than filtration. Filtration processes are generally to be avoided in handling hazardous substances, e.g., nuclear fuel reprocessing operations, in order to preclude the formation of aqueous effluents.

Pressure Treatment of Waste-Metal Mixture

J.R. May; U.S. Patent 4,280,921; July 28, 1981; assigned to Newport News Industrial Corporation describes a method for immobilizing, solidifying, and reducing the volume of powdered solid form waste material. The method includes blending waste material with powdered metal. The waste material is substantially anhydrous powdered solid material. The mixture of the waste material and powdered metal is compacted using a pressure of at least about 10 tons per square inch for sufficient time to provide a reduced volume, strong, solid product. The amount of the powdered metal employed is at least sufficient to solidify the waste material when the mixture is subjected to the above pressures.

The solidified product has a compressive strength in excess of 800 psi. The compacted solid product of the process includes: substantially anhydrous solid waste material, at least about 1.5% by weight of powdered metal, and optionally small amounts of lubricant which may be added for the purpose of reducing wear of pressing machine parts.

The waste material to be treated according to the process can be waste material or by-product from any source provided it is a substantially anhydrous powdered solid material. The powdered waste material is usually substantially of nonmetallic materials, but can include minor amounts (e.g., up to about 20%) of metals. The waste material to be treated in accordance with the process is usually in the form of particles having diameters of about 1 micron to about $\frac{1}{16}$ inch.

Examples of some particular metals suitable for the process include powdered iron, powdered nickel, powdered bronze alloys, powdered aluminum, and powdered steels such as powdered stainless steel. Powdered iron is preferred because of its relative availability and cost. The powdered metals employed generally have diameters of about 25 microns to about $\frac{1}{32}$ inch. Typical of the preferred compositions is Anchor MH-100, (Hoeganaes Corp), which has particle size distribution as follows:

U.S. Standard Mesh Sieve Analysis*	Distribution, %
+ 80 (177 microns)	1
– 80 + 100 (149 microns)	4
–100 + 140 (105 microns)	20
–140 + 200 (74 microns)	27
–200 + 325 (44 microns)	24
–325	24

*Microns also given

The preferred amounts of metal employed are from 1.5 to about 3% by weight and most preferred about 2 to about 3% by weight based upon the weight of the solid waste material.

The mechanical properties of the final product are not solely dependent upon the properties of the waste material but can be adjusted by the amount of powdered metal to be added thereto. Since no chemical process is relied upon for the initial immobilization of the waste material, end product properties obtained are not inherently totally dependent upon the variables of the waste material. Also, since no chemical reaction occurs between the waste material and the powdered metal, the leach resistance of the final product is not dependent upon the bonding agent used but is dependent upon either the container in which the material is placed and/or a particular property of any water resistant sealant which can be employed subsequent to the formation of the monolithic structure. Accordingly, the properties with respect to leach resistance can be tailored to particular specifications according to requirements of a specific user.

In addition, since the process results in a reduction in the volume of material to be finally discarded, the cost and environmental impacts are significantly reduced.

Some advantages of the process include the fact that the process results in volume reduction of the total amount of waste material to be immobilized whereas the methods employed require additional volume. In addition, since the processing equipment required according to this process is readily obtainable, such can be easily installed at a plant site and accordingly waste from such hazardous plants such as nuclear power plants, hazardous chemical process plants, and wood pulp plants would not necessarily require transportation to a dump site.

Encapsulation of Granular Material in Metal

D. Thiele; U.S. Patent 4,204,975; May 27, 1980; assigned to Kernforschungsanlage Jülich GmbH, Germany describe a process to assure the complete embedding of contaminated material within the metal used for encapsulation even when fine-grained radioactive or otherwise contaminated material of a carbide nature is to be encapsulated. It is a further object that the process and the apparatus for the purpose should be economical and capable of being simply performed and used.

Briefly, the metal in molten condition is sucked into a loose mass or accumulation of the material to be encapsulated and is then brought to solidification by cooling. It has been found particularly effective to carry out the process by putting the filling of the material to be encapsulated within a tube extending in both directions beyond the material filling a section thereof, in such a way that one of the free surfaces of the mass of material in the tube is limited by a partition through which molten metal may pass and the other free surface is either in contact with the metal or is at a small spacing from the surface of a body of the metal. The metal in the tube is heated above its melting point, and sucked into the mass of contaminated material by vacuum as soon as it melts.

It is particularly effective to utilize aluminum as the encapsulation metal for carrying out the process. In this case, it is necessary to carry on the process by first replacing the air with an inert gas.

Instead of utilizing aluminum, it is also possible to use other metals, preferably metals of low-melting point, such as lead, tin, zinc, copper or in special cases, also silver or gold.

Apparatus suitable for carrying out the method consists of a section of tube or pipe that can be closed off at its upper end by a cap with an air removal tube, the cap being removable for inserting the filling of contaminated material, as well as a solid block of the encapsulation metal. Near the lower end of the tube there is as a support for the filling, a grate for the material to be encapsulated, below which the interior of the pipe is connected through a connecting tube with a vacuum pump. Above the grate, as further support for the filling, there is provided a layer of a granular or lump-form material that does not combine with or otherwise react with the metal or the material to be encapsulated, forming a layer which allows the liquid metal to pass through but which is impervious to the material to be encapsulated. When aluminum is utilized as the encapsulating metal, it is desirable to provide the connecting tube below the grate to be equipped with means for supplying an inert gas to the encapsulating tube.

In cases when it is necessary to take account of the possibility of the granulate to be encapsulated floating to the surface in the operation of the process, it is desirable to provide a sieve within the tube section covering off the material to be encapsulated against the block of the encapsulation metal charged in the tube, the sieve allowing the liquid metal to pass through while preventing the contaminated material from floating to the surface of the liquid metal. In carrying out the process, in the apparatus provided for the purpose, it is observed in every case that the particles of the material to be encapsulated are fully encased in the metal and are caused to be embedded in a pore-free manner in the metal matrix. Figure 1.6 is a side view, mostly in cross-sction, of an apparatus for use in the process.

Figure 1.6: Apparatus for Embedding Radioactive Waste in Metal

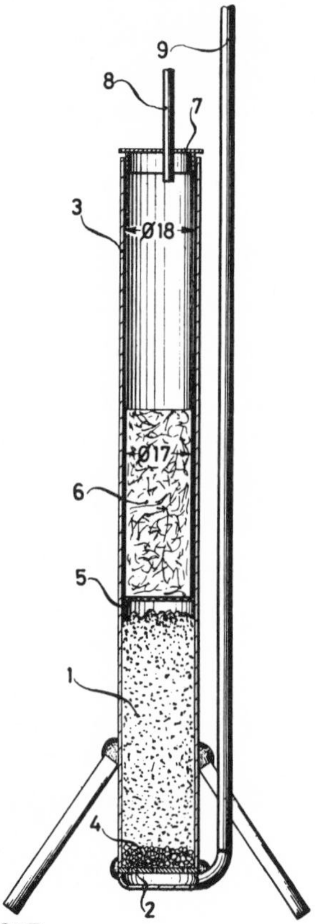

Source: U.S. Patent 4,204,975

MINERALS

Immobilization of HLW Calcine in Crystal Lattice

A.E. Ringwood; U.S. Patent 4,274,976; June 23, 1981; assigned to The Australian National University, Australia describes a process for immobilizing high level radioactive waste (HLW) calcine which comprises the steps of:

> (1) mixing the HLW calcine with a mixture of oxides, the oxides in the mixture and the relative proportions thereof being selected so as to form a mixture which, when heated and then cooled, crystallizes to produce a mineral assemblage containing well-formed crystals capable of providing lattice sites in which elements of the HLW are securely bound, the crystals belonging to, or possessing crystal structures closely related to crystals belonging to mineral classes which are resistant to leaching and alteration in appropriate geologic environments, and including crystals belonging to the titanate classes of minerals; and

> (2) heating and then cooling the mixture so as to cause crystallization of the mixture to a mineral assemblage having the elements of the HLW incorporated as solid solutions within the crystals thereof.

Preferably, a minor proportion of the HLW calcine is used in the mixture, for example, less than 30% by weight, more preferably 5-20% by weight.

According to a first exemplary method, the oxides and relative proportions thereof in the mixture of oxides are selected to form a mixture which can be melted at temperatures of less than 1350°C. In order to obtain such melting temperatures, these mixtures will generally be selected to form mineral assemblages including both silicate and titanate minerals. The mixture is melted with a minor proportion of the HLW calcine and allowed to cool. During cooling, the melt crystallizes to form the desired mineral assemblage and the HLW elements enter the minerals of this assemblage to form dilute solid solutions.

According to a second exemplary method, the oxides and the relative proportions thereof in the mixture of oxides are selected so that the mixture may be heated at a temperature in the range 1000°-1500°C without extensive melting of the mixture. While such mixtures may be selected to form assemblages including both silicate and titanate minerals, generally the mixture will be selected to exclude the formation of silicate minerals in the assemblage. Heat treatment of the mixture with a minor proportion of HLW calcine to a temperature in the above range without excessive melting causes extensive recrystallization and sintering, mainly in the solid state, and yields a fine grained form of the mineral assemblage in which the HLW elements are incorporated to form dilute solid solutions. The products of each of these methods, containing immobilized HLW elements, can then be safely buried.

The mixture of oxides comprises at least four members selected from the group consisting of CaO, TiO_2, ZrO_2, K_2O, BaO, Na_2O, Al_2O_3, SiO_2 and SrO, one of the members being TiO_2 and at least one of the members being selected from the subgroup consisting of BaO, CaO and SrO.

Preferably, the mixture comprises at least 5 members selected from the group, one of the members being TiO_2, at least one of the members being selected from the sub-group consisting of BaO, CaO and SrO, and at least one of the members being selected from the sub-group consisting of ZrO_2, SiO_2 and Al_2O_3.

If desired, in mixtures containing Al_2O_3, this component may be replaced partly or completely by the oxides of Fe, Ni, Co or Cr.

Preferably, the oxides and the proportions thereof are selected so as to form a mixture which, on heating and cooling, will crystallize to form a mineral assemblage containing crystals belonging to, or possessing crystal structures closely related to, at least three of the mineral classes selected from perovskite ($CaTiO_3$), zirconolite ($CaZrTi_2O_7$), a hollandite-type mineral ($BaAl_2Ti_6O_{16}$), barium felspar ($BaAl_2Si_2O_8$), leucite ($KAlSi_2O_6$), kalsilite ($KAlSiO_4$), and nepheline ($NaAlSiO_4$).

According to a preferred method of carrying out the process, the heat treatment (either melting and crystallizing, or recrystallizing in the solid state) is carried out under mildly reducing conditions, for example at an oxygen fugacity in the neighborhood of the nickel-nickel oxide buffer. This may be achieved by adding a small amount of a melt such as nickel to the mixture, or by carrying out the heat treatment under a reducing atmosphere, for example in a gaseous atmosphere containing no free oxygen and a small amount of a reducing gas such as hydrogen and/or carbon monoxide. As a result of the heat treatment under these conditions, molybdenum and technetium in the HLW are reduced to the tetravalent species Mo^{4+} and Tc^{4+} whereby they readily replace titanium Ti^{4+} in the hollandite, perovskite and zirconolite phases, thereby becoming insoluble and immobilized. Moreover, the volatility of ruthenium is minimized by heating under reducing conditions, while cesium enters the hollandite and/or leucite and kalsilite phases.

The table below sets out specific compositions according to two preferred embodiments of the process. The compositions of two alternative crystalline ceramic materials for HLW immobilization as disclosed in the prior art are given in Columns C and D for comparison.

	A	B	C*	D**
Mineral Structure	"Hollandite" Perovskite Zirconolite Ba—felspar Kalsilite Leucite	"Hollandite" Perovskite Zirconolite	Scheelite Cubic ZrO_2 Spinel Apatite Corundum Pollucite	Rutile Cubic ZrO_2 Metal $Gd_2Ti_2O_7$ Amorphous SiO_2 Pollucite
Radwaste (wt. %)	10	10	50	25
inert Additives (wt. %)	90	90	50	75
wt. %				
SiO_2	13	—	68	minor
TiO_2	33	60.4	—	~90
ZrO_2	10	9.9	—	—
Al_2O_3	16	11.0	11	minor
CaO	6	13.9	19	—
BaO	17	4.2	—	—
SrO	—	—	2	—
NiO	—	0.6	—	—
Na_2O	—	—	—	minor
K_2O	5	—	—	—

*Column C refers to "Supercalcine" [McCarthy, G.J. (1977) *Nuclear Techn.*, 32, 92].

**Column D refers to a ceramic developed by Sandia Corp. [Schwoebel and Johnstone (1977). ERDA Conf. 770102,101].

Immobilization of High Alumina- or Iron-Containing Sludge

A.E. Ringwood; U.S. Patent 4,329,248; May 11, 1982; assigned to The Australian National University, Australia describes a process for immobilizing high level waste (HLW) sludge containing high concentrations of Al, Fe, Mn, Ni and Na compounds which comprises the steps of (1) mixing the sludge with a mixture of oxides, the oxides in the mixture and the relative proportions thereof being selected so as to form a mixture which when heated at temperatures between 800° and 1400°C crystallizes to produce a mineral assemblage containing (i) crystals capable of providing lattice sites in which the fission product and actinide elements of the HLW sludge are securely bound, and (ii) crystals of at least one inert phase containing excess aluminum, iron, manganese, nickel and sodium, the crystals belonging to or possessing crystal structures closely related to crystals belonging to mineral classes which are resistant to leaching and alteration in the appropriate geologic environments, and (2) heating and then cooling the mixture under controlled redox conditions so as to cause crystallization of the mixture to a mineral assemblage having the fission product and actinide elements of the HLW sludge incorporated as solid solutions within the crystals thereof, and the excess aluminum, iron, manganese, nickel and sodium crystallized in at least one inert phase.

Preferably, the mixture of oxides which is added to the sludge to produce the desired mineral assemblage is comprised of at least three members selected from the group TiO_2, ZrO_2, SiO_2, Al_2O_3, CaO, at least two of the members being selected from the subgroup consisting of TiO_2, ZrO_2, and SiO_2.

Example: (a) A "high-alumina" sludge characterized by a mixture of fission products and actinide elements with excess oxides of Al, Fe, Mn, Ni, U and Na, possessing the following composition:

	Weight Percent
UO_2	3.5
Al_2O_3	50.3
Fe_2O_3	26.4
MnO	7.9
NiO	0.9
CaO	3.1
Na_2O	5.0
Fission products* plus actinides	~3.0

*Uranium has not been included with the remaining actinides. It is more appropriately classed with the inert components because of its very long half-life and correspondingly low alpha-activity.

is mixed with about 30% of TiO_2, ZrO_2, CaO and SiO_2, in proportions chosen so that when the mixture is heated, the added oxides combine with the sludge components to form a mineral assemblage consisting principally of hercynite-rich spinel + perovskite + zirconolite + nepheline. The heat treatment is carried out under controlled redox conditions such that most of the iron and nearly all manganese and nickel is maintained in the divalent state. The mixture is heated at a temperature of 1200°C for several hours and simultaneously subjected to a confining pressure using the conventional technique known as hot-pressing.

Alternatively, the mixture may be formed and sintered at 1200°C under the appropriate redox conditions without the application of pressure. The resulting product is found to be a fine grained, mechanically strong rock composed of the above minerals in which the HLW fission products and actinides are effectively immobilized. Actual compositions of the minerals in a rock produced in this manner are given below.

Compositions of Coexisting Mineral Phases in High-Alumina Sludge

	Nepheline	Perovskite	Zirconolite	Hercynite
SiO_2	41.5	–	–	–
TiO_2	0.2	53.4	29.5	5.8
ZrO_2	–	0.7	37.8	0.3
UO_2	–	0.2	13.9	–
Al_2O_3	35.9	2.4	1.1	48.2
Fe_2O_3	–	–	–	–
FeO	0.8	2.7	4.1	37.4
MnO	0.2	1.7	0.9	7.2
NiO	–	–	–	0.4
CaO	–	39.6	12.3	–
Na_2O	21.5	0.3	0.4	–
Sum	100.1	101.0	100.0	99.4

(b) In a modification of Example (a) above, the sludge is mixed with about 20-30% of the same oxides in proportions chosen to form a hercynite-rich spinel + zirconolite + nepheline mineral assemblage, and the mixture treated as above. A product physically similar to that of Example (a) is obtained with the fission products and actinides immobilized in the zirconolite phase.

(c) A high-alumina sludge as described in Example (a) is pretreated by washing to reduce the sodium content, mixed with about 20-30% of TiO_2, ZrO_2 and CaO in proportions chosen to form a hercynite-rich spinel + perovskite + zirconolite mineral assemblage and the mixture treated as above. A product physically similar to that of Example (a) is obtained.

(d) In a modification of Example (c) above, the sludge is mixed with about 20-30% of the same oxides in proportions chosen to form a hercynite-rich spinel + zirconolite mineral assemblage and the mixture treated as above. A product physically similar to that of Example (c) is obtained.

Backfill Composition

G.W. Beall and B.M. Allard; U.S. Patent 4,321,158; March 23, 1982; assigned to U.S. Department of Energy describe a combination of naturally-occurring minerals having sorptive capacity sufficient to retain hazardous elements present in nuclear wastes. The mixture contains phosphate mineral, sulfate mineral, ferrous mineral and clay mineral, in addition to quartz and montmorillonite, which provide thermal and structural stability and resistance to water transport, respectively.

Preferred backfill compositions are those comprising 55-65% by weight of quartz, 15-25% by weight of montmorillonite, 3-7% by weight of apatite, 3-7% by weight of olivine, garnet, magnetite or pyrite, 3-7% by weight of anhydrite and 3-7% by weight of attapulgite. The compositions are well suited for use in bedrock, salt beds, and the like. Most preferably, the ferrous mineral in these compositions is magnetite or olivine.

Nuclear wastes, which may be vitrified, are sealed in a conventional manner in canisters. A typical canister construction, consists of a thin chromium-nickel steel casing, a thick intermediate layer of lead and a thin outer layer of titanium. The canisters are placed in holes or tunnels in the earth, such as bedrock or salt beds, to the designed capacity of the repository. The repository can be kept open for inspection to assure that drainage and ventilation systems are functioning properly.

Prior to final abandonment of the repository, the backfill composition is used to fill holes around the canisters and vertical shafts. Conventional earth-moving and compacting equipment is used for filling the lower portions of the excavation; the higher portions can be filled by spraying.

Expansion of the backfill composition as the clay components swell owing to absorption of water ensures that the shafts, tunnels and holes will be completely filled with a material which is at least as impervious to harmful nuclides as the surrounding rock. In addition, the swollen backfill material prevents or reduces the effects of minor rock movement on the canisters.

STORAGE CONTAINERS

PROCESSES AND APPARATUS

Mixing System Within Storage Canisters

Radioactive waste is typically disposed of by packaging it in heavy-duty canisters. In order to maximize the security of such containers, they are made of shielding, often lead-filled material. Furthermore, the radioactive waste itself is normally mixed with a binder, normally a bitumen, so that even if a canister is broken open the radioactive waste contained thereby will not be able to escape or run off.

Normally the radioactive material is poured into the containers or introduced in fluid form. Thereafter, the binder is injected into the partially filled canister and a simple mixer head is inserted into the canister and rotated to mix the binder and waste together.

H. Baatz and D. Rittscher; U.S. Patent 4,235,739; November 25, 1980; assigned to Steag Kernenergie GmbH, Germany describe an arrangement where the canister, which is normally shaped as a body of a revolution, e.g., a cylinder, is held in a holder that can be tipped about a horizontal axis for orientation of the canister with its central axis horizontal and its fill hole opening horizontally. Furthermore, means is provided for rotating the canister about its central horizontal axis when thus tipped so that the waste and binder can be intimately mixed.

The waste is introduced through the fill hole into the container via a swivel coupling during rotation of the container about its horizontally oriented canister axis. What is more, a suction line can be connected to the interior of the canister at this fill hole via the coupling for placing the interior of the canister under a predetermined subatmospheric pressure so that the fluent radioactive waste can be sucked by the pressure differential into the canister for filling thereof.

Thus, with this system it is possible to first charge the binder into the canister while it is in the upright position. This can most simply be done by loading into the canister premeasured bags of cement, the bags being formed of polyvinyl alcohol, so that once the radioactive waste, which contains water or has water as a vehicle, enters the canister the bag will dissolve. Thereafter the canister is tipped on its side and connected to the swivel coupling through which the fluent radioactive waste is fed into the canister. In addition a vacuum line can be connected through this swivel coupling to the interior of the canister so that the interior of the canister can be placed under subatmospheric pressure. When the valve is open in the line to the fluent radioactive waste this subatmospheric pressure will suck a predetermined quantity of the waste into the canister. Simultaneous rotation of the canister during all of these operations will insure extremely intimate mixing of the binder with the radioactive waste. The amount of cement can easily be calculated in accordance with the known quantity of water to be admitted with the radioactive waste.

It is possible to operate in accordance with a system wherein the radioactive waste from a reactor tank can be fed into the canister through a pressurizing pump and wherein a vacuum pump can be connected to the canister. Furthermore the canister can be connected via a low-pressure or vacuum line back to the reactor tank. The system is set up so that the particulate radioactive material is carried as a suspension into the canister and allowed to sediment therein, the excess water being fed back into the reactor tank. In such an arrangement the granular cement is only added at the last stages.

In accordance with yet another feature of this process, mixing formations are provided inside the canister, normally in the form of inwardly directed vanes. As the canister is rotated these mixing formations insure excellent homogenization of the binder-waste mixture.

With this system radio active waste can be packaged in an extremely neat and simple manner. Evacuation of the interior of the canister before and during loading produces an almost perfectly porefree concrete formed of a mixture of cement and the radioactive waste, the latter normally being simple salt solutions. As a result of the intimate mixing a solid block is formed inside the canister so that even if the canister breaks open, the waste contained therein will not be messy and difficult to handle.

Remote Control Mixing and Drum-Handling Apparatus

Prior systems for putting radioactive materials into a drum or other container in general require that operators and maintenance personnel be exposed to radiation, even though such system may be intended to protect personnel. For example, the operators in many cases must go into areas containing radiation to open drums or close them or to insert nozzles in the drums or to handle the drums in storage. In some systems an operator may stand behind a shield wall, but must extend his arms into a radioactive zone, and expose his head to see, to connect pipes for feeding radioactive material. If spills occur, the operator must go into the radioactive zone to clean up spills. In prior operations where drums are stacked in multiple layers in decay storage areas, the operator must often go into such areas to place planks between the layers. Maintenance men must go into radioactive work areas to work on equipment requiring maintenance at intervals, such as conveying equipment, motors and switches. The total amount of radiation to which personnel can be safely exposed is limited by physiological reasons; therefore, personnel must be controlled as to their duties, and the radiation amounts to which they are exposed frequently checked to avoid their exposure to an excessive amount of radiation that can adversely affect health.

Moreover, in operation of the nuclear plant, if an emergency should arise correction of which would require exposure of operators or maintenance men to radiation during a time when all available men had reached their limits of radiation tolerance, a shut-down of the plant might be necessary or other adverse consequences might result because of lack of operators or maintenance men having safe radiation tolerances.

Moreover, prior systems do not in general provide desired close control to insure that proper amounts of radioactive material, cement, or water are put into the drum to insure proper solidification of drum contents. It is imperative to avoid improper loading of the drum or mixing of drum contents.

Moreover, prior systems can on occasion spill radioactive materials on the outside of the drums or on the floor. If the spill is on the drum, it is necessary to decontaminate the drum prior to shipment. If the spill is on the floor, then a certain amount of dust can be generated as the material dries. Such dust, which is radioactive, could find its way through the plant and thus make the plant unsafe because of radioactivity. Spilled materials also can collect in floor drains and clog them. Prior systems for putting radioactive material into drums in general have loaded drums in an open space, so there was no way of containing or taking care of the problems caused by spills of radioactive material.

Some previous systems have numerous operating mechanical parts requiring periodic maintenance such as motors and electrical switches in radioactive areas. Maintenance of such equipment can expose personnel to considerable radiation.

Previous equipment loads drums containing radioactive material onto trucks or casks in a haphazard fashion and thus not loading the truck or cask to full capacity would lose lading and could cause damage to the drums or drum enclosure.

Previous systems, because of loss of electrical power or air pressure or improper handling of the drum handling means, could topple a drum or cause irregularities in operation which could cause spillage of radioactive material. Previous systems did not provide for an accurate weight of resin to a weight of cement ratio in order to insure that the drum contents would be properly solidified with the most economical use of cement and with the lowest transportation costs. In prior systems that filter the resin in the drums to remove water, costs are understandably higher for the drum because of the added equipment contained therein.

Some prior systems mix radioactive resins, cement and water in a mixture outside of a drum. This involves exposure of considerable amounts of equipment to radioactivity and possibilities of considerable exposure of personnel to radioactivity. Moreover, the mixer must be cleaned after each use, which is difficult because the cement sticks to the mixer; moreover, the mixer will become radioactive and hence unsafe in time considerably shorter than the life of the plant, necessitating replacement expense. Some prior systems mix these materials in the drum; but if an open-top drum is used, considerable spillage occurs during mixing, and if the drum is rolled about its lengthwise axis to mix its contents, a core of poorly mixed materials is formed in the center of the drum.

Most, if not all, prior systems lack fail-safe features to prevent unsafe conditions in the event of failure of operations of any portion of the equipment.

A.J. Stock, D.E. Christofer and J.E. Brinza; U.S. Patent 3,940,620; February 24, 1976; assigned to Stock Equipment Company describe apparatus and methods for packaging fluent material such as dangerous or radioactive liquids, or slurries containing radioactive or otherwise dangerous particulate material without human handling or the necessity of personnel connecting pipes or the like, comprising means controllable by remote control for moving a container into any of a variety of preselected positions; means controllable by remote control for introducing into a container in a preselected loading position an essentially predetermined amount of such fluent material in which preferably the proportion of liquid to radioactive or other dangerous material is accurately predetermined; means controllable by remote control for closing the container; means controllable by remote control for agitating the container to mix thoroughly the contents thereof; and means to receive the closed container containing its mixed contents, such means being at a location away from the location at which the liquid is introduced into the container and the location in which the container is agitated. The last named means may be a storage location, or a location at which the container is loaded onto a vehicle for transportation, or both.

The apparatus may comprise a building including walls resistant to passage of radioactivity, which walls segregate from locations occupiable by personnel the location at which the container has radioactive material introduced and at which the container is agitated to mix its contents, and if desired a storage location.

The apparatus may comprise means for forming in a tank a dispersion of an essentially predetermined proportion of particulate material and liquid, then passing from the tank into a container an essentially predetermined amount of the dispersion. The apparatus preferably comprises remotely controllable means for closing the container; and remotely controllable means for agitating the container to mix its contents. The apparatus may comprise remotely controllable means for opening the container prior to introduction of fluent material.

The apparatus preferably comprises means for weighing the container before it is filled or after it has been closed, or both, to check on the amount of material introduced.

Introduction of Two Materials of Different Radioactivities into Containers

A.J. Stock and P.C. Williams; U.S. Patent 4,299,722; November 10, 1981; assigned to Stock Equipment Company describe a method for loading a container such as a steel drum with a quantity of one or more dry particulate solidifying agents and a fluent material such as dangerous or radioactive liquids or slurries containing radioactive or otherwise dangerous particulate materials that are mixable with the solidifying agent. The method ultimately provides a mass that is ultimately at least partially and preferably completely solidified within and substantially fills the container. The method includes the steps of introducing into the container a predetermined amount of the dry particulate solidifying agent that normally contains a substantial volume of air entrained between particles, and a quantity of the fluent material sufficient to at least partially and preferably completely fill the container, and then closing the container. The solidifying agent and the fluent material are then agitated in the container to cause the particles to mix with the fluent material and the liquid of the fluent material, thus removing the entrapped air from the solidifying agent and forming a first mixture of a volume considerably less than the initial volume of the contents of the container. The container is then opened and a second quantity of fluent material is introduced into the container that is sufficient substantially to fill the container and provide a predetermined proportion of total fluent material to solidifying agent, after which the container is closed. The first mixture and the second predetermined quantity of fluent material are then agitated in the container to provide another mixture which subsequently becomes at least partially, and preferably completely, solidified with no free liquid, in the container, and that fills the container to a considerably greater degree than if the second filling and agitating operation had not been carried out. More than two such operations of introducing fluent material can be carried out if desired.

In accordance with another aspect of the process, a first fluent material of a certain measured level of radioactivity and a second fluent material of another measured level of radioactivity different from that of the first fluent material are introduced into the container in amounts calculated to provide an advantageous intermediate level of radioactivity to achieve certain economies in transportation and storage of the container. Advantageously these fluent materials of different radioactivities are introduced in two steps, as described above.

Figure 2.1 is a flow diagram illustrating the process.

Figure 2.1: Flow Diagram of Process to Introduce Two Different Fluent Materials Into Containers

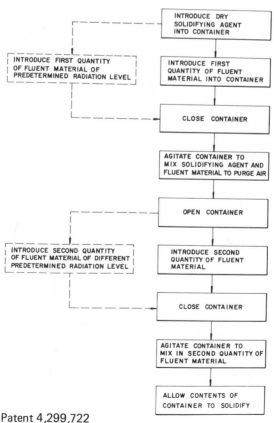

Source: U.S. Patent 4,299,722

Device for Transferring and Packaging Contaminants Within Lead-Tight Sheath

R. Goutard and R. Levardon; U.S. Patent 4,281,691; August 4, 1981; assigned to Commissariat a l'Energie Atomique, France describe a device for transferring and packaging of contaminants and especially radioactive products within a leak-tight sheath in such a manner as to ensure that the sequence of operations involved in placing the products within the sheath does not result in any contamination of the environment.

The device transfers a noxious product through a lock-chamber which is rigidly fixed to a leak-tight partition-wall and packages the product within a leak-tight sheath. The device essentially comprises a sleeve having a central cavity for establishing a communication between the lock-chamber and the sheath while ensuring tightness against outleakage to the environment, and a seal plug having a double wall and placed within the cavity for preventing the communication, the seal plug and the sleeve being intended to be cut simultaneously in order to seal-off the lock-chamber and the sheath separately by means of each wall of the seal plug.

According to a preferred embodiment, the sleeve is provided with an external channel at right angles to its axis and the seal plug is provided with an internal channel at right angles to its axis and located between its two walls, the channels being of substantially identical width and normally disposed in the same radial plane. The external surface of the seal plug can be rigidly fixed in leak-tight relation to the internal surface of the cavity formed within the seal plug over the entire periphery of the surface and over a distance extending on each side of the channels and measured parallel to the axis of the sleeve.

According to another distinctive feature of the process, each wall of the seal plug is provided with a gripping element disposed on the side remote from the sheath when the seal plug is in position within the sleeve.

According to a further distinctive feature of the process, the sleeve can be joined to the sheath in leak-tight manner at one end and is provided at the other end with an outer annular flange in which is formed a groove for receiving an annular seal which is capable of engaging in leak-tight manner with the internal wall of the lock-chamber. The sleeve can then be provided with an external rib having the same diameter as the annular flange and placed at a predetermined distance from this latter, the external rib being intended to be applied against latch-engagement means associated with the lock-chamber. Preferably in this case, the channel formed in the sleeve is disposed between the external rib and that end of the sleeve which is intended to be joined to the sheath.

The method for the practical application of the aforesaid device in order to remove noxious products contained in a cell and to transfer the products within a leak-tight sheath through a lock-chamber rigidly fixed to the partition-wall of the cell comprises the following successive steps:

(a) withdrawing the first sleeve into the interior of the cell while at the same time introducing the end of a second sleeve into the end of the lock-chamber remote from the cell, the second sleeve being closed off by a second seal plug and being rigidly fixed to a sheath, continuing the introduction of the second sleeve until an external rib is located within the interior of the lock-chamber and is maintained by latching means;

(b) removing the first sleeve and plug from the lock-chamber in order to open the second sleeve to the cell;

(c) withdrawing the second seal plug from the second sleeve into the interior of the cell in order to open the sheath to the cell;

(d) transferring the noxious products from within the cell into the sheath;

(e) inserting the second plug into the second sleeve;

(f) rigidly fixing the second seal plug in the second sleeve in a leak-tight manner; and,

(g) cutting the second sleeve and the second plug along a plane which passes between spaced walls on the plug to separate the sheath from the lock-chamber.

Placing Bitumen-Containing Wastes into Containers

In the bituminization processing of radioactive wastes the muds or concentrates are introduced into bitumen above 140°C by means of a multiple shaft extruder in which the water evaporates and the radioactive salts are mixed with the bitumen. The mixture of bitumen and waste leaves the worm shaft extruder under its own gravity force through an open channel into an available drum or waste container. The extruder acts as the evaporator and the mixture is at a temperature of 160° to 170°C when it leaves the extruder. Due to the low viscosity required, the temperature must be correspondingly high.

The particular drawback of this arrangement is that the bitumen-waste mixture is exposed to the atmosphere while at a high temperature, which involves great danger of combustion. The open discharge channel may also easily become clogged since the stream of bitumen and waste does not flow homogeneously and uniformly. This then requires intensive monitoring of the filling process.

W. Stegmaier, W. Kluger, H. Boden and F. Schneider; U.S. Patent 4,252,667; February 24, 1981; assigned to Kernforschungszentrum Karlsruhe GmbH and Werner & Pfleiderer, both of Germany describe a method in which the product temperature of the wastes mixed with bitumen is lowered considerably, compared to the processing temperature immediately before pouring the mixture into the drum or barrel.

The objects of the process are further achieved by provision of a discharging device provided for this purpose in a multiple worm shaft evaporator, which device is composed of an additional discharge housing attached to the housing of the evaporator to form an extension of the evaporator housing in the conveying direction of the multiple worm shaft extruder, and presenting a conveying bore in which the worm shafts are continued, the underside of the discharge housing being provided with a discharge opening and the walls of the discharge housing being provided with cooling channels.

An advantageous feature of the process resides in the provision of heating channels, in addition to the cooling channels, in the walls of the discharge housing, the heating channels and the cooling channels being disposed in concentric circles, and in making the conveying direction of the worm shafts or of the ends of the worm shafts, respectively, opposite to the main conveying direction, starting from the inner boundary of the discharge opening.

In this way, it is possible to advantageously lower the temperature and, despite the resulting increase in pressure, to achieve perfect positive discharge of the bitumen as a result of the pressure increase in the last stage. Cooling is advisably associated with the positive discharge or with the increase in pressure, respectively, as a result of the special design of the worm shaft ends, since otherwise there would exist the danger of clogging.

WASTE STORAGE CONTAINERS

Waste Storage Container with Metal Matrix

Nuclear waste storage containers containing a metal matrix for the nuclear waste are desired to provide greater impact strength for the waste container and to increase the thermal conductivity to prevent undesirably high centerline temperatures. Matrix fabrication has been effected by casting, although this process may prohibit the use of materials having high melting points which may be required to prevent melting of the matrix during storage from unforeseen temperature excursions or other accidents. Matrix fabrication by prior processes has generally been concerned with low temperature metals such as lead and aluminum.

Waste encapsulation technology includes the use of waste containers of such lengths as 15 feet and of from 1 to 2 feet in diameter. High-level waste particulate material to be stored in these containers may be only about 1 mm in diameter. In order to provide a matrix around these materials of such small particulate size, a very fluid metal, under perhaps considerable pressure, would be required to fully penetrate a large volume, such as a storage container, of small particles. The degree of difficulty in providing a cast matrix increases at a fast rate as the waste particulate material size decreases.

In addition to this drawback to casting, other drawbacks include the requirement for heated pipes or other heating elements to transport liquid metal into and around the cells in which the high-level waste is maintained, or the melting of large quantities of metals inside the cell in which the high-level waste is retained, with the associated problems of remote operation from the cell exterior to the cell interior.

Other pressing and forming type processes such as extrusion, mechanical or hydraulic pressing, swaging, high energy rate forming, etc., extensively complicate the container fabrication process, especially when required to be performed within a hot cell, and the use of this type of process to assemble several smaller units to form the large storage container would result in a loss of overall capacity. In addition, it is desirable to regain the size, shape, and integrity of the high-level nuclear waste particles, and, in some cases, to provide a protective cladding to these particles. The use of the pressing and forming processes described above would cause a degradation of these properties of the particles and of the cladding.

K.R. Sump; U.S. Patent 4,115,311; September 19, 1978; assigned to the United States Department of Energy describes a method for forming, and the product formed thereby, a storage container having a high-level waste disposed in a sintered metal matrix, comprising disposing a high-level waste into a storage container, thereafter disposing a metal powder into the container with the high-level waste, hermetically closing the port through which the high-level waste and metal matrix were fed into the container, and thereafter heating the hermetically sealed container containing high-level waste in the metal matrix to a sintering temperature to form a sintered metal matrix containing the high-level waste in the storage container.

It may be desirable, depending on the size and shape of the waste and matrix particles, to vibrate the container as the metal powder is fed into the container to obtain the desired matrix density such as by using a base-type vibrator. If the metal matrix powder is spherical, the advantage of vibration is reduced. The waste particles can occupy up to 50 to 70 vol % of the container, and the relative particle size and shape of the high-level waste particles will also cause the metal powder matrix density to vary from 35 to 65% of theoretical of the remaining volume. The void volume in the matrix is an additional asset and provides a desired result since this eliminates the need of leaving a space at the top of the container for any possible gas pressure caused by isotope decay and/or thermal outgassing.

Sintering of the metal powder may be done in a furnace at a temperature below the melting point of the matrix and below the previous fabrication temperature of the high-level waste particles. It is desirable to stay within the previously used fabrication temperature of the high-level waste particles to not adversely affect the physical properties of the particles or the cladding, which may be such as a coating of aluminum oxide over pyrolytic carbon on the particles. As an example, waste particles fabricated at 950°C and contained, embedded or enclosed in a copper matrix material could subsequently be gravity sintered at 700° to 950°C. Selection of the time, temperature and the like yields little or no shrinkage during gravity sintering. Employing gravity sintering, with or without vibration, consolidates the particles and matrix during heating and yields a sintered or diffusion bond at matrix contact points. The matrix stays in contact with the container wall and the individual matrix particles bond at their contact points and form a load-bearing structure and a good thermal conductivity path. For example, a 53% theoretical density stainless steel cyclindrical sample (without waste particles), initially 1¾ inches long and ⅞ inch diameter, sustained a maximum load of 59,600 lb and a 60% height reduction before failure. Since there is no pressure exerted upon the high-level waste particles during matrix fabrication, no degradation of the high-level waste particles, or if clad, of the cladding occurs.

Depending upon the materials involved, a stagnant cell atmosphere may be left in the can prior to gravity sintering, or the vessel storage container may be evacuated to a rough static vacuum of 50 microns, and thereafter sintered in the evacuated condition, or a dynamic vacuum may be applied during the sintering step which would have the added benefit of any outgas removal, or the chamber or storage container, after evacuation, may be backfilled with an inert gas such as helium or argon, to 1 atmosphere or less pressure prior to gravity sintering. Use of vacuum or inert gas improves the sintering process and may improve the subsequent mechanical properties of the materials because of a cleaner bond area between particles.

It should be noted that while the ease of providing a matrix for waste particles increases with the increased size of the particles, the process much more easily provides a matrix for small size waste particles, even as low as 0.5 mm in size, than prior art casting procedures.

There are several advantages to this process and the product of this process, including an economic advantage over cast matrix processes, because of less material and equipment required, a product that has superior thermal conductivity properties compared to solid glass or ceramic waste, and a product wherein the matrix provides strength.

A further advantage in high-level waste applications is that the matrix powders do not bond to the particles such that if there is an accidental fracture the particles themselves are not fractured but the matrix breaks around the particles, fines generated from fractured particles may become airborne or transported rapidly by other means.

The high-level waste particles that may be used in this process may be obtained from glass particle fabrication processes or ceramic particle fabrication processes where the waste form utilizes existing calcining processes with a compositionally modified waste liquid to achieve an improved or supercalcine waste form in which generally all of the radioactive atoms will be isolated in thermally and chemically stable phases. For example, the high-level waste particles may be chemical vapor deposition alumina and pyrolytic carbon-coated improved ceramic particles or supercalcine ceramic particles, and generally contain fission products as ceramic oxides or as glass modifiers.

The metal matrix powders that may be used are such as pure copper and its alloys, pure iron and its alloys, AISI Series 410, 304, 310 and 316 stainless steels, and superalloys such as Inconel or Hastalloy. A particularly good copper alloy matrix is manganese bronze having a nominal composition, expressed in wt %, of 57.5 copper, 39.25 zinc, 1.25 iron, 1.25 aluminum, and 0.25 manganese. The properties of the matrix materials that are desirable are a high melting point, good thermo-conductivity, good mechanical strength, good corrosion characteristics in salt, water, and/or air, and good oxidation resistance in air at operating temperatures.

Storage Assembly Containing Particulate Material

M.E. Lapides; U.S. Patent 4,326,918; April 27, 1982; assigned to Electric Power Research Institute, Inc. describes a fuel storage assembly which utilizes the same means for protecting the assembly against cooling malfunction while, at the same time, serving as a heat transfer medium, a neutronic poison and as an absorber/combining agent for any nongaseous material inadvertently released from the contained fuel rods.

The storage assembly is one which has housing means adapted for positioning underground and including a closed inner chamber for containing the nuclear fuel, e.g., the fuel rods. A thermally conductive member is located partially within the inner chamber and partially outside the housing means for transferring heat generated by the contained nuclear fuel from the housing chamber to the surrounding ground. Particulate material is located within the chamber and around the nuclear fuel contained therein. This material is selected so as to serve as a heat transfer medium between the contained nuclear fuel and the heat transferring member while at the same time standing ready to fuse into a solid mass around the contained nuclear fuel if the heat transferring rod malfunctions or otherwise fails to transfer the heat generated by the fuel out of the chamber in a predetermined manner.

In a preferred embodiment, the particulate material is borax. Figure 2.2 is a vertical sectional view of the assembly designed in accordance with the process. Referring to the figure, the overall storage assembly which is adapted for positioning underground is generally indicated by reference numeral (10). This assembly is shown including an elongated housing (12) having a closed bottom end (14), an opened top end (16) and an inner chamber (18). A cap (20) cooperates with the opened top end for closing the chamber.

Figure 2.2: Storage Assembly Containing Particulate Material

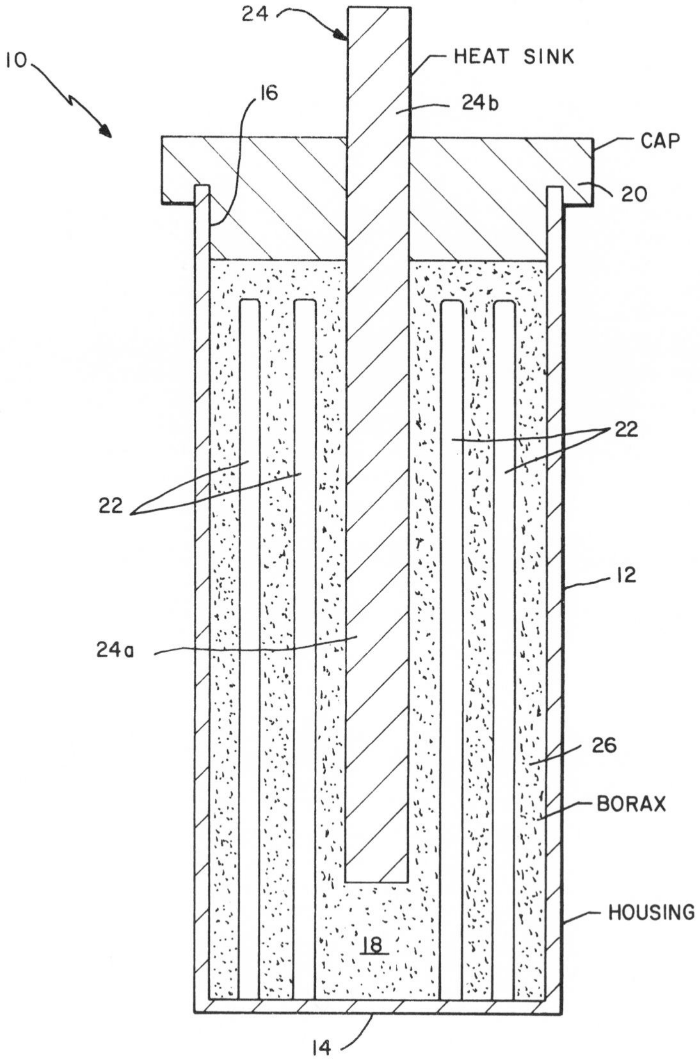

Source: U.S. Patent 4,326,918

Glass Storage Containers

P.B. Macedo, C.J. Simmons, D.C. Tran, N. Lagakos and J. Simmons; U.S. Patent 4,312,774; January 26, 1982 describe glass articles containing toxic solids and having high mechanical strength and high chemical durability to aqueous corrosion and having sufficiently low radioisotope diffusion coefficient values to provide protection to the environment from the release of radioactive material such as radioactive isotopes, nuclear waste materials, etc. and which are concentrated, immobilized and encapsulated therein and are suitable for burial underground or at sea. The glass articles are made by depositing the radioactive solids in a glass container followed by heating the container to drive off nonradioactive volatiles and to drive off nonradioactive decomposition products. The glass container may be made of porous glass and may or may not contain a porous or nonporous glass packing which can preferably be particulate or can be relatively large as a single or few glass rods.

The glass articles of this process have a composition characterized by a radiation activity illustratively above 1 mCi, preferably greater than 1 Ci, per cc of the article. (When highly dilute radwastes are treated pursuant to this process for the purpose of concentrating and immobilizing the radwaste for storage, the radiation activity of the resulting glass articles may not reach the level of 1 mCi/cc of the glass article and may remain below 1 mCi/cc when it becomes expedient for other reasons to collapse and seal the glass container. In concentrating and immobilizing radioactive materials in dilute radwastes, the glass container can be loaded up to 10 mCi/cc or more but usually is loaded up to 1 mCi/cc of the glass article.)

The radioactive material is in the form of radioactive solids that are sealed within the glass container. In one aspect, the amount of radioactive material contained in the glass articles is at least 1 ppb (part per billion based on weight), generally in solid form of at least 5, and preferably at least 10, of the radioactive elements listed hereinafter. Preferably, the glass articles should contain at least 75 mol % SiO_2, most preferably greater than 89 mol % SiO_2.

From the practical standpoint, the upper limit of radioactive material contained in the glass articles will be governed, to a degree, by such factors as: the concentration, form and type of radioactive material encapsulated in the glass article; by the volume fraction of pores, if any, in the glass container; by the amount, if any, of glass packing in the glass container; by the various techniques employed to encapsulate the radioactive material in the glass container; and other factors.

Radioactive materials which can be concentrated, encapsulated and immobilized in the glass container pursuant to this process include radioactive elements (naturally occurring isotopes and man-made isotopes exising as liquids or solids dissolved or dispersed in liquids or gases), in combined or uncombined form (i.e., as anions, cations, molecular or nonionic, or elemental form) such as rubidium, strontium, the lanthanides and actinides, cations and elements. Especially suitable are radioactive wastes from nuclear reactors, spent reactor fuel reprocessing, spent fuel storage pools or other radioactive waste producing processes. The glass in the final glass product can be vitreous throughout or partly vitreous and partly crystalline.

The process can be practiced in many ways. Illustratively, one facile yet highly effective way is to deposit the radioactive materials, e.g., radioactive nitrates, as a solid in a nonporous glass container such as a glass tube having at least one opening, and then followed by heating the container to drive off water and/or other nonradioactive volatiles, if present, and then to collapse the walls of the tube and seal it around the deposited radioactive solids. The heating step can be carried out in such manner that solids such as nitrates deposited in the tube decompose to provide nonradioactive gases such as nitrogen oxides, which are removed from the glass tube before sealing.

Alternatively, a nonporous glass container can be closed at one end and at least partially filled with a packing such as porous glass particles such as porous glass powder or tiny glass spheres or silica gel in particulate or other form. The fluid containing radioactive material is then poured into the container to fill the interstices between the glass particles followed by heating to drive off nonradioactive volatiles with or without decomposition of components such as nitrates in the fluids and ultimately to seal the glass container around the radioactive solids deposited on the glass particles and in the pores of porous glass particles, if present, contained by the container. In this instance, the contained glass particles provide surfaces on which the solids can be deposited and also act to control the volatilization to prevent eruption of fluid out of the tube during the heating step. The porous glass particles provide additional interior surfaces within the pores of the particles for deposit of additional dissolved solids in the fluids as well as external surfaces for deposit of dispersed solids.

In another embodiment, a nonporous glass container having open upper and lower ends can be filled with porous or nonporous glass particles which are held in the glass container by means of a porous structure such as glass wool or a porous glass disc or rod in the lower portion of the container to support the glass particles in the container. The fluid containing dissolved and/or dispersed radioactive solids is then poured into the upper or lower end of the container and passes through the bed of glass particles which act as a filter to remove dispersed radioactive solids from the fluid. The glass particle bed can contain glass particles having silicon-bonded cation exchange groups such as alkali metal oxide or ammonium oxide groups.

The porous cation exchange glass particles remove dissolved radioactive cations from the fluid. The fluid can be passed through one or more such beds using conventional techniques for multibed filtration and/or ion exchange until the fluid has been cleansed of radioactivity to the desired level. When the filtration-ion exchange glass particles become loaded or when, for some other reason, it is no longer desired to further utilize them, the beds and the container containing them can be heated to drive off water and/or other nonradioactive volatiles or gases such as decomposition products, e.g., nitrogen oxides, and then to collapse the pores of the porous glass particles containing the radioactive cations, to fuse the glass particles together thus entrapping radioactive solids and/or cations deposited on the inner and outer surfaces of the particles, and then to collapse the glass container and seal it around all of its contents to encapsulate the entire mass into a substantially solid leach-resistant structure suitable for long-term storage.

In still another embodiment, the glass container itself can be made of porous glass and the radioactive fluid is introduced into the interior of the container and caused to permeate through the pores of the glass from the interior walls to the outer walls of the glass container. The insoluble radioactive solids originally dispersed in the fluid are deposited on the interior wall of the container and the dissolved radioactive solids are disposed in the pores of the glass container where they can be deposited by various techniques such as those taught in U.S. Patent 4,110,096. The glass container can then be heated to drive off volatiles as described above, to collapse the glass container and seal it, thereby encapsulating the radioactive solids within the glass structure. Prior to heating, the outer wall surfaces of the container can be washed to remove deposited radioactive solids from the outer surface layer of the glass container so that ultimately a radioactive-free outer clad is provided after heating to collapse the pores and the container.

The nonporous glass compositions when used herein for the glass container and/or for the glass packing within the container are of any suitable type, but preferably are strong, durable, leach-resistant and chemical-resistant. Any glass composition having these properties can be used such as high silica glasses, for example, Vycor and Pyrex. Suitable glasses contain at least about 70%, preferably at least 80%, most preferably at least 93% silica.

The literature adequately describes the preparation of the porous silicate glass compositions. Suitable glass compositions which may be utilized as porous glass compositions in the above methods generally contain SiO_2 as a major component, and have a large surface area. In the practice of various embodiments of the process, the SiO_2 content of the porous glass or silica gel desirably is at least about 75 mol % SiO_2, and most preferably at least 89 mol % SiO_2. Such glasses are described in the literature, for example, in U.S. Patents 2,106,744; 2,215,036; 2,221,709; 2,272,342; 2,326,059; 2,336,227; 2,340,013; 4,110,093 and 4,110,096.

The porous silicate glass compositions can also be prepared in the manner described in U.S. Patent 3,147,225 by forming silicate glass frit particles, dropping them through a radiant heating zone wherein they become fluid while free falling and assume a generally spherical shape due to surface tension forces and thereafter cooling them to retain their glassy nature and spherical shape.

In general, the porous silicate glass can be made by melting an alkali-borosilicate glass, phase-separating it into two interconnected glass phases and leaching one of the phases, i.e., the boron oxide and alkali metal oxide phase, to leave behind a porous skeleton comprised mainly of the remaining high silicate glass phase. The principal property of the porous glass is that when formed it contains a large inner surface area covered by silicon-bonded hydroxyl groups.

It is preferred to use porous glass made by phase-separation and leaching because it can be made with a high surface area per unit volume and has small pore sizes to give a high concentration of silicon-bonded hydroxyl surface groups, and because the process of leaching to form the pores leaves residues of hydrolyzed silica groups in the pores thus increasing the number of silicon-bonded hydroxyl surface groups present. The porous silicate glass when used as packing may be in the form of powder as for use in chromatography columns or in a predetermined shape such as plates, spheres or cylinders.

It is preferable to utilize a glass composition in the container which will produce a clad or envelope that is low in leachable components such as alkali metals or boron. In the event that this is not possible or practical it is preferred to then insert the glass container before or after collapse into a second glass container which has a composition containing no or low amounts of alkali metals or boron or other leachable components. The absence of substantial amounts of leachable components in the outer glass clad obviously provides a more durable and leach-resistant product which is more suitable for burial. In those instances where a glass packing is used in the container, it is preferred that the composition thereof contain little or no leachable components, especially when the resulting products are intended for burial. It is most preferred that very high silica glasses are employed in both the glass container and the glass packing.

When it is desired to avoid cracking of the glass container in the case where a glass packing such as glass particles, spheres or a glass rod are disposed within the glass container, it is preferred to utilize as the container glass a glass which, after the deposition step, has a glass transition temperature of up to 100°C higher than the glass transition temperature of the glass formed from the glass packing and solids deposited in and on the glass packing. It is also preferred for this purpose to utilize as the glass container a glass which, after the deposition step, has a thermal expansion coefficient which is up to about $2 \times 10^{-6}/°C$ less than the thermal expansion coefficient of the glass resulting from sintering of the glass packing and solids deposited in or on the glass packing.

In determining the glass transition temperatures and thermal expansion coefficients, the amount and type of solids deposited in the pores of the glass container, when a porous one is used, and of solids deposited in the pores and on the outer surfaces of porous glass packing, when used, and of solids deposited on the outer surfaces of nonporous glass packing, when used, can have a considerable effect on the glass transition temperatures and thermal expansion coefficients and should be taken into consideration. It is also preferred to regulate the cooling of the composite glass container and contents, resulting from the depositions and sintering step, such that the rate of cooling is as nearly the same as possible throughout the composite glass container and contents. While cracking has been observed in certain instances, it has not prevented the achievement of the objects of this process, namely, the immobilization and isolation from the environment of radioactive solids from radioactive wastes containing such solids in dissolved and undissolved form.

Container with Deflectable Foil at Open End

R.A. Hay, II and H.E. St. Louis; U.S. Patent 4,234,447; November 18, 1980; assigned to The Dow Chemical Company describe a method for the disposal of low-level radioactive wastes by encapsulation of the radioactive waste within a hardenable resin. The steps of the method comprise: providing a disposable container and an agitator selectively disposable within the container and external to the container, the container having a generally open upwardly-facing end, adding to the container a predetermined quantity of hardenable resinous material and a radioactive waste, agitating the mixture of hardenable resin and radioactive waste until the radioactive waste is encapsulated within the hardenable resin and closing the open end of the container with a container cover during the mixing process. The container cover is maintained in a generally fixed relationship to the agitator and has disposable splash guard generally adjacent to the disposable container. Subsequently the agitator is removed from the container, the contents of the container permitted to harden and the container transported to an approved disposal site. The improvement comprises providing a generally annular membrane periphery partially closing the open end of the disposable container and permitting passage of the agitator therethrough, the membrane being deflected by and generally conforming to the splash guard. The membrane serves to reduce the amount of resin and radioactive material splashing directly on the splash guard.

Filter-Lined Containers

L.E. White and C.M. Gracey; U.S. Patent 4,058,479; November 15, 1977; assigned to Aerojet-General Corporation describe a package for storing toxic solids in which the stored solids are homogeneously dispersed in an immobilization material and also completely surrounded by a layer of the immobilization material. The immobilization material mixed with the toxic solids prevents dispersion of those solids if the package should be suddenly cracked or broken during transportation or other handling. The outer or surrounding surface area layer of immobilization material uniformly isolates the toxic material from the container wall and thereby prevents dispersion of that material by leaching of water-soluble chemicals, and the like, if the outer container deteriorates during the long-term storage.

The method comprises a tight mesh fabric bag inserted into a barrel or other suitable container. The bag is filled with particulate radioactive waste material and a nonradioactive liquid that mixes with the radioactive waste particles. The tight mesh of the fabric acts as a barrier to the solid waste material, but not to the liquid. The granular waste particles are thus held within the bag. A substantial quantity of liquid also stays inside the bag in mixture with the radioactive solids, filling the void spaces between particles and surrounding each particle. But, a portion of the liquid flows through the bag as it is filled to form a layer of nonradioactive material between the bag and barrel wall.

After the barrel is filled, the liquid material is solidified to provide a massive monolithic solid that contains radioactive waste immobilized by a solidification agent. The tight mesh bag is incorporated in the solidified mass by the solidification of the liquid contained internally, externally, and within the pores or weave of the tight mesh bag. It thus becomes an integral part of the overall solidified mass that adds strength and structural integrity, particularly to the outer portions of that mass. The tight mesh weave of the bag, and smaller diameter than that of the outer container, also exclude radioactive material from the layer of solidification material between the bag and outer container. This layer thus becomes a containment barrier that isolates radioactive material from the environment, and is also integral and contiguous with the material contained internal to that layer.

This surface area protective layer provides an effective barrier against leaching and dispersion of the radioactive salts and other materials. Its thickness can be easily controlled during formation of the storage package by selection of the sizes of the bag and barrel and the positioning of the bag within the barrel to provide an appropriate degree of protection to prevent impact or corrosion likely to rupture the outer container wall from penetrating to the radioactive material. Any break that may occur through both the outer container wall and this protective layer will expose only a minimum quantity of radioactive material to the environment because the protective layer does not separate from the contained radioactive mixture by shrinking, or otherwise, during solidification. Thus, only the material immediately beneath any break will be exposed. Even when ruptured, the bag and surface layer provide strength that causes the contained mixture to be less susceptible to breakup by water and other leaching agents.

Isostatic Pressing of Capsule

H. Larker; U.S. Patent 4,172,807; October 30, 1979; assigned to ASEA AS, Sweden describes a method for containing high-level radioactive waste in a body resistant to leaching by water. The method includes the steps of providing a mass containing radioactive substances and either a material which is resistant to leaching by water or a material which when heated forms a material resistant to leaching by water; enclosing the mass in a capsule; and then isostatically pressing the capsule at a temperature and pressure sufficient to form a coherent, dense body from the mass.

Examples of such resistant material include oxides of such kinds as are normally comprised in glasses of various kinds and different rocks, for example, SiO_2, B_2O_3, Al_2O_3, MgO, alkaline oxides, alkaline earth oxides, TiO_2, ZrO_2, Fe_2O_3 and Cr_2O_3, and further rocks existing in nature which are well known for their long-term stability, such as rocks composed of silicates, aluminates, phosphates and titanates. Preferred materials include aluminum oxide, titanium oxide, quartz and rocks existing in nature.

The mentioned resistant materials may also be added to materials which are brought into contact with the radioactive substances and in which the radioactive substances are then fixed. Thus, the resistant materials may be mixed with ion exchange materials before the ion exchange materials are brought into contact with the radioactive substances so as to reduce the management of radioactive substance. A suitable amount of resistant material incorporated in the particle mass may be 1 to 95% of the total weight of the particle mass and the incorporated material. The particles of the incorporated resistant material suitably have a size of less than 1 mm and preferably of less than 0.5 mm.

The capsule may, among other things, consist of sheets of tantalum, titanium, zirconium, alloys based on these metals such as, for example, zirconium based alloys sold under the trade name Zircaloy, steel, iron, nickel and further of quartz glass or boron silicate glass. The capsule material should be matched to the resistant material so that it has sufficiently high melting point for the capsule to fulfill its duties, and substantially the same coefficient of thermal expansion in such cases when the capsule is left to provide a reinforced containment. With quartz or titanium oxide as resistant material, quartz glass is preferred for use in the capsule, and with boron silicate glass as resistant material, a capsule of the same material is preferred. In certain cases it may be suitable to use a capsule of metal which is internally provided with a layer of quartz glass or boron silicate glass.

Between the capsule and the mass to be contained it may be suitable to arrange an intermediate layer of a resistant material such as any of the previously exemplified resistant materials. It may be particularly suitable to use a material of the same chemical composition as the mass but without radioactive isotopes as the intermediate layer. The particles in the material in the intermediate layer suitably have a size of less than 1 mm and preferably less than 0.2 mm. The intermediate layer may, for example, be applied as a layer of a thickness of a few millimeters or centimeters on the inner wall of the capsule.

The pressure during the isostatic pressing amounts to at least 10 MPa and is preferably between 50 and 300 MPa. The temperature is, of course, dependent on what materials are included in the particulate mass but is at least 700°C. A suitable temperature for particulate masses containing titanates, quartz or titanium dioxide is 1200° to 1300°C and for particulate masses containing aluminates and aluminum oxides is 1250° to 1350°C.

Self-Sealing Canister

W.J. Mecham; U.S. Patent 3,983,050; September 28, 1976; assigned to the U.S. Energy Research and Development Administration describes a method which eliminates many of the problems attendant upon the storage of radioactive calcine wastes in a metal canister for long periods of time in a moisture-containing environment such as open-air or water-tank storage. In this method, dry cement powder is added to the metal canister containing the calcined wastes so that the cement powder is in contact with the inner surface of the wall of the canister before the canister is sealed. Should the container wall fail and develop an opening to the environment, moisture from the environment will enter the canister, mix with the cement, forming concrete which will harden, seal the opening and prevent the escape of any radioactivity from the canister into the environment.

The process involves mixing cement powder with the solid calcined high-level radioactive waste in a ratio of 1 part cement powder to 3 to 10 parts by weight of calcine before the calcine is placed in the metal canister and sealed for storage in a moisture-containing environment.

It has been assumed that the wastes to be stored will be stable, dry, solid, high-level radioactive wastes. The reference design waste form is a calcine product from Light Water Reactor Uranium (PWR-2) fuels, typically packaged in a stainless steel canister, 12.75 inches outside diameter by 10 feet long, containing 6.3 cubic feet of waste from reprocessing 3.15 metric tons of fuel and generating approximately 3.25 kilowatts of decay heat at 10 years age.

The ratio of cement to calcine will depend upon the physical form of the calcine, i.e., whether it is a fine granular friable material such as sand or of a coarser form such as gravel. In general, a fine granular powder is limited to a ratio of 3 to 1, while a coarser granular form may vary up to a ratio of 10 parts to 1 part by weight cement.

The type of cement may be any of the ordinary Portland cements which have the ability to harden in a high-moisture environment such as where the canister has been placed in a water tank for storage. It may be desirable to use an expansive cement such as Atlas MX which will expand about 2% as it sets. Because of the high temperatures which the radioactive material in storage may reach, a cement having a high alumina content may also be considered.

The cement may be mixed with the calcine in various ways. For example, the cement powder can be added to the calciner to be mixed with the calcine as it is formed. This would enhance the incorporation of the oxide waste with the cement powder. The cement can also be mixed with the calcine as it is fed into the canister. This would give an even distribution of cement with the calcine within the canister. As another alternative, the cement powder can be inserted into the canister as an outer annulus, with the calcined waste in the center of the canister. This may be accomplished, for example, by inserting a cylindrical sleeve into the canister which is slightly smaller in diameter than the inner wall of the canister so that an annular space about 1 inch deep is formed between the sleeve and the wall, and filling this space with cement powder while filling the center of the sleeve with calcine. The sleeve can then be removed and the canister sealed for storage. This method has the advantage that the amount of cement necessary to provide the self-sealing feature for the canister would be minimized.

It is important that the inner surface of the wall of the canister be in contact with the cement powder so that a breach anywhere in the canister wall which would permit the entrance of moisture either from the cooling basin or from the air will result in the water mixing with and wetting the cement so that it can set, plug the breach and thus prevent the inflow of any more water.

The formation of cement under conditions of leak will, of course, not provide the usual proportioned well-mixed cement-water paste. However, a leak due to corrosion or to a crack in the canister will first open only a very small surface for the entry of water. Since the canister is sealed and the water is only at the pressure of the gravity head in the pool, the water will enter slowly. The temperature is favorable for quick setting of cement since the temperature of the outer wall of the canister is expected to be about 128°F. Thus, while the point adjacent to the breach in the canister wall may have too much water to form good cement, the inside portions of cement will not have excess water and a zone of setting proportions will encircle the hole. The leaching of the calcine will be reduced and the integrity of the calcine and of the container will be maintained for a longer time so that maintenance or removal of the canister can be carried out if desired.

Advantages of this method:

> (a) The properties and further accessibility of the dry calcine are not changed in the normal situation where the canister does not leak.

> (b) The amount of concrete formation, which may change the accessibility of the calcine, is limited to the amount of water leaked.

> (c) The formation of monolithic concrete involves the doubling of the volume of solid previously present as dry calcine, thus closing any void space in the calcine and forming a seal against any further water entry or exit of calcine or water.

> (d) This method is compatible with the formation of a concrete monolith of the calcine prior to canistering and thus should be deemed desirable for this type of interim storage of high-level wastes.

> (e) No expensive reagents and a minimum of equipment are required to carry out this method of the interim storage of high-level radioactive wastes.

Waste Canister Having Radial Fins

It has been proposed that high-level, long-half-life radioactive wastes be converted to a glass and stored within a container or canister in some location where the radioactivity cannot contaminate the environment. Conversion of the wastes to a glass is advantageous because of the great inertness and low solubility of a glass. However, a glassy waste has a very low thermal conductivity and a high heat generation rate. This combination causes very high centerline temperatures when the waste is stored in a cylindrical canister. An obvious solution to the problem would be to employ fins extending from the canister wall to the canister center to conduct the heat from the center of the canister to the walls. Unfortunately, such a simple solution to the problem is not practical since conventional fins produce local hot spots where they are attached to the wall of the canister. Such hot spots cannot be tolerated as they induce severe internal stresses in the canister.

J.B. Duffy; U.S. Patent 4,021,676; May 3, 1977; assigned to the U.S. Energy Research and Development Administration has found that fins centered in the canister so that the ends of the fins are spaced from the canister wall are effective to keep the centerline temperature in the glass filling the canister at an acceptable level without producing unacceptable hot spots in the canister wall. While the clearance has not been optimized, a one-half-inch gap would be suitable for a canister 12 inches in diameter having one-sixteenth inch walls formed of stainless steel. For canisters of differing sizes, wall thickness and wall material gaps of one-eighth to five-eighths inches would be suitable.

Figure 2.3 is a horizontal section of such a canister according to the description above.

Figure 2.3: Horizontal Section of Canister Having Radial Heat Dispersion Fins

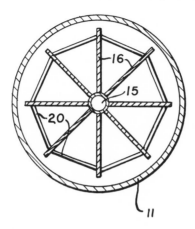

Source: U.S. Patent 4,021,676

Storage in Circulating Pipes

One method for the long-term storage of fission products is to store them as solutions or slurries in large tanks fitted with cooling coils to remove the decay heat and with means for circulating the fission product solution or slurry within the tank. However, it is difficult to ensure that sufficient circulation occurs within the large volume of liquid in the tank to prevent the accumulation of sediment on the tank walls and on the cooling coils which tends to reduce the efficiency of the cooling. As a safety measure there must be spare tanks available to transfer the fission product solution or slurry should any defect become apparent in the original tank. The capital investment in such storage tanks is large and it is, therefore, desirable to reduce the number of spare tanks which have to be provided.

A.L. Mills, J. Reekie and J.A. Williams; U.S. Patent 4,299,271; November 10, 1981; assigned to United Kingdom Atomic Energy Authority, England describe a storage installation for liquid radioactive material which comprises pipe circuits containing the liquid radioactive material, means for circulating the liquid radioactive material around the pipe circuits and means for circulating a liquid cooling medium over the external surface of the pipe circuits.

The liquid cooling medium may be in a tank and the pipe circuits may be immersed in the cooling medium in the tank. Alternatively, the liquid cooling medium may be passed through the annular gap between coaxial pipes, the inner one of which contains the liquid radioactive material.

The liquid cooling medium may be circulated from the tank or the annular gap between the coaxial pipes of the alternative pipe circuits described above to a heat exchanger by means of pumps. Should the pumps cease to operate, the temperature of the liquid radioactive material within the pipe circuits will rise because of the cessation of flow of the cooling medium. It is undesirable that the liquid radio-active material should boil within the pipe circuits. Additionally, as the temperature of the liquid radioactive material rises, the rate of corrosion of the pipe circuits by the liquid material therein also rises. To prevent boiling of the liquid radioactive material and to minimize the corrosion which could occur during a malfunction of the cooling medium, circulating pump secondary cooling systems are preferably provided.

In the case of pipe circuits which are immersed in the cooling medium in a tank, a condenser may be provided on the tank to prevent loss of any cooling medium should the temperature of the cooling medium be raised to a point at which evaporation of the cooling medium is occurring to a significant extent. The condenser is preferably an air condenser requiring no power input for its operation and it should be of such a size that no loss of cooling medium occurs even if the cooling medium boils.

The means for circulating the liquid radioactive material around the pipe circuits may comprise fluidic pump means, and such means may be operated by a pulsed-liquid column controlled by air pressure.

The cooling medium may be water but if a tank is used which is fitted with a reflux condenser as described above, a cooling medium having a boiling point in the range of 60° to 80°C is preferred so that the temperature of the pipe circuits does not rise to a point where the corrosion rate is excessive. Examples of cooling media which may be used include methanol, isopropanol, methylene chloride, carbon tetrachloride and other halogenated hydrocarbons such as those sold under the trade name Freon. In a situation where the pumps circulating the cooling medium are not operating, the latent heat of evaporation extracted from the pipe circuits as the cooling medium boils prevents excessive heating in the pipe circuits.

Seal for Storage Bore Hole

E.-P. Uerpmann; U.S. Patent 4,316,814; February 23, 1981; assigned to Gesell-schaft fur Strahlen-und Umweltforschung mbH, Germany describes closures for sealing storage bore holes which constitute the final disposal site of radioactive waste and a method of applying the closures.

The sealing closure is formed of at least one prefabricated body which is a metal and/or a dense ceramic material and/or cast steel and/or a lead alloy and which is arranged in the storage bore hole above the uppermost waste vessel in a close fit with respect to the bore hole wall.

The arrangement of prefabricated closure elements is manufactured with uniform standards in quality. Thus, extensive work in the vicinity of the storage bore holes may be dispensed with. The personnel are not exposed to any radiation since the sealing closures can be introduced into the storage bore holes by remote control. The proposed materials for the closure contain no water which could otherwise be decomposed radiologically by a γ-radiation. Since the abovenoted materials from which the sealing closures may be made are conventionally used as shielding materials (lead alloys and cast steel) or as a reactor building material (ceramic), their resistance to radiation is superior.

The thermostability of these materials is also of a superior degree; lead alloys will not melt under the conditions to be expected and the pressure resistance of ceramic and cast steel is sufficiently high for this purpose. Further, lead alloys are particularly advantageous since they are adapted to be deformed in a ductile manner and, therefore, provide an excellent seal. The handling of prefabricated bodies by remote control can be effected without difficulty. Cast steel is, similar to lead, a corrosion-resistant material. Dense ceramic is highly corrosion resistant and is widely used for conduits in chemical laboratories.

In summary, the particular advantages of the process are to be regarded in the configurational uniformity of the sealing closures, in a high degree of safety during installation and in the lack of water content in the material of the sealing bodies. All methods wherein the storage bore holes are filled either with ground salt or with cement require the presence of personnel in the vicinity of nonsealed storage bore holes. The quality of the storage bore hole closure can vary in these known methods and cannot be checked because of the exposure to large doses of radiation.

Sulfur-Containing Repositories

D.E. Schweitzer, C. Sastre and W. Winsche; U.S. Patent 4,269,728; May 26, 1981; assigned to the U.S. Department of Energy describe the storage of spent fuel in repositories in which sulfur is used as the storage medium.

The repositories used in accordance with the process may be of conventional design. For example, tanks or other suitable containers to be used for the storage of nuclear waste conventionally have diameters of from 6 to 18 feet. In most cases, containers with diameters of 15 or 16 feet are used.

The spent radioactive fuel elements from nuclear reactors are introduced into the abovedescribed repositories. Subsequently, these fuel elements are in their entirety surrounded by sulfur. Sulfur has the advantage that it is noncorrosive and is not subject to radiation damage. As a result, sulfur is an extremely stable material under radioactive conditions.

Since the sulfur is noncorrosive, it does not affect dissolution of the metallic parts of the fuel elements or the repositories. Thus, the sulfur itself does not represent a radiation hazard. Moreover, of course, in the event of a leak, the sulfur is not absorbed either into the ground or the atmosphere and since it does not contain radioactive salts in any event, the danger of a radioactive contamination is minimized.

In a preferred embodiment, the sulfur in which the fuel elements are stored is continuously kept at a temperature of more than about 112°C to maintain the sulfur in the liquid state. Accordingly, should a leak occur, the sulfur will immediately solidify at the leakage point where it is exposed to ambient temperatures below 112°C. As a result, it is ensured that the liquid sulfur will not escape from the container and the radioactive fuel elements are not exposed to the atmosphere.

In accordance with another feature of the process, the decay heat given off by the spent fuel is utilized to maintain the temperature of the sulfur at a level above the melting point of the sulfur. For this purpose, the fuel elements are spaced within the repository in which they are stored sufficiently close so that the temperature in the sulfur which surrounds the fuel elements is kept at a level of at least 112°C throughout the repository and the sulfur is constantly maintained in the molten state.

The method may be implemented in a variety of ways. Generally, the spent fuel elements can be introduced into a cylinder which is usually made of metal. The sulfur is then added to the cylinder in either the molten or solid particulate form so as to completely surround the fuel elements. If the sulfur is in the form of a solid particulate, the cylinder may be heated to make the sulfur molten. Thereafter, it may be necessary to add additional particulate sulfur in order to make certain that the elements are totally immersed. Also it may be desirable to continue the heating of the molten sulfur for a period of time sufficient to make certain that there are no air bubbles entrapped in the molten sulfur. Thereafter, the molten sulfur can be allowed to solidify about the elements.

As noted above, depending on the placement of the elements, when the individual cylinders containing the elements are placed into the final repository, they may be spaced at an appropriate distance so that any decay heat given off by the elements may be sufficient to keep the sulfur in the molten state.

Encasing of Waste Barrels in Leachproof Sheath

E.-P. Uerpmann; U.S. Patent 4,222,889; September 16, 1980; assigned to Gesell-schaft fur Strahlen-und Umweltforschung mbH, Germany describes methods for permanently encasing radioactive wastes solidified with binders or so encasing radioactive waste containers themselves such that resistance against water or aqueous salt solutions is assured by the creation of a closed sheath of inactive, hydrophobic and waterproof material which is essentially nonporous, i.e., having tight pores.

The method comprises:

> applying a bottom layer of a synthetic resin or a dissolvable spacer material on the bottom of the inside of a waste barrel;

> applying a lateral spacer layer inside the barrel along the walls to occupy the space of the lateral portion of a sheath to be applied around the waste;

> filling the wastes together with binder into the barrel up to or below the height of the lateral spacer layer and permitting the wastes and binder to set; and then

> filling the spaces occupied by the lateral spacer layer and, if present, the dissolvable bottom spacer material, as well as the space in the barrel above the waste and binder, with a synthetic resin containing a solvent which dissolves the spacer material of the lateral spacer layer and, if present, the dissolvable bottom spacer material, whereby the synthetic resin forms a sheath completely surrounding the wastes and binder, which sheath has a firm and tight mechanical bond of its sides with its top and bottom. The lateral spacer layer can be a preformed spacer inner container which can be inserted into the barrel.

These objects are also attained by providing another method for encasing radioactive wastes in a closed sheath, the wastes being solidified and accommodated in a barrel and the closed sheath to be resistant to water and to aqueous solutions of natural mineral salts and to leaching, which comprises:

applying a bottom layer of synthetic resin or dissolvable spacer material on the bottom of the inside of an outer sheath;

placing a waste barrel which contains the wastes in the outer sheath, with the outer sheath extending over the entire bottom and side walls of the barrel beyond the top thereof and the barrel resting on the bottom layer;

applying a lateral spacer layer along the outside walls of the barrel between the barrel and the outer sheath to occupy the space of the lateral portion of an inner sheath which is to be applied around the outside of the barrel; and

filling the space occupied by the lateral spacer layer and, if present, the dissolvable bottom spacer material as well as a disc area defined by the top of the barrel and the portion of the outer sheath which extends beyond the barrel with a synthetic resin containing a solvent which dissolves the spacer material of the lateral spacer layer and, if present, the dissolvable bottom spacer material whereby the synthetic resin forms a sheath completely surrounding the barrel, which sheath has a firm and tight mechanical bond of its sides with its top and bottom.

In practicing the above methods, it has been found to be of particular advantage to use polyester resin or epoxy resin as the synthetic resin, styrene as the solvent and foamed polystyrene as the material for the bottom spacer layer, for the lateral spacer layer and for the preformed inner container. Likewise, the barrels may be placed onto blocks of synthetic resin serving as spacers.

For any of the methods of this process, the thicknesses of the bottom layer, the lateral layer and the top layer should be 0.5 to 2.0 cm and preferably 1.0 to 2.0 cm. A preferable concentration of the solvent in the synthetic resin would be 15% styrene.

The particular advantages of the solution of the process include the introduction of a not previously used inactive layer of a waterproof and hydrophobic material around the waste. When provided in sufficient thickness, this layer assures that water or aqueous solutions of natural minerals will not come into direct contact with the radioactive wastes being permanently stored in geological formations. Consequently, the danger of leaching, possible chemical decomposition of the binder and spreading of radioactive materials is definitely reduced if not completely prevented until the wastes decompose to harmless values.

The process, and in particular the ease of its technological and industrial realization, offers great advantages for the security of permanent storage of radioactive wastes and also of chemical wastes in geological formations. It can also be applied to other methods of storage of other materials as long as appropriate consideration is given to the interactions among waste product, packaging material, inactive water-tight barrier layers and storage medium. Leaching and spread of radioactive or other dangerous wastes is decisively reduced.

Protective Barrier Material

F. Gagneraud; U.S. Patent 4,300,056; November 10, 1981 describes an inexpensive and very compact material offering excellent resistance to aggressive agents, to act as a barrier between a radioactive environment and an uncontaminated environment.

In its most general form, the process consists of placing between the radioactive environment and the outside environment, elements made of materials selected from the group of molten slags and scorias coming from making ferrous and nonferrous metals.

These elements, acting as antiradiation barriers, can be of various types such as, for example, walls or monolithic slabs of poured slag for putting up buildings sheltering the installations for fabrication or treatment of nuclear fuels; pipes of poured metallurgical slag for movement of the radioactive effluents in the treatment plants; poured slag blocks for putting up walls acting as radiation insulating barriers at storage sites; or containers in various shapes for insulating the entire volume to be buried or immersed, or in the open air.

According to a particularly advantageous embodiment, the process relates to surrounding metal containers, usually used to enclose solid or pasty radioactive wastes, with molten metallurgical scorias or slags, these containers optionally being provided with a protective layer of refractory concrete.

According to a preferred embodiment of this process, a waste container is wedged in an empty large-size external drum, then molten slag is poured in the empty space between the two containers, the operation being followed by the slowest cooling possible to avoid any cracking and to obtain a compact coating.

The covering around the metal waste container can be made from molten slag coming directly from its ladle or pouring trough at the output of the blast furnace or other production device. To improve and facilitate the work, the fluidity of the slag can be increased by lowering the solidification temperature by adding to the slag such substances as sodium borate, sodium carbonate, siliceous sand, iron oxides or other suitable products.

In case the insulating elements are made up of walls, slabs, or the like, they can be fabricated in the vicinity of the blast furnace or foundry and delivered prefabricated and assembled at the desired site. The joints of the components of the blocks and walls, generally of poured metallurgical slag, can be made of heavy granulated slag mortar, to which optionally is added granules of barite, cast iron or the like, the cement used preferably being made up of a cement with a high blast furnace slag content. According to a variant, it is possible to join the poured slag blocks with molten lead.

Various slags and scorias are suitable for use in the process. Of the most suitable, there can be cited as production sources blast furnaces and steel mills, foundries, installations making silicochromium and silicomanganese, lead-zinc smelters, copper smelting, manufacture of phosphorus by electric furnace, etc. However, it is advantageous to use molten products with a slight basicity index whose crystallization temperature is lower than that of the basic slags. These slags are most suitable for retaining a high degree of vitrification. Preferably, also for this purpose, slags can be selected that have considerable content of heavy metals (FeO, Cr_2O_3, MnO, NiO) coming from the ferroalloy industry or lead, zinc and copper metallurgy, because the presence of metal oxides increases the absorbing properties of the slags in regard to radioactive radiation.

WATER REMOVAL
AND CONCENTRATION PROCESSES

EVAPORATION

Prevention of Contamination of Heating Steam Line

One known method of treating a radioactive liquid waste produced from a nuclear power station comprises heating and vaporizing the radioactive liquid waste with heating steam so as to thereby concentrate the waste and reduce its volume.

More specifically, the features of this known method, and apparatus therefore are as follows. A radioactive liquid waste discharged from an installation such as, for example, a nuclear power station, is stored in a liquid supply tank, from which it is transferred by a liquid supply pump to an evaporation concentration device. This evaporation concentration device normally comprises a concentration vessel which is generally of the vertical type. A steam heater is installed in the concentration vessel or, alternatively, in a circulation circuit path including the concentration vessel and a circulation pump. This steam heater operates to heat the abovementioned radioactive liquid waste thus supplied to the evaporation concentration device, whereby the liquid waste is evaporated and concentrated to a specific concentration.

The vapor generated by the abovedescribed heating process is conducted to a condenser where the same is cooled and rendered into a condensate. On the other hand, the waste solution concentrated as described above is discharged into a storage tank for concentrated waste solution after the operation of the abovementioned evaporation concentration device and liquid supply pump has been terminated.

In the abovedescribed process, however, there is the possibility of the heating tube of the steam heater becoming corroded and being damaged or broken depending on the properties of the radioactive liquid waste being treated. If this damage should occur, since the pressure of the heating steam is higher than the internal pressure of the evaporation concentration device when this device is operating, there is little possibility of the heating steam line being contaminated. However, when the operation of the evaporation concentration device is terminated, the heating steam condenses within the steam heater and is replaced by the concentrated waste solution within the evaporation concentration device, whereby, when the operation of the evaporation concentration device is restarted, this concentrated waste solution is introduced into the heating steam line and gives rise to operational difficulties such as, for example, radioactive contamination of the entire heating steam line.

H. Irie, F. Tajima, N. Kuribayashi and K. Isozaki; U.S. Patent 4,208,298; June 17, 1980; assigned to Tokyo Shibaura Denki KK, Nippon Genshiryoku Jigyo KK and Toshiba Engineering KK, all of Japan describe a process for treating radioactive liquid waste of a system having an evaporation concentration device comprising a concentration vessel for containing radioactive liquid waste supplied thereinto, and a steam heater having steam inlet and outlet lines and operating to heat the radioactive liquid waste with heating steam so as to concentrate the same by evaporation. A sluice valve is installed in the steam outlet line for stopping fluid flow therethrough; a discharge line, for discharging radioactive waste which may have infiltrated into the heating steam tube path of the steam heater so as not to contaminate the steam in the steam outlet line, has at an intermediate part thereof, a drain valve disposed therein, and is connected at one end thereof to the steam heater; and means is provided for receiving radioactive waste solution discharged from the other end of the discharge line.

In accordance with the process when the pressure within the steam heater is less than within the concentration vessel, and the operation of the evaporation concentration device is to be resumed, the sluice valve is first closed and the drain valve is subsequently opened. As a result, any concentrated waste solution which has infiltrated into the steam heater is discharged through the discharge line and drain valve. After a predetermined operative period of time, the drain valve is then closed and the sluice valve opened, and consequently, noncontaminated heating steam is again permitted to flow through the steam line.

Preheating Distillate to Be Degassed

Evaporation installations are frequently employed for the purification of liquids, particularly radioactive wastewater. The distillate resulting from such installations are ordinarily conducted from the condensing stage thereof to a final degassing stage to remove the remaining volatile impurities therefrom.

In order to obtain an effective degassing effect, it is necessary that the temperature of the distillate to be conducted to the degassing stage be at or slightly below its boiling point. Since the temperature at the outlet point of the condensing stage is, for economic reasons, generally substantially below its boiling point, preheating of the distillate prior to the degassing is required.

Such preheating can be obtained in various known ways, all of which require substantial additional amounts of heat as well as a relatively large and expensive technical plant. It is known to preheat common water to be degassed using a heat exchanger heated by steam. The temperature of such water at the entrance of the heat exchanger is relatively constant. Therefore it is not difficult to control such a preheater. However, because of the great difference between the heat of condensation of water and the specific heat thereof, any slight unavoidable change in the amount of the condensing vapor leads to a strong change in the temperature of the distillate. Such strong temperature oscillations in the distillate cannot be totally removed and can only be diminished through highly sensitive and correspondingly expensive and disturbance-susceptible control arrangements. So it is much more difficult to control preheating of distillate than of common water. Moreover, such arrangement must obviously be laid out for the lowest possible temperature of the distillate, which creates additional expense.

G.E. Rajakovics; U.S. Patent 3,990,951; November 9, 1976; assigned to Vereinigte Edelstahlwerke AG, Austria describes an evaporation arrangement whose condensing stage receives input steam from the evaporating stage via a main pipe.

In an illustrative embodiment, the preheater includes a housing that contains a reservoir in which the distillate may be heated. The distillate is introduced into the reservoir by means of an outlet pipe of the condensing stage.

In order to effect the preheating of the distillate in the reservoir to approximately its boiling point, a portion of the input steam progressing in the main pipe from the evaporating stage to the condensing stage is shunted off by means of a branch pipe within the housing, such branch pipe extending to the reservoir to discharge the steam portion so that the latter can be condensed by the distillate in the reservoir; the resulting heat of condensation is effective to provide the required temperature rise of such distillate. Such branch pipe communicates with the main pipe upstream of a diaphragm member which blocks from the condensing stage the portion of the steam which is to be directed via the branch pipe to the reservoir.

The heated distillate is conducted to the degassing stage via an overflow pipe disposed in the reservoir at a predetermined height therein. The arrangements in accordance with the process are particularly advantageous since they employ only a predetermined small portion of the steam normally present in the pipe between the operating stages, so that no additional source of steam for preheating the distillate is required. Moreover, the arrangement is self-regulating because of the small temperature difference between the condensing vapor and the desired temperature of the distillate. The outlet temperature of the distillate therefore will be approximately constant and nearly independent of the inlet temperature of the distillate coming from the condensing stage.

Two-Stage Flash Vaporization Technique

Unfortunately, conventional evaporation techniques are limited in effectiveness by the fact that radioactive droplets of small diameter from the liquid to be purified are unavoidably carried along with the vapor from the evaporation stage, and serve to impose what has heretofore been an irreducible minimum concentration of radioactivity in the final product distillate.

G.E. Rajakovics, H. Gabernig and G.P. Klein; U.S. Patent 4,043,875; August 23, 1977; assigned to Vereinigte Edelstahlwerke AG, (VEW), Austria describe a process which employs at least one unit including first and second flash vaporization stages. The radioactive liquid to be purified is introduced to the first vaporization stage to be circulated, heated and flash evaporated. The vapor resulting from the flash evaporation of the liquid in the first vaporization stage is applied to a first direct condenser, illustratively a mixing condenser. A portion or even all of the enriched liquid obtained from the second vaporization stage is applied via a circulating pump to the first direct condenser to condense by contact the vapor from the first vaporization stage and form a mixture with this condensate. At least a portion of this mixture is then flash evaporated in the second vaporization stage. As a result, the second vaporization stage is effective to flash-evaporate components derived from the first vaporization stage and indirectly, via feedback, from the second vaporization stage itself. The second vaporization stage, together with the circulating pump, the first direct condenser, and the connecting conduits therebetween, form a "loop". For limiting the enrichment of radioactivity in the second vaporization stage and, thereby, improving the effectiveness of the process, a portion of enriched liquid is withdrawn from the loop.

Preferably, the purified vapor emanating from the second vaporization stage is further condensed in a second indirect condenser. A portion of the resulting distillate is routed out of the unit for external utilization, and an additional portion of such distillate is fed back to the loop, illustratively on the suction side of the circulating pump.

Enriched liquid obtained from the loop may be: cooled; discharged from the unit, possibly by dilution with separate liquid; combined with additional waste liquid, a portion of the resulting mixture, for example, may be discharged, another portion or even all of this mixture may be recycled to the first vaporization stage; evaporated, e.g., in the first vaporization stage or in a third vaporization stage, the vapor formed in the third vaporization stage may be condensed by condensing means, such as the first direct condenser, the second indirect condenser, or a third condenser; directed to another unit having first and second vaporization stages too; processed by any possible combination of the above.

Entrainment Separator Apparatus

Nuclear generation of electric power produces radioactive liquid wastes. Since the water component is by far the largest volume of such wastes, it is necessary to remove all traces of radioactivity before the water is discharged. One method is to concentrate the radioactive wastes by evaporation of the water and then decontamination of the water vapors by separation of any entrained radioactive liquid droplets. Attempts have been made to separate radioactive water droplets from steam by coalescing of the liquid on separator components. Such separators operate efficiently only when the steam flows within some predetermined narrow velocity range. If the steam velocity strays outside such a range, the separator permits so much radioactive water to escape that whatever receives the steam vapors downstream will be seriously contaminated by radioactivity. The necessity of avoiding such contamination puts serious constraints on the permissible variations in the operating capacity of the system.

A.N. Chirico; U.S. Patent 4,084,945; April 18, 1978; assigned to Ecodyne Corporation describes a method of disposing of a variable volume flow of aqueous radioactive waste, comprising: heating the waste until the water component vaporizes; passing the vapors and any entrained radioactive water droplets at a varying range of velocities serially through a plurality of water droplet separators having decreasing cross-sectional areas in the direction of vapor flow; the velocity at which the vapors and entrained radioactive water droplets pass through the water droplet separators varying with the variable volume of flow; one of the water droplet separators having a relatively small cross-sectional area being most effective in removing radioactive water droplets when the vapor velocity is at the low end of the range, and another of the water droplet separators having a relatively large cross-sectional area being most effective in removing radioactive water droplets when the vapor velocity is at the high end of the range; removing decontaminated vapors; returning the radioactive water droplets to the waste being heated; and removing concentrated radioactive wastewater for disposal.

Hot Air Drum Evaporator

R.L. Black; U.S. Patent 4,305,780; December 15, 1981; assigned to The U.S. Department of Energy describes an evaporation system for handling aqueous radioactive wastes. More specifically, the process relates to a disposable evaporation system using standard 30 and 55 gallon drums as containment vessels for an adiabatic saturation process wherein moisture is absorbed by hot dry air moving through cascading spray and over standing liquid.

In accordance with the features of the process, standard 30 and 55 gallon drums may be used to house the major component parts and to serve as long-term storage vessels for radioactive material. Within a drum, waste solutions, heated or unheated, pass over a series of vertically arranged, donut-shaped trays as cascading water sprays, thereby increasing the air water interface. A down-draft of heated air moving through the center of the drum removes water vapor in an adiabatic saturation process in which relatively dry air picks up moisture. An air distribution manifold delivers air at low velocity to the space between adjacent trays in radial flow. The humidified air is removed from the drum and passed through a double high efficiency filter to assure no radioactive particles escape in the process. Free liquid within the drum is recirculated by a pump and heated by the heating coils to speed the evaporation process. The drum is surrounded by a 2-inch layer of fiber glass insulation and is packed in a larger drum. The larger drum is encased in concrete providing a shield. A moisture absorbent is packed inside the unit and the unit is then sealed prior to shipment for storage.

Gas Transport Medium

K. Szivós, G. Lovass, L. Lipták, J. Hirling and O. Pavlik; U.S. Patent 4,040,973; August 9, 1977; assigned to Magyar Tudomanyos Akademia Izotop Intezete, Hungary describe a process and apparatus providing more simple possibilities, compared to the conventional ones, for the concentration of radioactive waste materials at temperatures below their boiling points, and for the final storage thereof.

These aims can be fulfilled when the radioactive waste materials are concentrated at the site of their final storage, and the concentration is carried out using a transport medium of 0.5 to 1.5 atmospheric pressure, the medium circulating in a closed loop circuit, and transporting both heat and vapor.

According to a preferred embodiment, steam is injected in a suitable manner into this transport medium flowing in closed circuit, then it is condensed together with the vapors of the waste liquor, transported by the medium.

According to a preferred method, the transport medium is tangentially injected into the vapor site through one or more inlet air ducts, thus the transport medium approaches the surface of the liquid containing the waste material to be concentrated in a vortex stream, thereafter the exhausted medium is led along the outer mantle of the container containing the waste material. Thereafter the vapor contained in the transport medium is condensed, the medium is reheated, and recirculated.

After the concentration step the concentrate is surrounded with a self-solidifying impermeable substance, preferably with bitumen.

The apparatus in which the above process can be carried out consists of an outer and an inner container, preferably made of a single double-walled container, in which the radioactive waste material to be concentrated is placed into the open-surface inner container, and a spiral gas-diverting channel, serving to remove the exhausted transport medium, is arranged in the space between the two containers.

According to another embodiment of the process, a single container can be used instead of the two container system.

The characteristic operative parameters of the transport medium (air or nitrogen) in the system may be adjusted to the following values:

Transport Medium

Inlet point temperature of evaporator (°C)	80 to 160
Inlet point temperature of condenser (°C)	40 to 90
Outlet point temperature of condenser (°C)	30 to 80
Specific flow-rate related to the water sur- face (m^3/m^2 h)	200 to 400
Wastewater temperature in the inner con- tainer (°C)	50 to 90

The evaporation rate (related to the water surface) varies to a great extent depending on the nature of the transport medium and on the actual values of the operative parameters. The approximate value of the evaporation rate may vary between 2 to 15 kp/m^2 h. Also, the decontamination factor depends on a plurality of parameters; its approximate value may vary between 10^5 to 10^6 (without purification of the vapor).

Depending on the composition of the starting liquid wastes, the liquid wastes can be evaporated using the apparatus so as to reach a solid content of 500 to 1,000 g/ℓ, thus the necessary storage volume, related to the volume of the starting liquid wastes, is far lower than in the case of other known apparatuses (with the exception of the IR-heated evaporator, where a similar concentration rate can be achieved). The apparatus can be operated economically while it can be heated using nonexpensive fuels (steam, hot water, etc.).

Evaporator Employing Washing Liquid

H. Mende; U.S. Patent 3,969,194; July 13, 1976; assigned to Luwa AG, Switzerland describes a method and apparatus which render possible reducing the concentration of the radioactive substance in the droplets entrained by the vapor out of a washing liquid in an economical manner.

The method is manifested in that the washing liquid is deflected a number of times at the floor in such a way that the washing liquid itself together with the droplets entrained by the vapor are uniformly admixed, and the washing liquid subjected to a constant increase or intake of the radioactive substance.

A further aspect of the process is the provision of apparatus for the performance of the aforesaid method which is of the type possessing an evaporator as well as a column connected therewith, the column being provided with a multiplicity of floors. Each floor has operatively associated therewith an infeed connection and an outfeed connection with associated infeed weir and outfeed weir. Further between the infeed weir and the outfeed weir there are arranged a multiplicity of guide walls, and the ends of the guide walls together with the inner wall of the column limit through-passages.

Due to the deflection of the washing or scrubbing liquid at the column floors there is achieved the result that the entire washing liquid is uniformly admixed thereat and thus there are prevented so-called dead zones. This admixing has the effect that in all of the washing liquid zones there prevails approximately a uniform radioactivity.

This effect renders possible a reduction in the quantity of washing liquid, resulting in a reduction in the required heat of vaporization. Reduction of the vaporization heat is of course of comparable significance to an increase of the economy of the method.

It has been further found that with the apparatus the drop in radioactivity from one floor to the other is increased in contrast to the heretofore known equipment. The reason for this resides in the fact that the droplets entrained by the vapor stream from one floor to the other possess a concentration of dissolved salts or suspended solid particles which on the average is less than that of the droplets produced in the known methods and apparatuses. Since the quantity of dissolved salts or suspended solid particles decreases per unit of volume, the radioactivity also decreases.

It is also found that at the floors of the known devices there are present zones with a very small radioactivity. One such zone is constituted by the central floor portion which extends from the infeed connection to the removal or outfeed connection. However, in this case the radioactivity in the marginal zones is so high that the droplets entrained from these zones by the vapor more markedly load on the average the next higher floor than the droplets which are entrained by the vapor streams with this method and apparatus respectively.

Addition of Activated Carbon, $KMnO_4$, $MnSO_4$ and $CaCO_3$

E. Schieder and R. Stoiber; U.S. Patent 4,303,511; December 1, 1981; assigned to Kraftwerk Union AG, Germany describe a method of purifying tenside- and detergent-containing contaminated wastewaters of nuclear power plant installations and other plants in which radioactive substances are processed, wherein the contaminated water is, prior to return to a water loop of the plant, passed through an evaporator and a mixed-bed filter.

The improvement comprises: (a) adjusting the pH of the contaminated wastewater to a value of 2.5 to 3 by the addition of sulfuric acid; (b) adding the following reactants to the contaminated wastewater: finely-divided activated carbon, $KMnO_4$, $MnSO_4$ and $CaCO_3$; (c) agitating the wastewater and reactants to effect intimate mixing of the reactants with the wastewater; (d) adjusting the pH of the mixture to a value of 8.5 to 9 by the addition of an alkaline material; and (e) settling the alkaline mixture to form a lower sludge layer and a supernatant water layer of reduced contaminant content which is subsequently passed to the evaporator and mixed-bed filter.

In accordance with the process, the wastewater is adjusted to a pH-value of 2.5 to 3 in a tank by means of sulfuric acid and is then reacted with finely-divided activated carbon, $KMnO_4$, $MnSO_4$ and $CaCO_3$. The mixture is mixed mechanically or by means of injected air for several minutes. The pH-value of the thus treated water is then raised, for instance, by ammonia or sodium hydroxide to 8.5 to 9 and is again mixed for several minutes. The supernatant water is further processed in known manner. The further processing involves feeding this treated water to an evaporator and the distillate thereof to a mixed-bed filter so that the filtrate leaving the latter can be returned to the loop of the plant from which the contaminated wastewater was obtained.

In a long series of tests involving the trial of various reagents, the combination of four agents, namely activated carbon, $KMnO_4$, $MnSO_4$ and $CaCO_3$ was found to give optimum results. The quantities of agents will of course vary with the quantities of contaminants. Excess quantities of agents, while generally not harmful, are wasteful. In practice, it was found that a water composition having 10 mg/ℓ cation-active, 5 mg/ℓ anion-active and 50 mg/ℓ neutral tensides, with a minor amount, a small fraction of a percent of oils, and fats can be adequately treated with 100 mg/ℓ activated carbon, 50 mg/ℓ $KMnO_4$, 50 mg/ℓ $MnSO_4$ and 50 mg/ℓ $CaCO_3$ and will serve as a guideline for the quantities of agents to be used.

Often wastewaters almost always show about the same tenside contents. In such instances the optimum quantities of agents can be readily determined by a few tests and the method thereby optimized also with respect to its economy.

The course of the purification process is summarized schematically in Figure 3.1. The wastewater containing the contaminants is usually at a pH value above 6, more generally above 7, i.e., on the alkaline side and is first placed in a collecting vessel which may be any suitable tank equipped with agitating means, for instance, with a stirrer or also with an air injection device. First, the wastewater is adjusted by the addition of sulfuric acid to a pH value of 2.5 to 3. Reaction with agents takes place best at that pH value. Subsequently, finely distributed activated carbon, $KMnO_4$, $MnSO_4$ and $CaCO_3$ are added in accordance with the quantity of wastewater and its contaminant content contained therein, and are thoroughly mixed for several minutes.

After this mixing process, the pH value of the mixture is raised to 8.5 to 9 by the addition of an alkaline material, for instance, ammonia or sodium hydroxide and again mixed for several minutes either mechanically or by injection of air. Sludge is formed in the process and slowly settles at the bottom of the collecting tank forming a lower sludge layer in the tank and an upper supernatant water layer which has a markedly lower content of tensides, oils and fats than the contaminated wastewater.

This settling procedure takes several hours. Subsequently, the sludge is drawn off to remove the waste and sent to the waste disposal system. The supernatant water is transferred from the collecting tank to the evaporator which is conventional equipment used in the treatment of water. The distillate released from the evaporator is then further purified in a mixed-bed filter of ion-exchange resins and also conventional, and the filtrate is returned to the loop of the plant. The sump remaining in the evaporator is likewise directed to the waste disposal system.

To illustrate the operation of this method, contaminated water containing about 10 mg/ℓ cation-active, about 5 mg/ℓ anion-active and about 50 mg/ℓ neutral tensides, oils and fats was acidified with sulfuric acid to a pH of 2.5 to 3 and then treated with 100 mg/ℓ activated carbon, 50 mg/ℓ $KMnO_4$, 50 mg/ℓ $MnSO_4$ and 50 mg/ℓ $CaCO_3$. The pH was raised to 8.5 to 9 and sludge allowed to settle at the bottom. The treated water contained only 0.6 mg/ℓ cation-active tensides, 1.0 mg/ℓ anion-active tensides and less than 1 mg/ℓ neutral tensides, oils and fats when transferred into the evaporator.

Figure 3.1: Purification of Tenside- and Detergent-Containing Wastewaters

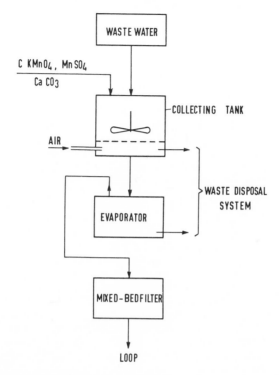

Source: U.S. Patent 4,303,511

Tests with very different tenside contents in washing waters of nuclear power plants have always shown an efficiency of considerably above 80%. The filtrates from the mixed-bed filter could not only be returned to the loop but could also be fed to the normal wastewater, as they had not yet reached the legal limits for wastewater applicable thereto.

It is advantageous to admix commercially available antifoaming agents to the wastewater, since this facilitates the mixing processes by inhibiting foam formation.

Iodine Retention During Evaporation

H.W. Godbee, G.I. Cathers and R.E. Blanco; U.S. Patent 3,920,577; November 18, 1975; assigned to The U.S. Energy Research and Development Administration have found that aqueous waste solutions from nuclear installations may be effectively concentrated for eventual storage of the radioactive contaminants, while being able to release about 90 to 99 volume percent of the waste solution back to the environment.

The first step in the process is to isotopically dilute the waste solution with nonradioactive iodine. The amount of nonradioactive iodine diluent added to the solution is at least 10^3 and preferably about 10^6 times the amount of radioactive iodine in solution. By isotopic dilution and the phenomenon of isotopic exchange, every chemical species of iodine within the solution should be composed of radioactive iodine and nonradioactive iodine in the same ratio as that of the active to the diluent isotopes in solution. The amount of volatile iodine species containing radioactive iodine is thus proportionately decreased because of the increase in the amount of those species containing nonradioactive iodine.

After isotopic dilution, the pH of the solution is raised to at least 9. Most of the volatile forms of iodine are in the higher oxidation states, e.g., +5 or +7. By maintaining the high pH, reduction of the iodine in the higher oxidation states to the iodide state is favored. The pH of the solution is adjusted by the addition of alkali or alkaline earth hydroxides, with sodium hydroxide being preferred.

As a further measure to insure that the iodine is reduced to the iodide state and to prevent the iodides from converting to free iodine, a reducing agent is added to the solution. Alkali sulfites or thiosulfates may be used as reducing agents in this step. Sodium sulfite and sodium thiosulfate are excellent reducing agents for use in this process with sodium sulfite in particular being preferred. The amount of reducing agent used in this step is an amount theoretically sufficient to reduce all zero and positive valence iodine species to the iodide state. This amount will, of course, vary depending on the particular type of waste and the amount of iodine present.

Having carried out the above steps, the thus treated waste solution is boiled to evaporate the aqueous phase and concentrate the radioactive as well as other contaminants.

It has been found that the combination of steps according to this process synergistically improves the iodine retention within an evaporator. If none of the above steps are taken, a decontamination factor of only about 50 to 100 is obtained. If only isotopic dilution is utilized a decontamination factor of about 5×10^2 is obtained. A similar decontamination factor is obtained utilizing isotopic dilution and pH control. However, when all three steps are carried out prior to evaporation, a decontamination factor of 10^3 to 10^4 is achieved.

Replacement Pump System

When the operation of a pump is vital to a system, duplicate pumps are frequently installed side by side. One such pump can always be maintained as a spare in top operating condition for immediate use should the working pump fail. In systems for concentrating aqueous radioactive waste by evaporation, a large volume of such waste must be pumped continuously through a waste concentrator vessel. If the pump fails, the system must be shut down until a new pump is put into operation. All operating components of a radioactive waste treatment system must be isolated within a shielded enclosure. Installing duplicate large volume pumps in such a shielded enclosure would disproportionately increase the size and cost of the system. Also, personnel who maintain the spare pump would have to enter the shielded enclosure frequently to check it out, and thereby expose themselves to radioactivity.

A.N. Chirico; U.S. Patent 4,246,065; January 20, 1981; assigned to Ecodyne Corporation describes a system in which a pump for a radioactive waste concentrator in a shielded enclosure is mounted on a wheeled vehicle that travels on tracks, and a spare pump is mounted on an identical wheeled vehicle outside of the shielded enclosure, adjacent the tracks. If the pump fails, it is wheeled out of the shielded enclosure, and the vehicle carrying the spare pump is put on the tracks and wheeled into the enclosure for quick connection to the concentrator.

Easier Maintenance of Evaporator System

It has been recognized in the operation of nuclear power plants that liquids accumulated from floor drainage systems and other internal reactor-related sources constitute a potential radiation hazard, and must therefore be handled and disposed of as radioactive waste material, typically by sealing the waste in concrete drums or barrels, and then burying it either on land or at sea. Because of the relatively high expense of the disposal process, it is highly desirable that the radioactive liquid waste be concentrated prior to being sealed in the containers for burial. To this end the waste may be first processed in one or more evaporator stages, wherein a large portion of the liquid is removed as harmless nonradioactive steam or water vapor leaving concentrated liquor for disposal.

Unfortunately, the concentrated waste liquor, which unlike primary water may contain suspended solids such as calcium, sodium and magnesium carbonates and sulfates in variable and unpredictable concentrations, is prone to boil and produce scaling in the heat exchanger and the other components of the conventional calandria and LTV natural flow evaporator systems heretofore used in the primary water concentrating process. This eventually reduces the efficiency of the evaporator to the point that its utility in the concentrating process is seriously impaired. While it is possible to remove some of the accumulated scaling by boiling out the evaporator system with a suitable solvent, it eventually becomes necessary to shut down and dismantle the system so that insoluble scales and plugged tubes can be cleaned by mechanical means. Because the liquor within the evaporator systems is highly radioactive and the systems are therefore normally installed in concrete vaults for shielding and are not easily accessible, dismantling prior-art systems for descaling or repair has been a time-consuming and expensive operation.

While forced circulation (FC) evaporators, where a pump or other mechanical circulation means is provided to circulate the liquor, overcome the scaling problem to a large degree by keeping the liquor circulating at a controlled rate with carefully controlled temperature gradients within the heat exchanger whereby boiling is restricted to the evaporation chamber, the use of these evaporators for concentrating radioactive liquids has been avoided in the past because of their use of a pump.

Heretofore, such pumps, which were arranged below the level of the liquid in the system and therefore necessarily required a shaft seal resistant to the passage of the fluid being pumped, required frequent servicing which could be accomplished only with great difficulty because of the surrounding radiation shielding and the high radiation levels of the liquor. Furthermore, the use of a pump was considered objectionable because it increased the volume requirements of the system and made necessary additional access passageways, thus significantly increasing the required radiation shielding around the evaporator.

W.E. Rushton; U.S. Patent 3,933,576; January 20, 1976; assigned to Whiting Corporation describes an evaporator for a nuclear power plant which minimizes scaling and the need for periodic maintenance, and a circulation pump for the evaporator system which can be serviced from one access plane without draining the system.

To attain the above objectives, an evaporator system is provided wherein the residence time of the radioactive waste in the heater tubes as well as the amount of heat transferred to the liquid during its passage through the tubes can be controlled, which permits minimization of scale formation. Furthermore, all components conventionally requiring servicing are arranged for convenient removal through an access port formed in the roof of the radiation chamber, and the arrangement and configuration of the components commonly requiring service is such as to minimize the need for frequent attention, particularly in the design and location of the troublesome pump shaft seal.

The evaporator includes an upright evaporation chamber providing a vapor head space above the intended liquid level and adapted to operate under atmospheric pressure. Conduit means interconnect the evaporation chamber with a pump chamber which supports on its removable head a pump drive from which is suspended a pump shaft and propeller, the pump shaft extending above the level of the liquid in the chamber. The pump operates to force the radioactive liquid waste from the pump chamber through second conduit means to an upright heat exchanger at a predetermined rate, the heat exchanger communicating by third conduit means with the evaporator chamber to form a closed loop.

The tubes of the heat exchanger and the steam chest which surround the tubes are secured to a removable head, thus permitting removal of the tube bundle and steam chest as a unit from the access opening in the chamber roof. A removable head is also provided on the evaporator chamber to permit removal of a demisting device.

The pump propeller is arranged in a channel of circular cross section which fits closely about the propeller tips in such manner that in operation the force of the fluid driven by the propeller also centers the propeller and eliminates the need for an immersed bearing. It is therefore possible by removing the horizontal pump mounting plate or head to remove the motor, the pump gear box, the pump shaft, the propeller, and the pump shaft seal as a unit through an overhead access opening. Furthermore, since the shaft seal is not required to withstand hydrostatic pressure or appreciable vapor pressure, it is capable of extended operation without maintenance or replacement.

DRYING AND SOLIDIFICATION

Treatment of Wastewater from Uranium Ore Preparation

Uranium, which principally finds application as fissionable material or nuclear fuel in modern atomic power plants, is obtained from uranium ore which at most contains only a fraction of one percent uranium.

In order to extract the uranium, the ore is subjected to a chemical treatment after some preliminary filtration and other mechanical steps. During such chemical treatment, which employs fresh water and in which gross quantities of auxiliary chemicals (e.g., acids during the ion exchange or adsorption operations for removing certain radioactive components) are used up extraordinarily large amounts of radioactively contaminated amounts of wastewater are generated. The wastewater emitted from installations of this type is in the form of aqueous solutions of inorganic salts, and also contains residual uranium and elements resulting from the decomposition of uranium, such as radium, polonium and lead. The treatment of such wastewater to negate its dangers represents a difficult problem for each installation.

Two principal methods of such treatment have been commonly employed. In one, which has been used where natural water courses or streams occur in the vicinity of the installation, the wastewater is further diluted with water and discharged into the stream, usually after some pretreatment via ion exchange or adsorption. Such treatment, of course, leads to radioactive contamination and poisoning of the water course.

In the second commonly used treatment method, employed where there are no water courses in the vicinity of the chemical treatment installation, the wastewater is collected in a natural, open settling tank. The spent ore is collected in the bottom portion of the settling tank, while the remainder of the wastewater which forms the top layer, is fed back to the installation for reuse.

The principal disadvantage of this latter scheme is that its life is limited, since the concentration of salt in the wastewater gradually climbs to an unacceptable limiting value beyond which a reuse of the water is no longer possible. Additionally, such settling tanks are expensive to maintain and, of course, exhibit the danger that the salts accumulating in the lower ground layers can leak out, leading again to contamination or poisoning.

V. Klicka, J. Mitas and J. Vacek; U.S. Patent 3,988,414; October 26, 1976; assigned to Vyzkumny ustav chemickych zarizeni, Czechoslovakia describe a safe and economical treatment technique for the wastewater which avoids these problems and operates in an improved way to recover the wastewater constituents for further reuse. This technique involves concentrating the wastewater through evaporation, and crystallization of the least soluble salt component from the resulting solution. This component is removed from the mother liquor and supplied to a suitable external utilization apparatus. The remaining mother liquor is fed back (preferably after the residual uranium is removed therefrom) to the installation as a substitute for the acids employed for the adsorption portion of the treatment, while the condensate distilled from the wastewater is fed back to the installation as a substitute for the fresh water employed therein.

The crystallization of the least soluble salt component is accomplished by subjecting the concentrated solutions to further evaporation, or to cooling, or to simultaneous cooling and vacuum evaporation.

The salt component crystallized after the first evaporation can, if desired, be redissolved to form a second concentrated solution, followed by recrystallization to obtain a salt component of higher radioactive purity.

Among the advantages of the process are its safety of operation and its economy of heat, water and chemicals, particularly during the effective feedback of steam condensate and the mother liquor. Also, it permits the effective use of installations in regions without large water courses, while avoiding the necessity of constructing spacious settling tanks with their attendant high maintenance expense and problems in removing the resultant radioactively contaminated products.

Additionally, the cost savings resulting from the effective utilization via feedback of the decomposition or disintegration products of the wastewater approximately make up for the cost of the processing of the wastewater. Finally, through the use of multistage arrangements in each evaporation station and through the employment of the heat of condensation to run the crystallizer, the steam supply requirements can be significantly reduced.

Two-Cylinder Dryer

H. Queiser, O. Meichsner and D. Erbse; U.S. Patent 4,314,877; February 9, 1982; assigned to Kraftwerk Union AG, Germany describe an apparatus for drying radioactive wastewater concentrates from evaporators.

The apparatus comprises: (a) a feed tank equipped with a stirrer and with a jacket for the passage of a heating medium to heat the tank; (b) an inlet opening in the tank for the introduction of wastewater concentrates from evaporators; (c) a discharge opening near the bottom of the tank and a return opening near the top of the tank together with connecting conduits and an interposed pump to form a circulatory loop for circulating wastewater concentrate from and to the tank; (d) a bleeder line having one end connected to the circulatory loop for bleeding wastewater concentrate, and having a line length shorter than the loop; (e) a pump connected to the bleeder line to regulate the flow of concentrate therethrough; and (f) the bleeder line having its other end connected to a two-cylinder dryer with a scraper associated with the cylinders to remove dry material from the cylinders.

In accordance with the process, the concentrate is preheated to 60°C and prior to drying it has a solids content of at most 20%. This preheated concentrate is fed to the dryer in a quantity of 10 to 20 ℓ per m^2 useful cylinder surface per hour for dwelling times of 7 to 18 seconds at cylinder temperatures between 160° to 210°C. The dry product produced by the method has low residual moisture of only 1 to 5% and therefore, excellent properties with respect to further processing. It can easily be taken off the cylinders and then worked with known embedding materials such as cement, bitumen, or plastic with which it is preferably filled into the customary 200 ℓ barrels and shipped for ultimate storage.

However, it is also well suited for interim storage without binder because it is only slightly hydroscopic. This means that the end product can be stored as a dry powder under the exclusion of air for extended periods of time until it can be removed according to the most recent ultimate storage requirements, possibly together with other residues in powder form in economically advantageous quantities and advantageously with respect, for instance, to permissible activity, abovementioned binders, etc.

The low residual moisture, which is decisive for the storability and has heretofore been neglected in this respect, can be achieved due to the preheating, the pH value and the underpressure at very specific cylinder temperatures which depend on the composition of the concentrate. For boiling water reactor evaporator concentrates and pressurized water reactor evaporator concentrates with organic components, a temperature of about 160°C applies, and for pressurized water reactor evaporator concentrates, a temperature of 180° to 210°C. It was ascertained that at such high temperatures (above 180°C), in conjunction with pH values of 3 to 8, boron salts with little crystal water content are produced which are far less hygroscopic than in other drying processes. The sodium metaborate content of the concentrate must be set to at most 10% by weight, especially by the addition of calcium chloride, because then the crystal form desired for the dry product is obtained with certainty.

Extensive investigations with a two-cylinder dryer for determining the operating parameters for the different types of concentrate to be powderized showed, on the basis of tests with other types of dryers, that, to obtain a dry product at all and particularly one with especially low hygroscopic properties, the parameters batch size, layer thickness on the cylinder, dwelling or residence time on the cylinder, usable cylinder surface, dry component content of the concentrate, the temperature of the cylinders and of the concentrate, and the pressure in the cylinder dryer are important. The parameters should fulfill, speaking generally, the following relation:

$$\frac{Q \times t_v}{F_N \times s} \times \frac{t_W}{t_M} \times p \times C_M = 15 \text{ to } 635 \times 10^{-5}$$

where

t_M = concentrate temperature (°C)
p = residual moisture of the product (%)
C_M = content of dry material in the concentrate (%)
Q = charging quantity (m^3/h)
t_v = dwelling time (h)
F_N = usable cylinder surface (m^2)
s = layer thickness (m)
t_W = cylinder temperature (°C)

It is advantageous to set the pH value to 3 to 8, depending on the concentrate composition, and an underpressure, i.e., subatmospheric or reduced pressure of 0.1 to 0.8 bar. The dwelling or residence time which determines the amount of heat supplied, is obtained for commercially available cylinder dryers at cylinder speeds of 2 to 5 rpm. The content of dry material (solids, dry basis) in the concentrate is preferably 15% by weight. However, the content of filterable solids should be at most 5% by weight. These limits can be maintained by admixing, for instance, filter concentrates or sedimentation with elutrition.

Inert Carrier Drying and Coating System

R.D. Sheeline; U.S. Patents 4,246,233; January 20, 1981 and 4,119,560; October 10, 1978; both assigned to United Technologies Corporation describes a system wherein a solution (which term includes both true solutions as well as dispersions) of liquid solvent and a solid solute is introduced into a hot inert carrier to cause the solvent to flash leaving dried solute in the inert carrier in the form of dispersed solid particles.

The inert carrier carrying the particles then flows to a second station where a binder for the particles is introduced to coat the particles by preferential wetting and then the coated particles coalesce so they can be readily separated from the inert carrier by gravity in a separation stage. As used herein the terms "preferential wetting" or "preferentially wetted" describe that condition which exists when the solid particles have a greater affinity to be wetted by the liquid binder than by the inert carrier. The existence of this condition is readily determinable since the liquid binder can actually be observed to displace the inert carrier as it flows around and coats the solid particle. Further, if this condition does not exist the process does not function in that the particles do not get coated and the result is a suspension of binder in the carrier and a suspension of particles in the carrier.

In general, preferential wetting will usually exist when the carrier is nonpolar and the binder and particles are polar or vice-versa, for example, although this may not be 100% predictable. The existence of the condition in specific systems can be verified by placing the materials in a Teflon or other nonsticking container at the operating conditions and shaking. If coalescing occurs as a separate phase, preferential wetting exists.

This process is useful whenever it is necessary to remove the solvent from a solution and/or encapsulate the dried, solid solute and in its most general application the following criteria must be met:

(1) The solid solute should be insoluble in and nonreactive with the inert carrier.

(2) The binder should be insoluble in and nonreactive with the inert carrier so that it is capable of forming a separate phase in the carrier.

(3) The binder should be a liquid at the operating condition but capable of solidifying, either thermoplastically or through a chemical reaction, upon removal from the system.

(4) The inert carrier should be a liquid with a relatively low vapor pressure to permit its continued reuse without extensive recovery operations.

(5) The particles should be preferentially wetted by the binder.

The binder may be any suitable polymeric material or cementitious material such as polyethylene, polypropylene, polystyrene, phenolics, cellulosics, epoxies, polyesters, acrylonitrile-butadiene-styrene (ABS), urea-formaldehydes, and others. The general characteristics of the binder are that it be relatively fluid at the temperatures of the process, be capable of encapsulating the particulate material by preferential wetting and be capable of hardening into a solid mass on curing or on cooling to ambient conditions.

For drying and coating aqueous solutions such high boiling liquids as paraffinic hydrocarbons, silicone fluids, phthalates, commercial heat transfer fluids such as Therminol or Dowtherm, high molecular weight alcohols, high temperature liquid polymers and others are suitable carriers and the previously listed polymers are suitable binders. This list is merely exemplary since an almost infinite combination of materials can be employed within the selection criteria set out above.

In a typical system the dried and coated end product may be between 65 and 75% particulate material such as sodium sulfate and 35 to 25% binder. The actual composition for any particular system may vary greatly.

It has been found that as the particle size of the particulate material is increased a higher solids loading can generally be obtained. The particle size distribution can be controlled by appropriate selection of the temperature of the evaporator, with higher temperatures yielding generally smaller particles and lower temperatures yielding generally larger particles. Another factor affecting particle size is average residence time of crystals in the evaporator. With longer residence times the recirculating particles contact fresh droplets of solution and can grow.

Example: An inert carrier drying and coating system was designed to process one gallon per minute of 20% aqueous sodium sulfate radwaste solution employing a dimethyl silicone oil as the inert carrier and a glycidyl ether, such as Epon (Shell Chemical Co.) as the binder. Hexahydrophthalic anhydride is used as the curing agent. The product cures in 3 hours at 300°F. The system was designed with a nominal operating temperature in the evaporator of 300°F. The inert carrier is recirculated through the heat exchangers at a high rate of approximately 125 gallons per minute and the temperature is increased to 320°F by 150 psi steam flowing through the heat exchanger. In the processing of the 20% sodium sulfate solution at a rate of 60 gallons per hour (120 lb/hr Na_2SO_4 and 470 lb/hr H_2O), binder is fed into the inert carrier through the jet mixer at the rate of 34.2 lb/hr and the coated particles removed in the separator.

The epoxy resin used is a solid at ambient temperatures and liquid at the 300°F operating temperature of the system. It forms a thermoplastic solid mass of sodium sulfate encapsulated in epoxy resin upon removal from the separator and cooling. The same resin system can be formed into a permanent solid by the addition of 5.8 lb/hr of curing agent and maintaining the removed product at 300°F for three hours. This produces approximately 1.2 ft^3 per hour of cured, dried, coated 75% Na_2SO_4. This cured product is stable at temperatures far higher than 300°F and significantly enhances the inherently low leach rate of the system. A comparison of the coated product with a conventional sodium sulfate-cement mixture shows a leach rate 3% of the cement leach rate.

Monitoring of Water Content

The treatment of aqueous solutions and suspensions of radioactive waste by drying and solidification has a problem in obtaining good solidified bodies. That is, no good solidified bodies are obtained only by compressing the powders to make the solidified bodies, unless the powders having a water content below a definite limit content are employed. The limit content is called "limit water content", which is usually a few percents, but is determined on the basis of strength, etc. required for the solidified bodies.

The aqueous solutions and suspensions of radioactive waste are continuously fed to the drier, and powders are formed. The resulting powders have not always a constant water content, depending upon the properties of the aqueous solutions and suspensions of radioactive waste and operating conditions of the drier. Sometimes, powders having a water content far above the limit water content are formed. No good solidified bodies are obtained from such poor powders only by direct compression, and thus a further step is necessary to take the resulting powders having the undesired water content, from the drier.

M. Hirano and S. Horiuchi; U.S. Patent 4,234,448; November 18, 1980; assigned to Hitachi, Ltd., Japan describe a method and an apparatus which comprises measuring a water content of powders between a step of drying and pulverizing aqueous solutions and suspensions of radioactive waste and a step of compressing and solidifying the resulting powders, and eliminating the powders when the measured water content of the powders is above a limit water content, from the system of treatment by washing without passing through the step of compressing and solidifying.

The present method for treating aqueous solutions and suspensions of radioactive waste comprises a step of drying and pulverizing aqueous solutions and suspensions of radioactive waste, a step of transferring powders formed in the step of drying and pulverizing to a hopper, a step of measuring a water content of the powders, and a step of pelletizing the powders, wherein a further step of introducing a washing solution into the hopper at the bottom when the measured water content of the powders fails to satisfy a predetermined condition, thereby dissolving the powders, and then discharging the washing solution is provided. The predetermined condition means a water content below the limit water content.

As a means for removing the powders from the hopper, an ordinary powder transportation means such as pneumatic transportation, etc., would be available, for example, in the case of dry powders, but is not applicable to powders of high water content. Thus, when the water content of powders is above the limit water content, a means of dissolution and washing is employed in this process. However, in that case, it is not suitable merely to introduce the washing solution to the hopper at the top and wash out the powders, because the washing solution cannot permeate into the powders. That is, dissolution of the powders can be readily carried out by introducing the washing solution to the hopper at the bottom, causing natural stirring in the hopper.

This process provides an apparatus for treating aqueous solutions and suspensions of radioactive waste, which comprises a drier, a pelletizer provided downstream of the drier, and a hopper and a water content-measuring device each provided in a conduit from the drier to the pelletizer, wherein an inlet for introducing a washing solution is provided at the bottom of the hopper. The term "the bottom of the hopper" means not only a bottom end of the hopper, but also a lower part near the bottom end, because sometimes the inlet for introducing the washing solution cannot be provided at the bottom end of the hopper owing to a design problem. The term "the bottom of the hopper" covers the bottom end and also the part near the bottom end and means a region where a stirring action is attained by introducing the washing solution.

Furthermore, a gas pipe can be connected to the inlet for introducing the washing solution to the hopper at the same time to inject air, etc., to the hopper at the washing to intensify the stirring action, or inject dry air, etc., after the washing to dry the system.

A further feature of this process relates to a treatment of a waste regenerating solution for desalters and resins, which takes most of the aqueous solutions and suspensions of radioactive waste generated in the atomic power stations. More than 80 to 90% of the powders formed by drying the waste regenerating solution for desalters and resins is comprised of sodium sulfate, Na_2SO_4. It is known as a physical property of sodium sulfate that sodium sulfate becomes very hard when deposited from its solution or when its powders are admixed with cold water.

Sodium sulfate has a maximum solubility of 33% at about 32°C. The solubility is slightly decreased in a temperature range above 32°C, but considerably decreased at a temperature range below 32°C. It is also known that precipitates formed above 32°C take the form of Na_2SO_4, whereas those formed below 32°C have water of crystallization in the form of $Na_2SO_4 \cdot 10H_2O$. For example, it was found by experiments that, when two-fold volume (based on volume of Na_2SO_4 powders) of city water at about 15°C was poured onto the Na_2SO_4 powders to dissolve the powders, very hard crystals were formed and dissolution of the powders was difficult to make, whereas, when the same volume of water at a temperature above 32°C, for example, hot water at 60°C, was poured onto the Na_2SO_4 powders, no crystals were formed, and the powders were converted to a slurry state.

That is, this process is further characterized by using water above 32°C, for example, hot water at 60°C or higher, as the washing solution, when most of the solid material of the aqueous solutions and suspensions of radioactive waste is comprised of Na_2SO_4.

Solid Particles from Droplets

K. Knotik, P. Leichter, E. Proksch and H. Huschka; U.S. Patent 4,203,863; May 20, 1980; assigned to Nukem GmbH, Germany describe a process for the production of solid particles from bioinjurious waste, e.g., radioactive concentrates, wherein a liquid, and in a given case binder-containing waste is divided into drops and supplied energy in a gaseous medium and the liquid is evaporated and solid particles are formed from the drops consists essentially of supplying carrier free energy to at least partially evaporate the liquid of the drops, in a given case under reduced pressure, and in a given case forming with a binder solid particles from the drops.

Consequently, it is therefore possible according to the process to divide into drops liquid waste mixtures, if necessary together with organic and/or inorganic binders, as, e.g., natural or synthetic glue, synthetic resins or their components, silicates, e.g., sodium silicate or potassium silicate, borates, e.g., sodium borate or potassium borate, etc., by means of a nozzle or the like and to change these drops into solid particles at reduced pressure or at normal pressure in a gaseous, nonturbulent atmosphere during the falling to the bottom by supplying carrier free energy.

An evaporation of the solvent can be caused by the supplying of energy or there can be attained a removal of the binder. In the use of synthetic resins as binder there are particularly suited as binders phenol and urea resins, e.g., phenol-formaldehyde, cresol-formaldehyde, resorcinol-formaldehyde and urea-formaldehyde. These are employed in the form of their components, e.g., phenol, resorcinol and/or urea or melamine or in the form of precondensates of the dissolved starting materials and dissolved therein. As hardening agent there is added an aqueous formaldehyde solution (e.g., a 40% solution) and in a given case a catalyst, for example, a mineral acid, e.g., hydrochloric acid or sulfuric acid, shortly before dropping the starting mixture out of the nozzle. Also by emulsifying in hydrophobic materials, as for example, bitumen, polyethylene, polystyrene, etc., there can be produced with heating of the drops a protective coating of the particle surface.

A further possibility is to coat the surface of the dropping drops, for example, by supplying a synthetic resin emulsion based on a silicone or similar compound in the dropping through a concentric double nozzle. Among others, for example, the waste solution can be mixed with a phenolic resin component and be brought together with a reactive gas, in this case, for example, formaldehyde vapor. There is formed a surface resin layer. According to an additional form of the process the energy is supplied by radiation, particularly IR rays. This type of energy carrier permits a particularly simple apparatus solution which is of particular significance directly in connection with the working up of bioinjurious waste since thereby maintenance and repair work can be held especially low.

According to a further form of the process, the energy is supplied by microwaves. The frequency region with aqueous solution is varied between 8 to 40 giga Hertz (gHz) depending on the size of the drops. By these means an especially simple heating of the drops can be attained, even in the interior of the drops so that there is caused a uniform and quick escape of the liquid.

The energy for drying the drops can additionally be supplied by electron beams, gamma rays in which case on the one hand there is present a dependence on the binder and on the other hand on the construction and length of the apparatus which can be between two and eight meters. The range of the energy spectrum is arranged according to the type of liquid, the binder added and according to the size of the drops.

It has proven especially advantageous to work with a drop size of approximately 0.5 to 5 mm. With this size of drops on the one hand it can be safely established that the solid drops obtained do not dust and on the other hand that solid, discrete particles are obtained which are particularly suited for a further working up.

Example; A radioactive aqueous waste concentrate solution which contained per liter of concentrate the ingredients below and radioactive impurities was divided by a nozzle into drops between 0.5 and 5 mm and subsequently converted by drying into solid particles.

	Grams
$NaBO_2$	200
KNO_3	10
$CaCl_2$	10
Na_3PO_4	5

It is not necessary to add a binder to this waste concentrate solution since the borate acts as such. The nozzle is arranged at the upper end of an essentially tubular shaped apparatus in which the drops fall to the bottom. A slightly reduced pressure prevails in the inside so that no toxic materials can go out of the apparatus and simultaneously the evaporation of the water is made easier. Air serves as the gaseous medium. The energy was supplied by an infrared (IR) emitter whose maximum output was between 3 and 6 μ. The use of an IR laser is also possible. The steam is condensed at a suitable condenser. The finely divided, dust-free particles collected at the lower end of the apparatus are present in discrete form and are outstandingly suited for embedding, for example, in bitumen, concrete, synthetic resin or the like.

Compressing Ion Exchange Resin into Pellets

Used granular ion exchange resin and used filter aid at a high radioactivity level are stored in a slurry state in a tank in a nuclear power plant. When the granular ion exchange resin and the filter aid are stored in the slurry state, a wall of the tank has a risk of corrosion. When the used granular ion exchange resin and filter aid are accumulated solely in the tank during the operating time of the nuclear power plant, the tank must be large. Thus, these radioactive solid wastes are placed in drums, and solidified by cement in the drums. However, in such a treatment, it is not possible to put a large amount of the radioactive solid wastes in the drum, and a large amount of solidified radioactive solid waste is inevitably produced.

To reduce the amount of the solidified wastes, the used granular ion exchange resin is made into powder in a centrifugal film drier in the same manner as for the regenerated waste solution, and the resulting powder is shaped into pellets. However, pellets cannot be obtained only by compressing the powder of granular ion exchange resin.

E. Ga, K. Chino, M. Kikuchi, A. Oda and S. Horiuchi; U.S. Patent 4,268,409; May 19, 1981; assigned to Hitachi, Ltd., Japan describes a process to shape pellets of mixed powder or various kinds of radioactive materials generated from radioactive material-handling facilities.

This process is characterized by making the granular ion exchange resin and the filter aid generated from radioactive material-handling facilities separately into the respective powders, mixing the powder of the granular ion exchange resin with the powder of the filter aid as a binder, and shaping the mixture into pellets, where a mixing ratio of the respective powders for shaping pellets is preferably such that the powder of the granular ion exchange resin is in not more than 90% by weight, the balance being the powder of the filter aid.

This process is further characterized by making the sodium sulfate generated from the radioactive material-handling facility into powder, mixing the powder of sodium sulfate as a binder in the two kinds of powders, that is, the powder of granular ion exchange resin and the powder of filter aid, and shaping the mixture into pellets, where preferably pellets containing X% by weight of the powder of sodium sulfate, Y% by weight of the powder of granular ion exchange resin, and Z% by weight of the powder of filter aid are shaped, wherein Y and Z are obtained according to the following formulas:

$$Y = 90 - 21.8(X/10) + 1.6(X/10)^2 + 0.44(X/10)^3 - 0.045(X/10)^4$$

$$Z = 100 - (X+Y)$$

Dewatering Directly in Storage Drum

H. Queiser and O. Meichsner; U.S. Patent 4,274,962; June 23, 1981; assigned to Kraftwerk Union AG, Germany describe an apparatus for treating various radioactive concentrates having a liquid component (suspensions, salt solutions), which concentrates are present separately in a nuclear processing plant from evaporation systems, resin bead ion exchange filters and from at least one further separating stage having, for example, mechanical filters, sedimentation basins and/or powdered resin ion exchange filters, comprising dewatering the filter concentrates containing suspended solids in a filter-cake-producing filter, dewatering the concentrates from the evaporation system, wholly or in part, directly in a transporting and storage drum to the dryness required for storage, and conducting at least one of the various concentrates, at least in part and at least for a time, directly without dewatering, into a storage drum where it is mixed with binders and converted to solids by hardening.

In addition to the technical advance realized by the solution of this process, the process has the further advantage that part of the concentrates and/or residues is initially fixed to a binder directly in the final storage drum with simultaneous dilution of the radioactive substances. This also produces a great internal shielding effect to reduce the radiation dose energy at the outside of the drum.

In preferred embodiments, this process enables the storage drum capacity to be utilized better. Further, the liquid to solid ratio in the individual concentrates can be set and various concentrates can be mixed with each other to set the overall liquid to solid ratio of the concentrate added to the storage drum. In addition, the concentrate to binder ratio can be set. The setting of these ratios makes it possible to obtain a desired or permissible radiation dose energy value in the storage drum.

Drying of Boric Acid Solutions

Boric acid is used in nuclear reactors, in particular in pressurized water reactors, as a neutron absorber. Indeed, the setting of the available reactivity in these reactors can be affected by varying the boric acid content of the cooling fluid.

In order to change the reactivity using boric acid, it is therefore necessary to bleed off some coolant or to inject boric acid into it.

The solution containing boric acid extracted in a bleeding operation is either recovered or concentrated by evaporation before carrying out an insolubilization treatment. This insolubilization is imperative as far as the storage of the concentrates is concerned since they are contaminated with radioactive substances.

The insolubilization processes presently used for such concentrates consist in embedding concentrate in cement, plastics or bitumen, as the case may be, after drying of the concentrate.

Direct embedding in cement or plastics can give rise to homogeneity problems as regards the product to be stored. For that reason, it is preferable to dry the concentrates before embedding them.

However, the drying of concentrates containing boric acid is not simple. Indeed, it is noted that drying equipment is easily jammed by setting of the product and that the dried product does not have a physical form suitable for embedding.

It has already been suggested to add lime when drying the concentrates which prevents jamming of the drying equipment and improves the physical form of the dried product. Indeed, lime converts soluble borates into insoluble calcium borates but too much lime must be added in order to obtain suitable drying which leads to an excessive increase in the quantity of product to be stored.

Furthermore, the boron decontamination factor associated with the drying is not very high. It is generally of an order of magnitude ranging from 1,000 to 3,000. By boron decontamination factor is meant the ratio of the boron concentrations in the concentrate and evaporated water.

J.-P. Cordier and M. Vandorpe; U.S. Patent 4,086,325; April 25, 1978; assigned to Belgonucleaire, SA, Belgium describe a process which makes it possible to decrease the quantity of product to be stored and to increase the boron decontamination factor, through evaporation of the concentrates after addition of lime, wherein an oxidizing agent is added to the concentrate.

The oxidizing agent can be hydrogen peroxide, a permanganate, a chromate and, in general, any oxidizing agent compatible with the boric acid and lime concentrate. The use of hydrogen peroxide is preferred because it is a strong oxidizing agent which, in addition to oxygen atoms, only introduces water molecules into the concentrate.

The quantity of oxidizing agent to be added is determined by the stoichiometry of the conversion reaction which will take place, i.e., the formation of perborates. Thus, for example, when using hydrogen peroxide, one mol of hydrogen peroxide will be added per mol of soda present in the concentrate. It should be noted that borates are generally present as sodium borates.

The addition of an oxidizing agent has multiple effects on the concentrates containing boric acid and/or borates.

Indeed, by oxidizing the concentrate, the soluble borates are converted to insoluble perborates which makes it possible to decrease the quantity of lime to be added.

Furthermore, it has been noted that perborates do not have a tendency to polymerize during drying which prevents the formation of agglomerates. In addition, perborates are insoluble in water and therefore undergo little entrainment with steam thus increasing considerably the boron decontamination factor which can easily have an order of magnitude ranging from 20,000 to 40,000. The process is illustrated by the following example.

Example: A solution to be dried contained 12% boric acid. After neutralization with sodium hydroxide to a pH ranging from 9 to 10, a quantity of lime equivalent to 60% of the quantity of boric acid was added. The sodium borate was then oxidized to the perborate by adding H_2O. The molar quantity of hydrogen peroxide added was equal to the molar quantity of soda present in the solution. Drying gave a dry product at the rate of 1.30 kg of dry product per kg of boric acid. The bulk specific gravity of the bulk product was 0.70 for a residual humidity of less than 10%. The actual specific gravity of the dry product was of the order of 1.90.

A similar solution treated with lime only required the addition of 1 kg of lime per kg of boric acid which resulted in the production of at least 1.5 kg of dry product per kg of boric acid.

OTHER CONCENTRATION PROCESSES

Adjustment of pH and Centrifuging

T. Sondermann; U.S. Patent 4,269,706; May 26, 1981; assigned to Reaktor-Brennelment Union GmbH, Germany describes a method for decontaminating process wastewater, especially from the manufacture of nuclear fuel, containing radioactive isotopes as contaminants, including (a) adjusting the pH of the contaminated process water to a value of about 5.8, (b) adding a reagent selected from the group consisting of CaO and $Ca(OH)_2$ to the contaminated process water of about 5.8 pH in an amount sufficient to raise the pH of the water in the range of up to about 8.5 to about 10, (c) agitating the mixture of the process water and the reagent for at least 5 minutes to effect intimate contact therebetween and produce a suspension in the water of solids containing radioactive contaminants, and (d) centrifuging the water containing the suspended solids to separate the solids containing radioactive contaminants from the water from which radioactive contaminants have been removed in the solids.

Absorption Followed by Volatilization and Purification

H. Mallek, W. Jablonski and W. Plum; U.S. Patent 4,277,362; July 7, 1981; assigned to Kernforschungsanlage Jülich Gesellschaft mit beschrankter Haftung, Germany describe a method which makes it possible, even with the accumulation of different solvent wastes, to separate the radioactive materials in a safe way from the solvent portions, thereby concentrating the radioactive materials in a relatively small volume. The method should be economical, and should be carried out in an entirely automatic operation.

The method of treating radioactively contaminated solvent waste is characterized primarily in that the solvent waste is supplied to a material such as peat, vermiculite, etc. This material effects the distribution or the dispersion of the solvent and absorbs the foreign substances found in the solvent waste. Air or an inert gas flows through the material in order to pick up the solvent portions which are volatile as a result of their vapor pressure. The thus formed gas mixture, which comprises air or inert gas and solvent portions, is purified in a known manner by thermal, electrical, or catalytic combustion of the solvent portions.

For absorbing the radioactively contaminated solvent waste, any material can be used which offers the solvent a greatly increased vaporization surface and can absorb the solvent. Such a material guarantees that the air or inert gas stream flowing through the material will be extensively saturated with solvent vapor, thus taking with it a large proportion of the solvent. In this connection, in order to increase the vapor pressure of the solvent, it may be expedient to heat up the material which effects the distribution of the solvent waste to 30° to 50°C.

In contrast to the known methods, according to which the contaminated solvent waste is burned directly, with this method the temperature provided for vaporizing the solvent is very low. Thus, an advantageous further feature of the method is possible with a series of solvents, such as alcohols, esters, and ketones, and consists in that the foreign substances are bound by added materials, such as ion exchange material and the like, which are supplied to the material which absorbs the solvent waste. As a result of the thereby obtained increase in the decontamination factor, this method is even more effective.

Since the gas mixture which is formed is extensively saturated with solvent vapors, it is expedient to add air to this gas mixture prior to the purification by combustion of the solvent portion to avoid explosions. The air supply, especially with fully automatic operations, is expediently controlled in such a way that the concentration of the solvent portion in the gas mixture is not more than half of the lower concentration limit of the combustible gas or vapor in the gas mixture at which the gas mixture can be exploded by being heated.

The method is advantageously carried out with an apparatus which is characterized primarily by a sealable or closable receptacle provided for receiving the material which effects the distribution of the solvent waste. The receptacle has supply lines for supplying the solvent waste and the air or inert gas into the lower portion of the receptacle where the material is contained. The receptacle also has a pipe connection for carrying off the gas mixture out of the top portion of the receptacle. The apparatus is further characterized by a device for purifying the gas mixture. This device is located above the pipe connection which carries off the gas mixture and is connected with the receptacle. To heat up the material which is located in the receptacle for absorbing the solvent waste, a device is expediently provided which may for example, comprise a further receptacle which is filled with water and is provided with a heating device. The first receptacle is located inside of this latter receptacle.

Pursuant to a further embodiment of the apparatus, the conduit provided for the supply of air or inert gas projects from above into the receptacle and empties at the bottom of the receptacle. In this way, if a distributing head is provided at the lower end of the conduit which projects into the receptacle, and if the conduit which projects into the receptacle is rotatably mounted and is connected with a drive for turning the conduit, then the absorbent material for the solvent waste can be stirred up. In this manner, a uniform distribution of the radioactive foreign substances in the absorbent material is achieved.

To mix air in the gas mixture which leaves the absorbent material, another conduit is provided which is connected with the conduit provided for carrying off the gas mixture. A valve for dosing or adding measured amounts of the air which is to be mixed in is expediently located in this new conduit. This valve is controlled by a device for regulating the quantity of air to be mixed in, so that the concentration of the solvent portion in the gas mixture provided for combustion can be kept below the intended concentration.

It may be expedient to provide a valve in the air or inert gas supply line. This makes it possible, by corresponding dosing of the air or inert gas introduced into the absorbent material, to adjust the solvent concentration in the gas mixture which is supplied to the combustion apparatus to a value which is suitable for an efficient combustion. In such a case, it is expedient to add a constant amount of air to the gas mixture prior to entry of the latter into the device for combustion.

After cooling the exhaust gas obtained during the combustion of the solvent portion, this exhaust gas is additionally purified before being released into the atmosphere. For this purpose, the known wet cleaning or washing method can be used, or filters, for example activated carbon filters, can also be used. The absorbent material, after the radioactive foreign substances contained therein are concentrated, is expediently provided for final storage in a known manner.

CHEMICAL TREATMENTS

ION EXCHANGE

Sand Pack Containing Barium Salts

R.P. Learmont; U.S. Patent 4,054,320; October 18, 1977; assigned to the United States Steel Corporation describes a process used in the in-situ leaching of valuable minerals such as uranium, whereby a leaching solution is injected into the mineral-bearing formation, permitted to remain in contact with the formation to effect the solubilization of desirable mineral values therefrom, and then withdrawn either from the original injection well or a nearby production well. Upon removal from the production well, the mineral-bearing solution is filtered aboveground prior to chemical extraction of the valuable minerals. A serious problem has arisen in that highly radioactive radium ions have been found to accumulate, both in the circulating leach solution and in various aboveground equipment resulting in serious waste disposal problems. This method utilizes a sand pack containing barium salts, which ion exchange with the radium ions so that the latter are prevented from reaching the surface.

To make the use of such a sand pack practicable, it is desirable that the barium-containing, ion-exchange material be employed in an amount at least sufficient to decrease the radium ion concentration of the pregnant solution to a level (a) which is no greater than about 10% of the level of the unexchanged pregnant solution, and (b) desirably to that of U.S. regulations, which specify a maximum permissible concentration in water of 3 picocuries per liter. Preferably, the sand pack radial dimension, measured outward from the outer surface of the well screen will be greater than about 10 cm so as to better insure that the radium ion concentration will be decreased to a level no greater than 2% of the unexchanged pregnant solution.

Barium-Salt-Containing Sorbent

K.J.A. Peeters and N.L.C. Van De Voorde; U.S. Patent 3,896,045; July 22, 1975; assigned to Belgonucleaire SA, Belgium describe a method for extracting radioactive ions present in tracer quantities in a contaminated liquid, in whch the liquid is contacted with a sorbent in a sulfate-containing medium selected from the group consisting of a barium salt, and a barium salt mixed with a metal ferrocyanide.

Ions such as Na, Ca, Mg, Fe, Al, thus are not extracted, while the radioactive ions are extracted from the liquids to be purified.

The sorbent is preferably from 85 to 98% barium salt which has an excess of barium cations and from 15 to 2% of a metal ferrocyanide in a sulfate-containing medium.

Purification of Waste Product Gypsum

P.H. Lange, Jr.; U.S. Patent 4,146,568; March 27, 1979; assigned to Olin Corporation describes a process for reducing the radioactive contamination in waste product gypsum which comprises:

(a) admixing waste product gypsum containing radioactive contamination with dilute sulfuric acid containing barium sulfate at an elevated temperature to form an acid slurry, having a solid component comprised of a fine fraction and a coarse fraction,

(b) separating the fine fraction of solids from the coarse fraction,

(c) whereby the fine fraction predominates in the barium sulfate and the radioactive contamination, and

(d) whereby the coarse fraction predominates in gypsum of reduced radioactive contamination.

In a preferred embodiment, the acid slurry is cooled, the acid is separated from the solid component, and the solid component is washed with water prior to separating the fine fraction from the coarse fraction, as described more fully below.

In the first step for purifying the waste product gypsum in accordance with the process, a dilute sulfuric acid is used to digest the gypsum. Generally, an aqueous sulfuric acid having a concentration in the range from about 10 to about 50% H_2SO_4 by weight is employed. It is preferred to employ an acid having a sulfuric acid concentration in the range from about 15 to 35% sulfuric acid by weight. However, more dilute or concentrated solutions may be employed if desired.

Sufficient barium sulfate is added to the slurry of gypsum and sulfuric acid to provide a barium sulfate concentration in the resulting slurry in the range of from about 0.02 to about 1.0% by weight and preferably from about 0.03 to about 0.3% by weight. Excess barium sulfate may be employed, but it is generally unnecessary since such an excess adds to the cost of carrying out the process without significantly improving the reduction of radioactive contamination. Barium sulfate is preferably added as a concentrated sulfuric acid solution containing from about 0.10 to about 10% by weight of barium sulfate.

The concentrated sulfuric acid contains from about 90 to about 100% by weight of sulfuric acid. Fuming sulfuric acid may be employed to dissolve the barium sulfate, if desired. The concentration of the sulfuric acid used to dissolve the barium sulfate should not be less than about 90% by weight, since precipitation of the barium sulfate may occur, which will increase the time of reaction and thereby reduce the efficiency of the process. If desired, solid finely divided particles of barium sulfate may be added to the acid slurry of gypsum, but this procedure will unduly extend the reaction time. In another embodiment, more expensive barium sulfate may be formed in situ by adding a reactant such as barium chloride to the dilute acid slurry.

The proportion of gypsum, in terms of anhydrous calcium sulfate, in the aqueous slurry of sulfuric acid and barium sulfate generally ranges from about 50 to about 500 grams per liter, and preferably from about 150 to about 350 grams per liter.

The resulting aqueous slurry of gypsum, sulfuric acid and barium sulfate is heated to an elevated temperature, generally to at least about 60°C, and preferably to a temperature in the range from about 80°C to the boiling point. If desired, super-atmospheric pressures may be employed. In a preferred embodiment, the barium sulfate is added to the slurry after the temperature has been raised to within the above-identified range in order to take advantage of superior absorptive properties of freshly precipitated barium sulfate. After the desired elevated temperature is obtained, the slurry is agitated for a sufficient period to effect solubilization of radioactive contaminants, followed by absorption and/or coprecipitation of radium sulfate crystals with barium sulfate crystals in finely divided form. Generally, this digestion period ranges from about 1 to about 2,000 minutes, and preferably from about 5 to about 250 minutes.

After the reaction of the acid slurry has been completed, the slurry is preferably cooled, generally to a temperature below about 60°C, and preferably in the range from about 15° to about 40°C. If desired, the hot acid slurry may be washed without a separate cooling step. The acid slurry with or without prior cooling is subjected to a solid-liquid separation step, such as filtration or cyclone separation, and the clarified acid is recovered. It may be recycled, after reconstitution, for use in reacting with additional impure waste product gypsum or used in other parts of the fertilizer process.

The solid component of the acid slurry, after separation of the clarified acid, is then washed, for example, by slurrying water and subjected to a solids separation step wherein the coarse fraction of purified gypsum crystals are separated from the fine fraction of crystals of barium sulfate and radium sulfate. This solid separation step is preferably carried out by wet screening, but can also be accomplished in a cyclone separation apparatus of other suitable size classification techniques known in the art. If desired, the solids from the acid digestion step may be washed and dried and stored prior to the solids separation step.

Generally, the separation of the fine fraction and the coarse fraction is made at a size of at least about 150 microns, and preferably at about 80 microns. The fine fraction is generally less than –150 microns, preferably less than –80 microns and more preferably less than –50 microns to achieve the improved results of the process.

Example: An aqueous sulfuric acid solution (2.97 liters) having a concentration of 28% was admixed with a concentrated sulfuric acid solution (30 milliliters) containing dissolved therein 3 grams of barium sulfate at a temperature of 105°C. Waste product phosphogypsum from a wet process phosphoric acid plant (750 grams containing a radium concentration of 26 picocuries per gram) was agitated with the solution of sulfuric acid and barium sulfate for a period of 12 minutes, while maintaining the temperature of the resulting slurry in the range of 98° to 105°C.

At the end of this period, the slurry was cooled at a temperature of 30° to 35°C and then filtered. The filter cake was washed with 6.2 liters of water and the solids were dried at a temperature of 60°C. The weight of the dry solids was 695 grams.

Fifty grams of the dried solids was slurried with 1 liter of water and then wet screened on a 200 mesh (74 micron) screen. The solid gypsum material retained on the 200 mesh screen was collected and dried at 60°C. Analysis of the purified gypsum product showed that it had a radium concentration of 1.42 picocuries per gram. This purified gypsum material was suitable for use in preparing plaster, gypsum wallboards, and other construction materials.

Disposable Apparatus for Removal of Impurities

J.A. Levendusky; U.S. Patent 4,107,044; August 15, 1978; assigned to Epicor, Inc. describes an apparatus for the purification of water containing radioactive impurities wherein the apparatus is disposable upon the occurrence of the active capacities of the filter element and demineralization materials being exhausted.

The disposable filter apparatus for the removal of radioactive dissolved and undissolved solids from a fluid comprises:

(a) a fluid tight vessel, the fluid tight vessel including a wall having an upper wall portion and a lower wall portion, the fluid tight vessel defining a chamber containing a bed of demineralization materials for removing the dissolved solids from the fluid;

(b) an inlet pipe extending through a first opening formed in the upper wall portion, the joint defined by the external surface of the inlet pipe and the first opening being closed such as to rigidly secure the inlet pipe to the vessel and to establish a fluid type seal therebetween;

(c) an inlet manifold rigidly secured to the inlet pipe within the chamber, the inlet manifold being in fluid communication with the inlet pipe;

(d) dispersion means rigidly secured to the inlet manifold, the dispersion means for introducing fluid evenly into the bed of materials, the dispersion means being in fluid communication with the inlet manifold to cooperate therewith in accommodating the introduction of contaminated fluid through the inlet pipe and the inlet manifold for dispersion into the bed of material;

(e) an effluent pipe extending through a second opening formed in the upper wall portion, the joint defined by the external surface of the effluent pipe and the second opening being closed such as to rigidly secure the effluent pipe to the vessel and to establish a fluid tight seal therebetween;

(f) an effluent manifold rigidly secured to the effluent pipe within the chamber, the effluent manifold being in fluid communication with the effluent pipe, and the effluent manifold being positioned within the chamber, adjacent the lower wall portion;

(g)   at least one drain line rigidly secured to and in fluid communication with the effluent manifold, each drain line comprising a pipe having a plurality of openings therethrough to
place the interior of the pipe in fluid communication with
the bed of demineralization materials;

(h)   filter means disposed on at least one of the dipersion means
and the drain line for removing the undissolved solids from
the fluid; and

(i)   a shielding means disposed around the outer surface of the
vessel, the shielding means capable of precluding the emission of radiation from the vessel to the surrounding environs,
and the shielding means defining a casing for the fluid tight
vessel whereby, upon exhaustion of the demineralization
materials and means for removing undissolved solids from
the fluid, the entire structure may be removed to an author-
ized dumping site and buried thereby precluding the necessity for removing the exhausted, highly radioactive means
and materials for removing the dissolved and undissolved
solids from the filter apparatus.

Figure 4.1 is a plan view of apparatus structured in accordance with this process.

Figure 4.1: Apparatus for Removal of Radioactive Impurities

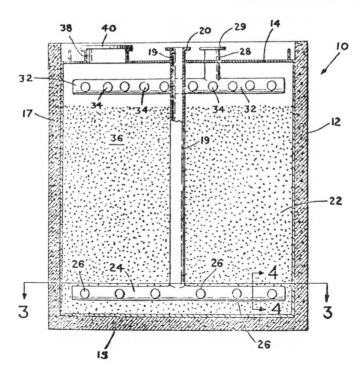

Source: U.S. Patent 4,107,044

Mycelia of Fungi

H. Prochazka, K. Stamberg, R. Jilek, P. Hulak and J. Katzer; U.S. Patent 3,993,558; November 23, 1976; assigned to Ceskoslovenska komise pro atomovou energii, Czechoslovakia describe a method of separation of stable and unstable isotopes from liquid waste based on the application of sorbents of suitable porosity and shape with functional groups forming chelates and exchange groups. According to this process, mycelia of fungi are used, namely materials formed on the basis of proteins, hexosamines, polysaccharides, nuclear acids, and materials of the type of chelate resins. Mycelia of fungi of strains, for instance, *Penicillium chrysogenum, Mycelium sterilium, Aspergillus ochraceus* and the like, stiffened by resorcinol-formaldehyde, urea-formaldehyde or by some other resin, proved to be suitable for this purpose.

An advantage of the method is the fact that one type of sorbent is used for separation of all mentioned stable and unstable isotopes of elements of fission and corrosion products, which sorbent can be, according to need, regenerated only by a suitable process. There are substantially three possible alternatives of application, for instance, of a stiffened *Mycelium penicillium chrysogenum:*

(a) Without application of activation or regeneration, that is, following a fundamental working cycle sorption-elution-sorption-elution and so forth, which is suitable for the majority of fission products with the exception of I-131, and of anionic forms of nuclides generally, for instance, Cs, partially also of Sr and the like.

(b) With application of a single stage regeneration called a "sorption-precipitation method." The fundamental working cycle is composed of the operations of sorption-elution-regeneration-sorption and so forth, whereby the regeneration is introduced according to need, for instance, to each third to fifth cycle.

(c) With the application of a two or more stage activation or regeneration, where the biosorbent is converted in the first stage into a cycle of a cation forming a precipitate (Fe^{2+}, Zn^{2+}, Cu^{2+}, Mn^{2+}, Fe^{3+}, Al^{3+}, and the like) or their mixtures (Zn^{2+} + Al^{3+}, Ag^+ + Fe^{3+} and the like). In the second stage a solution acts on the biosorbent, the solution containing an anionic component forming a precipitate (for instance, OH^-, CO_3^{2-}, $Fe(Cn)_6^{4-}$ and the like).

Thus the sorbent is prepared in such manner that the biosorbent is in the first place a supporting skeleton (it has the required properties, namely porosity, sorption of electrolytes and the like) and the created precipitate is the proper sorbent having properties similar to analogous types of synthetic ion exchangers on the basis of hydrated oxides, insoluble ferro- and ferricyanides and the like.

It is known from technical papers (G.B. Amphlett: *Inorganic Ion Exchangers,* Amsterdam, 1964) that these types of synthetic ion exchangers have sorption properties for a number of nuclides, particularly for alkali metals and alkali earth metals (see nuclides with longest half lives: Cs-137 33 years, Sr-90 19.9 years). The fundamental working cycle is substantially the same as in alternative (b) including the periodically introduced regeneration process according to need.

The three alternatives of application of, for instance, a stiffened mycelium of fungi of the strain *Penicillium chrysogenum,* as indicated in (a) through (c) above enable the separation of stable and unstable isotopes which are present both in the cationic or anionic state. The regeneration of the biosorbent [see (b) and (c) above] converts the resulting sorption properties towards an increase of the selectivity to individual elements or groups of elements by means which are technically easy to apply and are economically advantageous.

The alternatives mentioned, (a) through (c), can also be used for a single sorption process (for instance, for highly active waste), where it is undesirable to wash out the retained radioactive material back into the elution solution, but it is left in the solid phase. In such a case, the biosorbent can be burnt in a suitable arrangement at temperatures of about 500° to 800°C, and the volume of the contaminated solid phase can be thus substantially reduced prior to storage in suitable storage places for radioactive waste.

Biosorbents of the type R have been prepared by reinforcing a dry mycelium of the strain *Penicillium chrysogenum* with resorcinol-formaldehyde resin. Biosorbents of the type M have been prepared by reinforcing the same mycelium with urea-formaldehyde resin.

Example 1: Into wastewater, containing a mixture of fission and corrosion products, that is, isotopes of, for instance, Mo, Mn, Fe, Co, Sr + Y, Ru + Rh, Cs, Ce, La, Pm having an overall activity, for instance, on the order of 10^{-2} Ci/liter and having a pH value within the neutral range, a biosorbent of the type R (defined above) was added in a volume amount of 1 to 1,000 and the mixture was intensively stirred for 20 minutes. After separating the solid and liquid phase, the overall activity of the wastewater was reduced to 10^{-4} Ci/liter.

Example 2: Wastewater as used in Example 1 was allowed to pass through a pressure column with a solid biosorbent bed of the type M at a specific load of the column 10 to 20 volumes to a volume-hour. By passage through the column a reduction of the overall activity of the wastewater was achieved to 10^{-4} Ci/liter for the first 300 to 500 volumes per volume of the bisorbent.

Mixture with Ion Exchange Resins

O. Meichsner and H. Queiser; U.S. Patent 4,033,868; July 5, 1977; assigned to Licentia Patent-Verwaltungs-GmbH, Germany describe a method for processing contaminated wash water which will be very efficient and economical resulting in savings of operating agents and energy, and a reduction of the size of the apparatus required.

This is accomplished by mixing the contaminated wash waters with ion exchange resins in a mixing vessel to form a suspension and then feeding the resulting suspension into a filter.

The method thus is a combination of ion exchange and filtration. The usually predominant portion of ionic, radioactive substances is removed from the wash water in the mixing vessel by the ion-exchange resin. On the other hand, the solid matter in the wash water is removed from the wash water by being retained on the filter when the suspension of wash water and ion-exchange resin is fed to the filter. The wash water leaving the filter is thus freed of both ionically dissolved radioactive contaminants as well as radioactive solids suspended in the wash water. Generally, no ionic cleaning occurs in the filter, and the ion-exchange resin in the suspension serves as a filtering aid on the filter for mechanically filtering solid substances from the suspension.

The mixing of the wash water with the ion-exchange resins is distinguished from typical ion exchange processes where wastewater usually is processed in a straight-through operation, that is, the water is continuously passed or flowed through a bed of ion exchange resin and removed therefrom. In contrast, in this process, the water and ion exchange resin are mixed to form a batch suspension and this suspension is left for a period of time so that the ion exchange resin and water contact each other for a substantial time and a thorough ion exchange is effected. By mixing the resins which cause the exchange of ions with the wash water, optimally required contact periods for the exchange of ions can be set. This permits a much better utilization of the total exchange capacity of the resins than would be possible in the conventional straight-through operations. Usually, a contact period of 15 to 20 minutes is sufficient and optimal to achieve the desired ion exchange. The mixing vessel contains no filter plates or the like, but is simply a container in which ionic cleaning takes place.

In accordance with a preferred embodiment, waste ion exchange resins which can no longer be used in straight-through operation or which are not as yet fully spent are used to treat the waste wash waters. Waste ion exchange resins can be used in this process because of the long contact time of the resin with the wash water in the mixing vessel.

The decontamination effect realized in the method for wash waters from nuclear power plants can be equated to the evaporation method previously used for wash waters since decontamination factors of from 10^3 up to 10^5 and residual activities of 10^{-5} to 10^{-6} Ci/m^3 are realized.

The method is particularly suitable for use in connection with a nuclear power plant which operates with boiling-water reactors. Powdered resin used in such a plant for cleaning radioactive condensate is employed only to a very low level of its useful (volume) capacity, since high quality condensate is required. Such powdered resin yields a large volume of sediment or slurry (mud), due to an electrostatic effect; the solids in this mud are only about between 5 and 10% by weight.

Purification of Ion Exchange Resins

K.-H. Neeb and H. Richter; U.S. Patent 4,118,317; October 3, 1978; assigned to Kraftwerk Union AG, Germany describe a method of freeing from radioactive corrosion products and from conditioning substances stemming from the coolant loop of a nuclear reactor, ion exchanger resins spent in the separation of the reactor which comprises thoroughly rinsing the resin with deionate so as to release suspended radioactive substances from the resins, separating the released suspended radioactive substances in a mechanical filter, rinsing with diluted mineral acid the resin previously rinsed with deionate so as to release radioactive cations and anions therefrom, adsorbing the cations and anions in an adsorber substance fixed on an inorganic carrier, and separating the conditioning substances from the mineral acid with which the resins have been rinsed and forming chemical compounds therewith for reuse.

In accordance with other modes of the method, the mechanical filter is formed of inorganic filter material, the adsorber substance which reacts with the nuclides is selected from the group consisting of ferro-ferricyanides and Ag compounds, the conditioning substances comprise boron and lithium, the mineral acid is nitric acid, and the chemical compounds formed are compounds thereof.

In accordance with a further mode of the method, the rinsing steps are effected with a rinsing agent circulating in the coolant loop of the reactor.

The method thus makes possible a simplification of waste disposal of the spent ion exchanger resins; a concentration of the radioactive corrosion products from the coolant loop of the nuclear reactor into a narrow space and thereby preparation thereof for ultimate storage; and recovery in addition, of the valuable conditioning substances for the reactor coolant. A slight amount of radioactivity that may yet accompany these recovered substances offers no interference, from a practical stand-point, since the extent thereof is certainly less than that of the coolant loop during operation.

Conditioning of Contaminated Ion Exchange Resins with a Base

M. Buchwalder, R. Perrin and D. Thiéry; U.S. Patent 4,122,048; October 24, 1978; assigned to Commissariat a l'Energie Atomique, France describe a process permitting an effective conditioning of contaminated ion-exchange resins no matter whether they are anionic, cationic or a mixture of two and no matter to what degree they are spent.

According to the process, the contaminated ion-exchange resin or resins are brought into contact with a basic compound in a sufficient quantity to block the active sites of the cationic resin or resins, the thus treated ion-exchange resin or resins are in-corporated into an ambient temperature-thermosetting resin and the latter is cross-linked.

The basic compound which serves to block the active sites of the ion-exchange resin or resins comprises either a metallic hydroxide such as soda, ammonia or lime, or a metal salt such as aluminum chloride, sodium acetate, sodium citrate or sodium oxalate, or an amine such as pyridine. This basic compound can either be used in the form of an aqueous solution with a molarity of 0.1 to 10 M or as it is.

In the case where it is desired to condition a mixture of cationic and anionic ion-exchange resins, it is necessary to choose for the pretreatment a basic compound whereof the nature and the concentration in which it is used are suitable for both types of ion-exchange resins. Thus, for example, to pretreat a mixture of two-thirds cationic resin and one-third anionic resin by means of soda, it is necessary to use a soda solution whose concentration is 5 to 10%, that is to say the pH is preferably 6 to 12.

The thermosetting resin into which is incorporated the contaminated ion-exchange resin or resins following their pretreatment by means of a basic compound can ad-vantageously be constituted by a polyester resin such as a glycol-maleophthalate-based resin mixed with styrene.

In this case the conventional compounds necessary for ensuring the copolymeriza-tion of the styrene with the polyester and the control of the crosslinking time are used, i.e., a catalyst such as methyl-ethyl-ketone peroxide or benzoyl peroxide (in a proportion of 1 to 2% by weight of catalyst based on the resin), an accelerator such as cobalt naphthenate or dimethylaniline (in a proportion of 0.1 to 0.2% by weight based on the resin), reaction controlling agents such as retarding agents (catechol-based compound known as "NLC 10") and moderators (α-methyl-sty-rene).

Thus, the polymerization reaction started by the active radicals of the peroxide activated by the accelerator bring about the grafting of the styrene molecules onto the glycol-maleophthalate chains, followed by an arrangement in a three-dimen-sional network.

According to the process, it is also possible to use as the thermosetting resin an epoxy resin mixed with an appropriate hardening agent (amine or organic acid). It is also possible to use a phenoplast resin.

Preferably one part of pretreated ion-exchange resin is incorporated into one part of thermosetting resin. More specifically the process is performed in the following manner:

The first embodiment of this process consists of passing a solution of the basic compound over the contaminated ion-exchange resins located in a column. After passing the solution of the basic compound into the column, the thus pretreated resins can optionally be washed. They are then suction-filtered, incorporated in identical proportions in a thermosetting resin and finally the thermosetting resin is crosslinked.

This first embodiment has the advantage that during the passage of the solution of the basic compound over the contaminated resins, there is a continuous exchange during the displacement of the solution of the basic compound along the column and consequently a maximum effectiveness with regard to the blocking of the active sites of the ion-exchange resins. However, according to this embodiment the solution of the basic compound may extract certain of the radioelements, particularly cesium which were fixed to the ion-exchange resins, so that this solution which has become radioactive must be conditioned in a random manner.

A second embodiment of this process comprises mixing the contaminated ion-exchange resins with a solution of the basic compound in a container, whereby the contact time must be about two hours. Then, following the optional washing of the thus pretreated ion-exchange resins, they are suction-filtered and incorporated in equal proportions into a thermosetting resin. Finally, the thermosetting resin is crosslinked. This second embodiment has the advantage of being very simple to perform. However, it should only be used for conditioning contaminated resins which do not have a too high radioactivity (the integrated dose remaining below 10^9 rads).

The pretreatment operation by means of a basic compound can also be perfomed by mixing the compound as it is with ion-exchange resins. This is particularly the case when using lime or sodium oxalate.

Example: A mixture of two-thirds cationic resin "Duolite ARC 351" and one-third anionic resin "Duolite ARA 366" (marketed by the Diaprosim Co.) filled with ^{60}Co is successively treated with a basic compound constituted by soda (Example 1), ammonia (Example 2), pyridine (Example 3), sodium acetate (Example 4), sodium citrate (Example 5), aluminum chloride (Example 6), lime (Example 7) and sodium oxalate (Example 8). This mixture of ion-exchange resins is in the form of moist grains (55% humidity).

The table below summarizes the conditions of this pretreatment according to the nature of the basic compound used.

In the table the basic compound quantity is in parts by weight for treating 100 parts by weight of ion-exchanger resins. The concentration of the basic compound solution for neutralizing the active sites is in g/ℓ.

Example Number	Basic Compound . . . . (parts by weight)		Concentration of Basic Compound Solution (g/ℓ)
1	Soda	30	50
2	Ammonia	75	500
3	Pyridine	60	100 to 500
4	Sodium acetate	70	70 to 100
5	Sodium citrate	75	100 to 200
6	Aluminum chloride	60	300
7	Lime	15	*
8	Sodium oxalate	50	*

*Directly mixed with the ion-exchange resin.

The mixture of the thus pretreated ion-exchange resins is then incorporated into a glycol-maleophthalate-based polyester resin mixed with styrene in the following proportions: 50 parts by weight of pretreated resins and 50 parts by weight of polyester resin to which is also added 1.5% by weight of catalyst based on the polyester resin and 0.2% by weight of accelerator based on the polyester resin.

After between 30 and 60 minutes a solid homogeneous block is obtained having the following characteristics: leaching $\cong 2 \times 10^{-6}$ cm per day—110 days; compression behavior $\cong 100$ kg/cm^2.

Volume Reduction of Spent Radioactive Ion Exchange Material

E.W. Tiepel, P.K. Lee, A.S. Kitzes and D.L. Grover; U.S. Patent 4,008,171; February 15, 1977; assigned to Westinghouse Electric Corporation describe a process which substantially reduces the volume of radioactive resins required to be encapsulated for disposition by removing all of the slurry water as well as the intrinsic water from the resins. The resins are subjected to a vacuum filtration process which removes the free water and the remaining wet resins are then exposed to a vacuum environment, where superheated steam injected into the evacuated resin container acts to remove the intrinsic water. Since the removed radioactive free water is sufficiently pure to permit recycling in a hold-up tank while the dried resin, which shrinks about 50% during treatment, is discharged to a steel or other drum suitable for burial according to conventional practices.

Figure 4.2 is a schematic showing of the system used for reducing the volume of ion exchange resins.

Figure 4.2: System for Volume Reduction of Spent Ion Exchange Resins

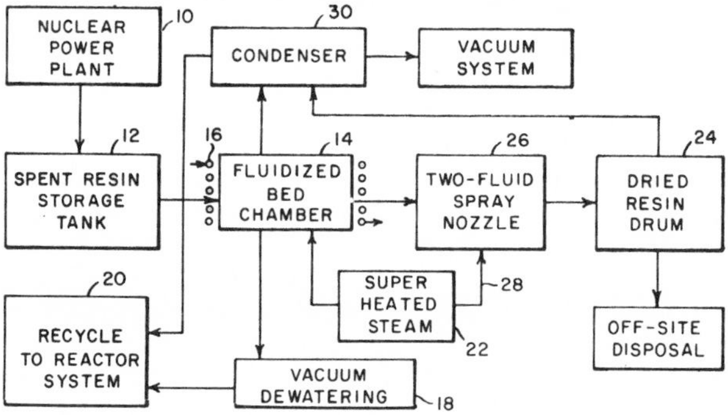

Source: U.S. Patent 4,008,171

The abovedescribed process exhibits several important advantages since the process affords substantial reduction in resin disposal and operating costs while simultaneously minimizing release of radioactivity to the environment or atmosphere. The process does not generate any other contaminated waste streams, other than the water from the steam and resin which can be sent directly to the reactor liquid waste treatment system, because the process described herein and such treatment systems are all contained within closed loops.

The water from the dewatering steps is of a quality that is acceptable for direct recycle to the reactor make-up tanks through a polishing demineralizer. Moreover, the components may be sized smaller than those used at commercial reactor plants and mounted on vehicles of different types and sizes. These vehicles with the installed components may then be used to service nuclear reactor plants which may desire such a disposal service.

E.W. Tiepel, C.K. Wu and A.S. Kitzes; U.S. Patent 4,053,432; October 11, 1977; assigned to Westinghouse Electric Corporation describe a related process which substantially reduces the volume of radioactive organic ion-exchange materials while minimizing the amount of radioactive release to the off-gas system.

The ion-exchange material is removed from the ion-exchangers, dried to a moisture content less than 50% by weight, and inserted into a fluid bed reactor. A carrier gas is inserted into the fluid bed reactor, and the ion-exchange material is heated. The heat thermally decomposes the ion-exchange material, producing an effluent gas containing the volatile decomposition products. The carrier and effluent gases are removed from the fluid bed reactors, and after the ion-exchange material has thermally decomposed, the insertion of carrier gas is stopped, and an oxygen-containing gas is inserted in the reactor. The remaining ion-exchange material is burned with the oxygen-containing gas, and a final volume reduction of approximately 20:1 from the original ion-exchange material settled bed volume is obtained.

The effluent gas is supplied to an afterburner, where it is combusted with air or oxygen, passed through a filter to remove any entrained solids, passed through an absorption material to remove any acid gases or radioactive species, passed through a high efficiency particulate absolute filter to remove any fine dust present, and expelled to the atmosphere. The final gas phase composition consists of carbon dioxide, water, nitrogen, and oxygen.

Containment of Radioactive Spills

C. Calmon; U.S. Patent 4,056,112; November 1, 1977 describes a process which provides for a relatively rapid containment and removal of radioactive spills by a chemical absorptive means which results in the handling of much smaller volumes of liquid in a much more effective and safer manner than any presently known method.

The process utilizes ion-exchange compositions having cationic groups either in salt or base form, and/or anionic functional groups either in salt or acid form, such materials never having been used up to now for the purpose of containing and removing radioactive spills. Such materials have been used for purifying liquid waste containing particulates and dissolved ionic species, wherein solid ion-exchange materials are placed in a purifying tank or column and act as filter and exchange media, the degree of efficiency thereof depending on particle size. In such purifying processes the ion-exchange compositions are normally shipped in wet form because the particles in dry form, when wetted, tend to crack or break, in which case such materials cannot be used in purifying column operations due to high head losses as well as possible losses due to back-washing. In all such purification processes the liquid waste which is to be purified must be appropriately contained and brought to the wet ion-exchange material which is confined in the tank or column usually fixed in a specific area.

In accordance with this process, however, ion-exchange compositions are used for containing and removing liquid radioactive spills at the site thereof so that, where time for action thereon is limited, no time is lost in trying to pick up and carry the spill to a remote location for purification. In accordance with the process, for spills of liquid radioactive material, a dry ion-exchange resin is used. Such dry ion-exchange resins comprise one or more suitable monomers crosslinked with an appropriate crosslinking agent to form a matrix for the ion-exchange group or groups as is well known. The crosslinking thereof can be arranged to enhance the liquid absorption characteristics thereof so that when the ion exchange resin is deposited over the radioactive liquid spill and over the regions surrounding such spill, large volumes of liquid spill are readily absorbed and are effectively confined to the spill region, while the radioactive ionic species becomes simultaneously chemically attached to the exchanger.

The ion-exchange resin is selected so that its ability to provide an ion-exchange with radioactive ions, such as strontium, cesium and plutonium is very high. Thus, the soluble radioactive ions are picked up and held tightly to the solid, dry ion exchanger resin through exchange for nonradioactive species. For example, when using a sodium resin exchanger with strontium-90, the exchange operation can be described as follows:

Sodium Exchanger + Strontium-90 → Strontium-90 Exchanger + Sodium

Because of the desire to avoid handling large volumes of liquid, it is recognized that it is of the greatest importance in a liquid spill to absorb substantially all the radioactive liquid solution as quickly as possible. Accordingly, the capability of the ion-exchange resin to absorb large amounts of such liquid is suitably enhanced. Standard dry ion-exchange resins can hold at best 45 to 60% of their total volumes as water. However, this amount of water pickup can be greatly increased if the crosslinking in the ion exchange resin is appropriately arranged. The existence of a relationship between the absorption characteristics of an ion-exchange resin and the amount of crosslinking agent that is known, is discussed, for example, in the text "Ion Exchanger Properties and Applications," K. Dorfner, Ann Arbor Science Publishers, Inc. 1972. Such relationship, as determined by the mol percentage of the crosslinking agent used, can be seen from the table below with respect to an exemplary ion-exchange resin material comprising sulfonated copolymers of styrene and a divinylbenzene (DVB) crosslinking agent.

Mol % DVB in the Ion-Exchange Resin	Ratio of the Volume of Liquid Absorbed to the Initial Dry Volume of Resin
0.25	50
0.5	22
1.0	10
2.0	7
3.5	5
10	<2

Thus, for example, if 5 gallons of dry ion-exchange resin having 0.5 mol % of DVB as a crosslinking agent is used for the absorption of a radioactive spill, approximately 110 gallons of such spill would be so absorbed. For practical applications the crosslinking may range from about 0.25 to about 25%.

While such absorption characteristics of the ion-exchange resin are dependent upon the crosslinking therein, the ion-exchange capacity of such a resin primarily depends on the dry weight of the resin and is not much affected by the crosslinking thereof.

In order to increase the rate at which absorption of the liquid can be achieved, i.e., to increase the kinetics of the ion-exchange operation, the resin can be made in relatively fine sizes so as to increase the surface area thereof. The sizes of the resin particles may lie within a size range from about –16 mesh to about +200 mesh, such particles already being generally manufactured and available within a range from about –16 mesh to about +50 mesh. Such particles provide relatively quick absorption action and therefore prevent the percolation or rapid spreading of the radioactive spill.

However, resin particles as small as +200 mesh can be used to provide rapid absorption so long as effective methods for retrieving the particles for disposal are used. As can be seen by the above table the use of a relatively low crosslinking resin together with the use of relatively small particle sizes permits relatively high absorption in comparison to the initial dry volume of the resin so that pickup of the radioactive spill is greatly facilitated.

Thus, in accordance with the process, when a liquid radioactive spill takes place, the dry ion-exchange material is suitably deposited over the spill region and over the regions surrounding the spill region, particularly in the direction of flow thereof, so as to absorb the solution and to confine the liquid to the spill area and, thus, prevent its spread or its percolation or absorption into the surface on which the spill has taken place. In this way the ion-exchange products act effectively as sponge materials and, in addition, due to the ion-exchange operation, permit the radioactive nuclides simultaneously to be held in the ion-exchange resin through an ion-exchange operation.

Ion-exchanger sheets, in cloth, paper or foam forms, can be appropriately deposited at a location whre a spill is likely to take place, as a precautionary measure prior to an actual spill, e.g., on trays or in containers used to store or transport radioactive liquids so that spills can be immediately taken care of without the loss of time needed to move the ion-exchange material from a storage place to the spill area. Many liquids or dangerous products are shipped in containers which are in turn placed, for example, in cartons containing foamed polymeric products for protection. In accordance with the process, such foamed products can be arranged to contain ion-exchange materials which can absorb both the liquid and the radioactive nuclides in case of breakage of the container with the carton. Such a packaging process makes the transportation of radioactive nuclide solutions much safer than has hitherto been possible.

ABSORPTION AND ADSORPTION

Impregnation of Activated Charcoal with Oxine

Oxine (8-hydroxyquinoline) was first prepared about 100 years ago and it was introduced as an analytical reagent in 1926. Many investigations have been carried out on the reagent and it has proven to be one of the most valualbe analytical re-agents available for the separation and gravimetric, volumetric and photometric determination of many metals. Oxine forms stable and water-insoluble chelates with many metals.

A method of removing radionuclides from the cooling and leakage-water from nuclear reactors using the properties of oxine as stated above is already known. The principle of the method is simple and comprises adding an oxine into the cool-ing-water of the nuclear reactor or leakage-water therefrom to form water-insoluble and stable complexes of the radionuclides contained therein and adsorbing them on activated charcoal. Many kinds of operations can be employed in the method, for example, an operation using activated charcoal on which oxine is adsorbed, a batchwise opertion in which activated charcoal is added after addition of oxine, or a continuous operation using an activated charcoal column. However, the operation using activated charcoal on which an oxine is impregnated is recommended.

The conventional method for preparing oxine impregnated activated charcoal com-prises dissolving oxine in proper solvents and passing the solution through a col-umn of activated charcoal or dipping the activated charcoal in the solution with stirring so that the activated charcoal adsorbs the solution and finally removing the solvents. However, the conventional method suffers from the defect that a large amount of organic solvent is required, because oxine is not readily soluble in water. In addition to that, organic solvents are particularly troublesome to handle and remove.

K. Motojima, E. Tachikawa and H. Kamiyama; U.S. Patent 4,222,892; Septem-ber 16, 1980; assigned to Japan Atomic Energy Research Institute, Japan describe a method for preparing an oxine impregnated activated charcoal comprising con-tacting solid oxine with activated charcoal.

The process can be performed either of two ways: (1) Reaction of an oxine with an activated charcoal in the air hereinafter referred to as method (1) is as follows: The boiling point and melting point at atmospheric pressure of oxine are approxi-mately 270° and 74°C, respectively, and the vapor pressure at around room tem-perature of oxine is not so high. Furthermore, since the molecular weight of oxine is relatively large (145.2), the diffusion rate of gaseous oxine is not so high. There-fore, the surface of solid oxine is covered and saturated with gaseous oxine and after equilibrium is established, the vaporization of oxine scarcely proceeds. How-ever, when activated charcoal exists in contact with solid oxine, the oxine vaporized is quickly adsorbed by the activated charcoal and removed from the surface of solid oxine. Accordingly, the equilibrium state between solid oxine and gaseous oxine is broken and oxine is rapidly vaporized and is adsorbed by the activated charcoal. Therefore, when a sufficient amount of activated charcoal for oxine exists in reac-tion system, all of the solid oxine is adsorbed by the activated charcoal. The reac-tion as stated above is illustrated as follows:

$$\text{oxine (solid)} \xrightarrow{\text{vaporization}} \text{oxine (gas) + activated charcoal} \xrightarrow{\text{adsorption}} \text{oxine impregnated-activated charcoal}$$

(2) Reaction of oxine with activated charcoal in water hereinafter referred to as method (2) is as follows: Oxine is readily soluble in organic solvents such as alco-hol, acetone, chloroform, etc., and is relatively soluble in acidic or alkaline aqueous

solution. However, it is only slightly soluble in water; that is, the solubility of oxine in water is approximately 0.05 g/ℓ. However, when activated charcoal exists in the reaction system, the oxine dissolved in water is rapidly adsorbed by the activated charcoal and removed from the reaction system. Therefore, when a sufficient amount of activated charcoal for oxine exists in the reaction system, all of the solid oxine is adsorbed by the activated charcoal. The reaction as stated above is illustrated as follows:

oxine (solid) ―dissolved→ oxine-dissolved in aqueous solution + activated charcoal→oxine adsorbed-activated charcoal

In accordance with method (1) or method (2) as explained above, an oxine can be quantitatively adsorbed by activated charcoal in its saturation amount regardless of the kind or shape of the activated charcoal. In the method of this process, the adsorption of oxine by activated charcoal proceeds relatively rapidly, but it does not proceed instantaneously. Therefore, if the proper means of mixing the reaction system is adopted according to the particle size of activated charcoal used in this process, homogeneous products can easily be obtained.

The preferred oxine useful in this process is needle-like or scaly crystalline powder.

The reaction conditions used such as temperature, pressure, etc., are not specifically restricted, except that temperatures above the melting point of oxine and temperatures below the freezing point of the oxine containing aqueous solution must be avoided. The method (1) can be carried out under vacuum or one or more inactive gases which do not react with oxine and activated charcoal. The method (2) can be carried out in the presence of solutions and/or solutes which do not react with oxine and activated charcoal. In accordance with the process, compounds having properties similar to those of oxine together with oxine can be adsorbed by activated charcoal; an oxine and a derivative thereof such as 8-hydroxyquinaldine can be simultaneously adsorbed by activated charcoal to improve the properties thereof.

Example 1: This example illustrates the reaction of oxine with activated-charcoal in the air. A twin-cylinder mixer was charged with 1,000 g of coconut shell activated charcoal of 12 to 32 mesh and 100 g of oxine crystalline powder and rotated at 40 rpm. After 30 minutes, the oxine added was completely adsorbed by the activated charcoal and an oxine adsorbed-activated charcoal (100:10) was obtained.

In the reaction of oxine with activated charcoal in the air, the gases (air components) adsorbed by activated charcoal are replaced with oxine to be released into the reaction system and therefore the pressure in the reaction system is slightly increased. However, since the increase in pressure in the reaction system is less than $1/50$-60 atmospheric pressure, the reaction of oxine with activated charcoal in the air can be carried out in a closed vessel.

It was determined that approximately 35% by weight of the oxine added had been adsorbed by the activated charcoal.

Example 2: This example illustrates the reaction of oxine with activated charcoal in water. Into a flask of 1,000 cc was put 100 g of coconut shell activated charcoal of 12 to 32 mesh, 500 cc of water and 10 g of oxine crystalline powder and shaken gently. After 30 minutes, the oxine was completely adsorbed by the activated charcoal. It was determined that approximately 35% by weight of the oxine added had been adsorbed by the activated charcoal.

Removal of Plutonium Using Chitin

G.L. Silver; U.S. Patent 4,120,933; October 17, 1978; assigned to the U.S. Department of Energy describes a process which comprises contacting an aqueous solution containing a radionuclide with chitin until a portion of the radionuclide is absorbed by or otherwise associated with the chitin and thereafter separating the chitin from the aqueous solution.

Chitin is the principal ingredient of the shells of certain crustacea and insects. It is a polysaccharide, long-chain, unbranched polymer of N-acetyl-2-amino-2-deoxy-β-D-glucopyranose units. Structurally, chitin is an analogue of cellulose wherein the hydroxyl groups on the C-2 carbon have been replaced by acetylamino or amino groups. Chitin does not occur in its pure form in nature but is usually associated with substantial quantities of protein and inorganic salts such as calcium carbonate. These impurities may be removed through a series of mild chemical separations resulting in a solid, flake, white to off-white product of about 99% purity. It should be noted that chitin is normally found in association with living organisms and would be expected to exhibit altered characteristics when removed from the ambit of biological processes.

Chitin may be used to decontaminate or extract radionuclides from aqueous solutions using any of the solid liquid contacting unit operations well-known in the art. One of the simplest of these is batch equilibration and is illustrated in the example. Those skilled in the art will recognize that other means of contacting and separating may produce higher levels of decontamination. These methods include multistage mixing-settling and countercurrent or stationary column operation. The purified aqueous stream and the contaminated chitin may be separated by well-known gravity separation or filtration techniques.

Example: Into several 100 ml flasks were placed equal samples of hexavalent plutonium of such amount that when diluted to the volumetric mark, the samples contained 13,000 disintegrations (d) per minute per milliliter of solution. Into each flask was weighed 387±5 mg of chitin in comminuted or flake form. The flakes were irregular white flakes having a surface area from about 1 to 10 square millimeters on each side. Equal volumes of sodium acetate/ammonium nitrate buffer solutions of various pHs were added to each flask and the solutions were diluted to 100 ml with distilled water. After standing with frequent shaking at substantially atmospheric pressure and room temperature for one week, an aliquot of liquid was withdrawn from each flask and counted in a liquid scintillation counter.

As can be seen from the table, the chitin was effective in removing from $\frac{5}{6}$ to $\frac{15}{16}$ of the initial radioactivity.

Flask No.	pH	Radioactivity (d/min/ml)
Original	NA	13,000
1	4.94	810
2	6.65	1,125
3	8.10	2,363
4	8.65	2,163
5	9.30	1,874
6	9.75	2,192
7	10.24	2,380

DENITRATION PROCESSES

Biological Denitrification

Waste solutions containing high concentrations of nitrate and nitrite ions represent a serious threat to the ecological balances which exist in nature. Accelerated eutrophication of lakes and streams is often caused by discharging conventionally treated waste effluent into the surface waters because these wastes contain quantities of nitrogen and phosphorus which can promote excessive algae production. Aside from being a major nutrient for algae production, nitrogen in the form of ammonia is toxic to aquatic life and can react with chlorine to form chloramines which are toxic to certain fishes. Water for livestock is considered unsafe at nitrate nitrogen concentration exceeding 10 ppm. These concentrations can cause methemoglobinemia, vitamin A deficiency, loss of milk production, thyroid disturbances and reproductive difficulties. Nitrite wastes are considered to be injurious to several species of fish at concentrations on the order of 5 ppm nitrite nitrogen. Complete denitrification (conversion of nitrate or nitrite to elemental nitrogen gas) prior to releasing wastes to surface waters is thus desirable.

While the prior art units are satisfactory for handling the low nitrate concentration levels of municipal and agricultural wastes, they have been found unsatisfactory for handling high concentrations of nitrate wastes. In fact very little effort has been directed towards removing nitrates from wastewaters containing concentrations of nitrate nitrogen in excess of 1,000 ppm. Wastewater streams containing this magnitude of nitrates are generated in fertilizer and explosive manufacturing operations. Large quantities of wastewater effluent containing high nitrate concentrations are generated in nuclear fuel processing operations and at uranium oxide fuel fabrication plants.

C.W. Francis and F.S. Brinkley; U.S. Patent 4,043,936; August 23, 1977; assigned to the U.S. Energy Research and Development Administration have found that high concentration nitrate wastes can be effectively denitrified by passing the waste with a source of carbon continuously in an upflow mode through a column packed with a support having attached thereto denitrifying bacteria. The nitrate waste, under the influence of the denitrifying bacteria attached to the column packing, is converted along with the carbon source to free nitrogen and carbon dioxide. While a cylindrical column can be used to carry out the process, a conical column is preferably utilized because of the support stability and greatly enhanced denitrification rates.

The process is applicable to nitrate wastes containing a concentration of nitrate within the range of 1 gram to about 15 kilograms per cubic meter and having a pH within the range of 5.5 to 9.0. The process operates optimally at concentrations within the range of 1 to 5 kilograms of nitrate per cubic meter and at a pH within the range of 6.4 to 7.0. Below pH 5.5 and above pH 9.0 biological denitrification proceeds very slowly. Its optimum pH range is 6.5 to 8.5. Thus, to ensure the most effective denitrification rates in a unit the influent pH should be at or near the lower pH value (6.5) as the pH increases during denitrification.

The packing material to which the denitrifying bacteria is attached is preferably anthracite coal particles having an effective diameter within the range of 2 to 3 mm. However, other packing such as polypropylene or ceramic rings or saddles may be used. Coal is preferred because its density (2.3 to 2.5 g/cm^3) allows the packing to act as a fluidized bed under maximum denitrification conditions in a conically shaped column. A small sample of stock culture of bacteria can be used to incubate a column for use in the process. Fresh soil (preferably a soil of medium pH and high organic content, 2 to 5% C) containing the bacteria is added to a solution of calcium nitrate containing approximately 1,000 grams of nitrate per cubic meter and 600 grams of methanol per cubic meter. The stock culture is allowed to grow anaerobically for a period of from 10 to 20 days with stirring every three to four days. This procedure produces a healthy culture of denitrifying bacteria. Column units are seeded with the above culture by recycling a solution containing the microorganisms, calcium nitrate and methanol. Preferably such a solution comprises about 0.006 molar calcium nitrate aqueous solution having about 600 grams of methanol per cubic meter. This solution is recycled through the column until a microbial population is established on the packing media. Generally about 1 to 2 weeks are required to establish a culture in a packed column.

Sinc the denitrifying bacteria require a carbon substrate to denitrify nitrate to elemental nitrogen gas as well as use for cell synthesis, a source of carbon must be provided along with the aqueous nitrate source. Many sources of carbon may be used in carrying out the process. Such carbon source may be a compound selected from the group consisting of glucose, malate, methanol, acetone, ethanol, acetate (neutralized acetic acid) or any other carbon source possessing a high biological oxygen demand (BOD). Methanol, however, is the preferred substrate for use in the process because of its availability and costs.

In carrying out the process, a satisfactory resident time for the reactants in contact with the bacteria must be selected. This is dependent upon both the volume of the column through which the reactants flow and the flow rate of the reactants. The maximum denitrification rates achieved using a conical column packed with anthracite coal has been about 1.2 to 1.4 grams of nitrate per second per cubic meter of initial anthracite coal packing. Because hydraulic residence time and bacteria are variable and difficult to determine accurately, rates of denitrification are most easily expressed in these terms. Denitrification kinetic data indicates that denitrification rates decrease at nitrate concentrations greater than about 5 kilograms of nitrate per cubic meter of solution. Thus, an appropriate size reactor can be computed by using 5 kilograms of nitrate per cubic meter of solution as a maximum influent concentration. A reactor volume (initial bed packing) capable of handling the required or desired removal rate of nitrate based on the 1.2 to 1.4 grams of nitrate per cubic meter per second should be used in computing column size.

The cross-sectional flow rate at the base of the conical column should be maintained at less than 2.3 $dm^3/m^2/s$. For example, to denitrify 5 metric tons of nitrate a day a volume of 10^3 m^3 of nitrate solution could be pumped at a flow rate of 11.6 dm^3/s into a conical column containing approximately 58 cubic meters of packing.

A great and unexpected advantage of the process is that trace elements which are present in the nitrate wastes are somehow retained within the column such that the wastes which are released to the environment are free of or appreciably reduced in the concentration of these metal ions. For example, ammonium nitrate solution containing 5 to 10 grams of uranium per cubic meter was lowered to less than 0.5 gram per cubic meter in one pass through a column packed with 11.12 cubic decimeters of anthracite coal. Zinc, copper, cadmium and plutonium will be similarly retained within the column. The exact reason for this retention within the column is unknown. However, it is believed that these cations are precipitated as carbonates or converted to insoluble phosphate minerals similar to apatite.

Reduction with Powdered or Gaseous Reactants

The process of *F.-J. Gattys; U.S. Patent 4,271,034; June 2, 1981; assigned to F.J. Gattys Ingenieurbüro, Germany* is concerned with a method for significantly decreasing the volume of secondary radioactive waste from denitration processes. Specifically, this is accomplished by using reactants in the form of powders or gases in dosed amounts for the denitration.

With this procedure, numerous advantages in terms of elimination of secondary waste, and reduction of reaction times and production costs can be realized.

The loss of the gaseous reactants, especially when paraformaldehyde is used, can be significantly decreased by conducting the discharge from the reacting vessel through an irrigation path containing Raschig rings or similar packing, and then through a reflux condenser.

More particularly, in this process, paraformaldehyde, instead of formaldehyde solution, and sodium biphosphate (NaH_2PO_4), instead of sodium nitrate and orthophosphoric acid solutions, are used. For continuous dosing, it is possible to use known dosing units with dosing throughputs of 300 g/h and up and a continuous manner of operation with a maximum deviation of less than $\pm 0.5\%$ of the adjusted nominal value, as measured over one minute.

The table below shows the balance of the solids and the water for a throughput of 200 m^3 radioactive solution, with an operating time of 6,000 hours per year:

		Solids (kg/h)	Water (kg/h)
(a)	Waste solution	11.2	28.8
(b)	Paraformaldehyde	—	3.2
(c)	NaH$_2$PO$_4$, 51.6 t/a	8.6	—
	Total 1	19.8	32.0
	Total 2	51.8	
		(38.0% by weight)	
	Evaporation	49.5	2.3
		(40.0% by weight)	

Water of crystallization and moisture in solids have not been considered (sodium biphosphate is the only sodium salt which has only 1 mol water of crystallization per mol of salt).

It can be seen that by a dosed addition of solids with the second manner of operation, practically the same result is obtained as in the evaporation with the first manner of operation. However, by using the second process as described above, the following advantageous results are obtained:

(1)   Reduction or even elimination of radioactive secondary wastes: (32.0–2.3) 0.6000 = 178 200 kg/a = 178.2 t/a. The reduction is approximately 92.8%. The complete elimination takes place in the next process step with a changed manner of operation.

(2)   Nitric acid of ~50% by weight which is ready for use can be recovered for dissolving the burnt fuel elements.

(3)   It is estimated that the denitration time can be cut in half or that the output of an existing denitration plant can be doubled.

(4)   A reduction of the production costs is obtained (vapor, cooling water, use of labor, etc.).

As further noted, significant decreases in the loss of gaseous reactants and more complete conversion of the reactants can be achieved by conducting the discharge from the vessel through an irrigation path and then through a reflux condenser.

By using such an irrigation path, the completeness of conversion, i.e., the destruction of nitrates with paraformaldehyde, can be increased from 75 to 80% to up to 85% and more. Simultaneously, the yield in paraformaldehyde, i.e., the percentage of the chemically reacted paraformaldehyde, increases from approximately 45% to up to 80% and more.

The intensity of the irrigation by means of the vapors which were discharged from the reaction vessel and, subsequently, condensed in the reflux condenser, advantageously is approximately 10 to 12.5 m³/h of condensate for each square meter of irrigation surface.

An even further acceleration and completion of the process can be obtained in this connection by the use of at least 0.25%, preferably 0.35 to 0.50%, potassium bisulfate relative to the amount of nitrate to be destroyed. As a result, a displacement reaction takes place in the denitration system which, on the one hand, displaces the free nitric acid from the nitrates and, on the other hand, hydrolizes the double cation compound to form the orignal salt. In this manner, the potassium bisulfate acting as the catalyst is restored.

$$KHSO_4 + NaNO_3 \rightarrow HNO_3 + KNaSO_4$$

$$KHSO_4 + NaOH \leftarrow HOH + KNaSO_4$$

Finally, the powdery reaction material can be introduced in the form of a paraformaldehyde suspension, wherein the waste solution taken from the reaction vessel and cooled down, preferably to 40° to 60°C, serves as a suspension medium. Stable suspensions with 30 to 35% by weight paraformaldehyde can be easily produced and can be returned into the reaction vessel by means of air lifts or pumps.

The following numerical examples shall serve to further illustrate the process.

Example 1: 1,440 g sodium nitrate ($NaNO_3$) were dissolved in 2,160 g desalted water. This solution was heated to 105°C and then treated for approximately 10 hours with 450 g of pneumatically introduced paraformaldehyde. In this process, the denitration was 79.5% and the yield in paraformaldehyde was approximately 45%.

Example 2: The procedure was the same as in Example 1, however, between the gas outlet of the reaction vessel and the reflux condenser, an irrigation path in the form of a glass tube which was filled with glass rings of 5 mm diameter was arranged. Only 300 g paraformaldehyde were used, while the remaining conditions were the same as in Example 1. An intensive flow of condensate from the vapors discharged from the reaction vessel was admitted to the irrigation path, namely, 10 to 12.5 m³/h of condensate for each square meter of irrigation surface. This resulted in a denitration of 95.8% and a paraformaldehyde yield of 81%.

Example 3: Another test was carried out in accordance with Example 2. However, 7.5 g of potassium bisulfate ($KHSO_4$) was added to the waste solution to be treated. In this case, despite a reduction in reaction time to 6.5 hours, an almost total denitration of 99.6% and an even higher yield in paraformaldehyde of 85.4% were obtained.

Addition of Urea

J.R. Berreth; U.S. Patent 3,962,114; June 8, 1976; assigned to the U.S. Energy Research and Development Administration describes a process for suppressing the evolution of noxious nitrogen oxides during solidification by heating of liquid radioactive wastes containing nitrates and nitrites which may be met by adding a slight excess of a stoichiometric amount of urea relative to the nitrates and nitrites present to the liquid waste, heating the waste containing the urea to at least about 130°C while bubbling carbon dioxide through the solution whereby the urea reacts with the nitrates and nitrites present in the solution, evolving elemental nitrogen, carbon dioxide and ammonia gas.

The amount of urea to be added to the liquid radioactive waste is dependent upon the quantity of nitrates and nitrites present in the solution. Thus, while one mol of urea will react with four mols of nitrate, one mol of urea will only react with two mols of nitrites, which are generally present as the sodium salts. In addition, it is preferred that a slight excess of the stoichiometric amount be added to the solution to ensure a complete reaction between the urea and the nitrates and nitrites in order to prevent evolution of any nitrogen oxides.

After addition of the urea to the waste solution, the solution should be heated to at least from about 130° to about 180°C to ensure complete reaction between the urea and the nitrates and nitrites present.

It is preferred, but not required, that carbon dioxide be bubbled through the solution to ensure that the reaction goes to completion and to sweep the evolved gases from the waste solution as they are formed.

The process is particularly useful for the destruction of nitrates and nitrites contained in neutralized and basic solutions. The process may also be useful to a lesser extent with acidic waste solutions where some destruction of the urea may take place before reaction by the urea with the nitrates and nitrites is complete.

The method for suppressing the evolution of nitrogen oxides can be used with several different methods for the solidification of liquid radioactive wastes. For example, the method can be used with the pot calcination process. In this process, a slight excess of the stoichiometric amount of urea relative to the nitrates and nitrites present is added to the waste solution and the solution is heated to from about 130° to about 180°C while carbon dioxide may or may not be bubbled through the waste solution. After the solution has been heated to dryness, with the evolution of elemental nitrogen, carbon dioxide and ammonia, the remaining solid is heated to from about 500° to about 700°C to calcine the solid and prepare it for storage.

The method may also be used in the fluidized-bed calcination process, where the liquid radioactive waste is sprayed onto a fluidized-bed calciner at a temperature of from about 400° to about 600°C wherein the urea reacts with the nitrates and nitrites present to evolve elemental nitrogen, carbon dioxide and ammonia and suppress the formation of nitrogen oxides. An advantage of the use of this method with fluidized-bed calciners is that the destruction of the nitrates and nitrites by the urea with the addition of proper additives such as hydrated alumina may prevent harmful agglomeration of sodium nitrate and sodium nitrite in the fluidized bed.

Example: An experiment on the destruction of nitrates and nitrites was tried using a simulated neutralized Purex waste solution having the following composition:

Constituent	Molarity
$NaNO_3$	3.0
$NaNO_2$	1.0
NaOH	0.3
$Fe(OH)_3$	0.2
Na_2SO_4	0.2
$NaAlO_2$	0.6
Na_2CO_3	0.3
MnO_2	0.2
NaF	0.02
$Hg(NO_3)$	0.001
Na_3PO_4	0.01
NaCl	0.01
NaI	0.001
$Mg(OH)_2$	0.01
$Ca(OH)_2$	0.003

A stoichiometric amount of urea relative to the sodium nitrate and sodium nitrite present was added to the waste solution, which is 4 molar in nitrate-nitrite and 6 molar in sodium. The solution was heated to dryness at temperatures not above 180°C with the evolution of N_2 and NH_3. Heating of the solid to 600°C liberated very minimal amounts of nitrogen oxides. Untreated solids heated to this temperature liberate copious amounts of nitrogen oxides. It was noted that the solid material was not as soluble as untreated dried waste solids.

Electrolysis of Waste Solutions

H. Schmieder and R. Kroebel; U.S. Patent 4,056,482; November 1, 1977; assigned to Gesellschaft fur Kernforschung mbH, Germany describe a method for preparing aqueous radioactive waste solutions for noncontaminating solidification and/or removal of such solutions. In the method, the total quantity of the various inorganic and organic substances is reduced by the destruction of nitric acid, nitrates and nitrites and formation of a waste gas mixture which is practically free of higher nitrous oxides. In accordance with this process, the radioactive waste solutions are subjected to an electrolysis current at such current densities at the anode and at the cathode that, in one process step, the substances of the group hydrazine, hydroxylamine, oxalic acid, oxalates, tartaric acid and tartrates are oxidized at the anode and the substances of the group nitric acid, nitrates and nitrites are reduced at the cathode.

The process produces gaseous oxidation and reduction products such as nitrogen, oxygen, and carbon dioxide, for example. In this process, the abovementioned substances hydrazine, hydroxylamine, oxalic acid, oxalates, tartaric acid and tartrates are completely destroyed in contradistinction to prior art processes.

The apparatus required for the electrolytic denitration is very simple and does not include diaphragms. For discontinuous operation, any conventional electrolysis cell structure can be employed in practice. In the experiment described in the example which follows, the cell was made of glass. For practical use, it is also possible to use the cathode material itself as the structure material for the cell. For continuous operation in a flowthrough cell, trough-shaped, elongated cells are recommended which are provided at various points with extraction devices for the various fission product fractions. The continuation of the oxidation-reduction processes is advisably monitored by checking the electrical conductivity of the solution.

The anode material of the cell can be platinum or a platinized metal with passivation properties. The cathode material of the cell can be titanium or graphite. The preferred platinized metals are platinized tantalum and platinized titanium. In principle, other metals, such as, for example, platinized zirconium and the like, can also be used, but if the surface of such anodes is damaged, corrosion will be too strong and the anode soon will be useless.

When a graphite cathode is used, difficulties or delays, respectively, which may occur at the start of the reduction reaction can be overcome by the addition of small quantities of metal ions from the group of copper, lead and titanium as starting aids to the waste solution. Additives which result in a concentration between 10 and 50 mg/ℓ in the waste solution are sufficient. Reduction at the cathode takes place at a current density of about 10 mA/cm^2 or more, and the surface ratio of the effective cathode surface to the effective anode surface lies in the range of about 0.1 to about 10. The ratio of the electrode surfaces can be varied over a much wider range, for example, the ratio can be less than 0.2 or more than 10.

For practical use, however, surface ratios in the abovementioned range of 0.1 to 10 are most appropriate. The upper limit of the current density at the cathode should be in accordance with economical considerations, for example, about 100 mA/cm^2. The length of the treatment is not relevant, because the surface of the electrode determines the rapidity with which the destroying reactions run and thus the length of time.

The advantages of the process are that no addition of neutralizing and/or reduction chemicals is required, thus reducing costs and apparatus involved as well as space requirements. Further, the waste gases contain practically no nitrous gases and the process is dependable and easily controllable.

A further advantage of this process is that it is possible to fractionate the fission products. In the experiments with fission products, it has been found that the noble metals, such as Ag, Pd, Ru and Rh, for example, in the still strongly acid solution are cathodically reduced to metal and precipitated while the remaining fission products remain in solution at these acid concentrations. Under certain circumstances, it will be possible in this way to indirectly shorten the intermediate storage periods for the waste solutions.

Example: This example illustrates the treatment of a simulated aqueous waste solution in an electrolysis cell. The waste solution serves as the electrolyte in the cell.

The cathode of the cell was made of titanium and had a surface area of about 80 cm^2. The current density at the cathode was a constant 50 mA/cm^2. The anode of the cell was made of platinum. The current density at the anode was about 250 mA/cm^2.

The volume of the simulated waste solution used in the cell was 90 ml. The waste solution had a 1.15 molar concentration of nitric acid and contained 0.1 mol per liter N$_2$H$_4$. Inactive noble metals Ag, Pd, Ru and Rh were added to the waste solution to simulate fission products. The waste solution was kept at a constant 20°C. At the beginning of the electrolysis process, precipitation started of the inactive noble metals Ag, Pd, Ru and Rh. The nitrite concentration at no time exceeded 100 mg/ℓ. Gaseous reaction products which were found at the cathode were mainly hydrogen and nitrogen, as well as small quantities of NO and N_2O. Gaseous reaction products which were found at the anode were mainly nitrogen and oxygen, as well as small quantities of NO. At acid concentrations below 0.5 molar, NH_3 could be determined. After about 15 minutes, hydrazine could no longer be determined in the solution. After about 140 Ah/ℓ, the hydrogen ion concentration had dropped from 1.15 mols per liter to 0.1 mol per liter. The drop in hydrogen ion concentration was linear.

MISCELLANEOUS PROCESSES

Chemical Digestion of Low Level Waste

C.R. Cooley and R.E. Lerch; U.S. Patent 3,957,676; May 18, 1976; assigned to the U.S. Energy Research and Development Administration describe a process whereby combustible solid waste material containing low level solid nuclear wastes are chemically digested by reacting the combustible solid waste material with concentrated sulfuric acid at a temperature within the range of 230° to 300°C and simultaneously and/or thereafter contacting the reacted mixture with concentrated nitric acid or nitrogen dioxide whereby the carbonaceous material is oxidized to gaseous byproducts and a low volume residue.

The process may be conducted batchwise or by incremental additions of solid waste material and nitric acid or nitrogen dioxide. The low volume residue may be further processed by separating the noncombustible solids from the resulting aqueous solution, neutralizing and drying the residue. Apparent volume reductions on laboratory scale runs of up to 160 have been achieved, but depend on the amount of inorganic chemicals in the material being digested. Advantageously, very little acid is consumed since it can be recycled or recovered and reused indefinitely, it being used only as a chemical combustion media to release combustion products at a controlled rate and at a low temperature. The final residue or "ash" is inactive and suitable as a noncombustible material for shipping and storage.

Example 1: 100 grams (~320 ml) of mixed waste material (i.e., 15 g tygon tubing, 15 g neoprene rubber, 15 g polyethylene, 15 g latex tubing, 15 g latex rubber gloves, 15 g plastic vial and 10 g plastic tape) were added to one liter of concentrated sulfuric acid at ~270°C. The mixture was allowed to digest for one hour, then concentrated nitric aicd was added to the mixed solution at the rate of 25 ml every 30 minutes. A volume of 380 ml of concentrated nitric aicd was necessary to change the solution from opaque black to a clear yellow. A fine grey residue remained undissolved.

The solution was cooled to room temperature and filtered. The separated residue weighed 3.7 grams and occupied about 2 ml, giving an apparent volume reduction of 160.

The residue was insoluble in water, acetone, dilute sodium hydroxide, nitric acid, ethyl alcohol, and carbn tetrachloride. It was not combustible and converted to a fine white ash when heated to 1200°C (indicating the presence of inorganic matter rather than organic).

Example 2: Exactly 250 ml of concentrated sulfuric acid containing five volume percent concentrated nitric acid were heated to 270°C in a closed flask connected to a condenser. Polyvinyl chloride (PVC) and concentrated nitric acid were simultaneously added to the hot acid, polyvinyl chloride at a rate of 16 grams per hour and nitric acid at a rate of 30 ml per hour. Nitric acid addition was continued for 15 minutes after each 16 gram batch of PVC was added in order to complete the oxidation. The experiment was stopped after nine 16-gram batches of polyvinyl chloride had been added. Final conditions were as follows: 144 grams of polyvinyl chloride digested, 320 ml of nitric acid added, 185 ml additional sulfuric acid added to maintain the starting volume of 250 ml, 490 ml of condensate collected. The sulfuric acid was cooled to room temperature and filtered. Approximately 4.3 grams (~5 ml in volume) of residue were filtered from the solution representing a weight reduction of 39 and a volume reduction of 33. The final residue was insoluble in water, acetone, nitric acid and sodium hydroxide.

Off-gases from the process include: CO, CO_2, Cl_2, HCl, NO_x, etc. Sulfur dioxide (SO_2) evolution is greatly suppressed by maintaining a nitric acid-rich system.

R.G. Cowan and A.G. Blasewitz; U.S. Patent 4,313,845; February 2, 1982; assigned to the U.S. Department of Energy describe a similar system for chemically digesting low level radioactive solid waste material wherein the solid waste is reacted with concentrated sulfuric acid at a temperature within the range of 220° to 330°C and simultaneously the reacting mixture is brought in contact with concentrated nitric acid or nitrogen dioxide. The improved system comprises an annular vessel constructed to be substantially filled with the concentrated sulfuric acid. The waste material is introduced into the annular vessel and the nitric acid or nitrogen dioxide is added to the sulfuric acid while the sulfuric acid is reacting with the solid waste. Means are provided for mixing the solid waste within the sulfuric acid so that the solid waste remains substantially fully immersed.

During the reaction, the off-gas and the product slurry residue are removed from the annular vessel. In one preferred form the means for mixing includes an air lift recirculator wherein mixing is provided by air used to oxidize the off-gases and the nitric acid or nitrogen dioxide used to oxidize the carbon slurry residue. In another preferred form the vessel is constructed to retain the heat of the exothermic chemical reaction to substantially maintain the reaction temperature within the range of 220° to 330°C.

Removal of Silicon from Incinerator Ash

Plutonium contaminated material, e.g., paper, plastic, wood, rags, ion-exchange resins and other burnable material is generally incinerated and the plutonium thereafter removed or recovered from the resulting ash. Generally, the ash contains large amounts of silicon, primarily in the form of silica (SiO_2). The process used for recovering plutonium involves dissolution of the ash in a nitric-hydrofluoric acid solution. An inherent problem with this process is that some of the silicon is converted to fluosilicic acid (H_2SiF_6) and silicon tetrafluoride, which compounds hydrolyze and form an objectionable gelatinous, sticky silica polymer which coats components of the plutonium recovery system such as filters, piping, valves, pumps and ion-exchange columns. This results in an increase in maintenance requirements, increase in shutdown time for maintenance, increased costs for recovery of the radioisotope, and other like problems.

R.G. Auge and R.P. DeGrazio; U.S. Patent 3,882,040; May 6, 1975; assigned to the U.S. Energy Research and Development Administration describe a process whereby radioisotope contaminated materials such as paper, plastic, wood, rags, ion-exchange resins and other burnable material may be incinerated to form incinerator ash which is thereafter contacted with NOF·3HF to react with the silicon present in the incinerator ash and convert it to silicon tetrafluoride gas in accordance with the following equation:

$$SiO_2 + NOF \cdot 3HF \rightarrow SiF_4 + NO_2 + \text{various oxides of nitrogen}$$

The silicon tetrafluoride gas which forms may be removed through suitable means such as an exhaust blower. The radioisotope contaminant may be such as plutonium, americium, cerium, and uranium.

The NOF·3HF as discussed above is preferably in a from about 0.75 to about 1.0 ml proportion of NOF·3HF with a from about 3.2 to about 5.0 ml proportion of water per gram of incinerator ash. A preferred concentration is about 10 ml of NOF·3HF with about 50 ml of water which may be contacted with about 15 grams of ash.

The resultant mixture may be appropriately heated, such as on a hot plate or other heating apparatus, to from about 75° to about 100°C and preferably at about 80°C and, with stirring, may be maintained at temperature for a period of from about 75 to about 100 minutes and preferably for about 90 minutes. The silicon tetrafluoride gas which was generated during this heating may be removed by suitable exhaust means such as an exhaust blower. If desired, the remaining ash may then be cooled, filtered and dried. Filtration may be accomplished through well-known means such as by using No. 30 Whatman filter paper. Drying may be accomplished by heating the residue under a vacuum in an oven at a suitable temperature such as about 120°C for about 2 hours.

In an example illustrating the efficiency of this process, 15 grams of ash (containing 1.935 grams of plutonium) were contacted with a mixture of 50 ml of water and 10 ml of NOF·3HF, heated to a temperature of about 80°C and maintained at that temperature for about 90 minutes, and thereafter the remaining ash was filtered and washed with water. The precipitate or ash was then dried in an appropriate manner by using a vacuum type oven. The weight of remaining ash after drying was 11.5 grams (containing 1.495 grams of plutonium) which was a weight reduction of 23.3%. There was a slight loss (0.44 gram) of plutonium from the ash between or during the reaction and processing due to some dissolution of plutonium and loss in the washing cycle. This plutonium was recovered in subsequent solution ion-exchange recovery processes. In this example, the silicon content before contact with the NOF·3HF was greater than 10% and after contact was 2.3% thereby giving a minimum of 77% silicon removal from the incinerator ash as silicon tetrafluoride gas.

Incinerator ash as used herein is generally of a particle size of from about 0.001 to about 0.5 inch diameter. As is well known in the art, when contacting a granular material with a liquid, maximum surface area presented for contact is desirable and, as such, the smaller size particles of incinerator ash for contact with the NOF·3HF would be preferred. Stirring or agitation of the solution may be accomplished by any well-known means such as the use of a magnetic stirrer.

Further removal of silicon from incinerator ash may be accomplished by recycling the incinerator ash through subsequent stages of this process. The number of times that this process is repeated will be dependent upon the percent removal of silicon from incinerator ash desired. Further, the volume of water, or the like wash solution, used in washing the ash in the filtration step is preferably minimized in order to maintain as small as possible the amount of solution contacting radioactive material.

HEAT TREATMENTS

CALCINATION OR COMBUSTION OF RADIOACTIVE WASTE

Minimization of Chloride Volatilization During Calcination

A typical problem which arises in the fluidized-bed calcining of the many types of waste is the fouling of the fluidized bed by particle agglomeration due to the presence of sodium nitrate. Sodium nitrate does not decompose but melts and exists in a molten state between 305° and 833°C which includes the normal range of calcination temperatures. Therefore, it is present in a molten state and can cause agglomeration of the bed particles and consequent fouling of the fluidized bed.

Volatilization of various corrosive components, such as fluorides and chlorides, presents problems downstream from the fluidized bed in the off-gas cleanup system. Consequently, it is desirable to minimize fluoride and chloride volatility.

D.C. Kilian; U.S. Patent 3,954,661; May 4, 1976; assigned to the U.S. Energy Research and Development Administration describes a process whereby liquids containing sodium, nitrate, and chloride ions are calcined to solids in a fluidized-bed calciner while minimizing volatilization of chlorides and preventing agglomeration of the bed particles by molten sodium nitrate. Zirconium and fluoride are introduced into the liquid containing the sodium, nitrate, and chloride ions and ½ mol of calcium nitrate per mol of fluoride present in the liquid mixture is added. The combined mixture is then calcined in a fluidized-bed calciner at about 500°C, resulting in a high bulk density calcine product containing the chloride and thus minimizing the chloride volatilization. In a specific embodiment, intermediate-level liquid radioactive wastes are combined with zirconium fluoride radioactive waste and ½ mol calcium nitrate per mol of fluoride present is added to the combined mixture. Preferably, 3 parts zirconium fluoride waste are blended with 1 part intermediate-level waste.

Cryolite Formation During Calcination

J.A. Wielang and L.L. Taylor; U.S. Patent 3,943,062; March 9, 1976; assigned to the U.S. Energy Research and Development Administration describe a method for the fluidized-bed calcination of sodium-containing liquid radioactive wastes which will not result in agglomeration of the fluidized-bed particles.

In accordance with the process, aluminum and a fluoride are added to liquid radioactive wastes containing significant quantities of sodium or sodium compounds, which wastes are to be solidified for long-term storage as a solid by calcining the wastes in a fluidized-bed calciner. The aluminum and fluoride are added to the liquid wastes prior to the spraying of the wastes in the calciner. Preferably, the aluminum and fluoride are added to the wastes in the form of solid AlF_3 compounds or in the form of hydrofluoric acid and aluminum metal, although addition of the aluminum and fluoride in other forms is also possible. In the fluidized bed the sodium in the wastes combines with the aluminum-fluoride to produce cryolite (Na_3AlF_6 or $3NaF \cdot AlF_3$). Fluoride addition displaces the nitrate during the calcination process while the aluminum complexes the fluoride in the aqueous phase. The reaction for the main reactants can be expressed:

$$3Na^+ + Al^{+3} + 6F^- \rightarrow Na_3AlF_6$$

The more complete reaction can be written:

$$6NaNO_3 + 4AlF_3 \rightarrow 2Na_3AlF_6 + Al_2O_3 + 6NO_2 + \tfrac{3}{2}O_2$$

The cryolite formation during calcining allows the sodium waste to be solidified to hard granular particles without fluid-bed agglomeration.

Since significant quantities of fluoride will be present and fluoride volatility will result in very undesirable corrosion problems, it is preferred that aluminum and sodium be present in excess over that required for cryolite formation. An excess of about 10% is believed to be sufficient to eliminate any possibility of fluoride volatilization in the fluidized bed. When excess aluminum and sodium are contained in the solutions introduced into the fluidized bed, the excess aluminum will be converted to aluminum oxide, while the sodium will form sodium nitrate. Therefore, the amount of excess must be kept sufficiently low such that the amount of sodium nitrate formed remains sufficiently small that it does not pose a problem with bed agglomeration. An excess of about 10% has been found to cause no problems with bed agglomeration and is therefore preferred.

The addition of fluoride and the formation of cryolite, which is a fluoride-containing compound, raise other considerations which must be taken into account in devising a satisfactory process for solidifying these wastes. The off-gases of the calciner must be freed of undesirable pollutants in the off-gas cleanup system, and routinely are scrubbed in order to remove any particulate matter which may be given off from the calciner. It is likely, and preliminary studies have shown, that some cryolite will be contained in the off-gases of the calciner and will be dissolved downstream in the off-gas scrub solution. When the cryolite dissolves in the scrub solution, fluoride could volatilize and cause corrosion problems in the off-gas cleanup system. Therefore, aluminum is added to the scrub solution in the off-gas cleanup system in order to complex any fluoride released from cryolite dissolution, thereby preventing fluoride volatility.

While the aluminum could be added to this solution in many forms, it has been found that aluminum nitrate is a convenient form in which to introduce the Al^{+3} ion. The aluminum nitrate is added in amounts sufficient to complex any fluoride released from dissolution of the cryolite. Consequently, corrosion problems from fluoride volatility are eliminated.

Addition of Aluminum to Waste Containing Zirconium and Halides

It is known that adding calcium nitrate to waste solutions before calcining the solutions will suppress the volatility of the fluoride to acceptable levels which can then be removed from the calciner off-gas by scrubbing equipment. However, the addition of calcium nitrate to the waste has little suppressive effect upon the chloride which, although present in the waste solutions in only relatively small amounts builds up in the fluidized bed of the calciner over a long period of operation, so that the quantity, in time, becomes significant. The addition of calcium nitrate to the waste solutions also results in the formation of a gelatinous solid. This solid, which is a hydrated calcium fluorozirconate, clogs transfer piping and calciner spray nozzles and generally disrupts calciner operation by increasing down-time for cleanup. The substitution of magnesium nitrate for calcium nitrate has been tried, and although it eliminates the formation of gelatinous solids while maintaining fluoride volatility suppression at acceptable levels, it has insufficient effect upon chloride volatility. When magnesium nitrate is added to first cycle waste, a calcine is produced which is very soft and breaks easily into fines during fluidized bed operation, plugging and bridging calciner off-gas and transport systems.

B.J. Newby; U.S. Patent 4,164,479; August 14, 1979; assigned to the U.S. Department of Energy has found that, by adding aluminum to the waste solutions to increase the aluminum to fluoride mol ratio, before adding the calcium nitrate to the solution, the before enumerated problems are substantially reduced or eliminated.

The method therefore consists of adding aluminum to the zirconium-fluoride waste solution containing zirconium, fluoride and chloride prior to adding calcium nitrate, in an amount sufficient to establish an aluminum to fluoride mol ratio of at least 0.27, whereby the quantity of gelatinous solid formed by the subsequent addition of calcium nitrate to the solution is substantially reduced and the volatility of the chloride during calcination of the waste solution is suppressed. The method further consists of adding aluminum to a blend of 3 parts zirconium-fluoride waste and 1 part second cycle waste prior to adding calcium nitrate, in an amount sufficient to establish an aluminum to fluoride mol ratio of 0.32, thereby reducing gelatinous solids formation and chloride volatility during calcination of the blend.

To demonstrate the effect of the addition of calcium and aluminum on the amount of gelatinous solids formed in the first cycle waste and in the blend, experiments were run in which varying amounts of ions were added to the wastes. In the table, the rate of filtration of solids after calcium nitrate or aluminum nitrate plus calcium nitrate had been added to the wastes is used as a measure of the gelatinous nature of the residue; the less the filtering time, the less the gelatinous nature of the solid. In each case 30 ml of homogenized slurry is sucked through a sintered glass filter (that has never been used before) having a 14 micron porosity by a vacuum pressure of 17 inches of mercury. The results are given in the table below.

Waste	Ca to F Mol Ratio	Al to F Mol Ratio	Filtering Time	Residue (g from 30 ml waste)
1st cycle Zr-F waste	0.55	0.21	25 min	4.6
1st cycle Zr-F waste	0.55	0.27	5 min 10 sec	2.6

(continued)

Waste	Ca to F Mol Ratio	Al to F Mol Ratio	Filtering Time	Residue (g from 30 ml waste)
1st cycle Zr-F waste	0.55	0.40	1 min 5 sec	0.55
3 vol 1st cycle Zr-F waste*	0.7	0.28	3 min	2.4
3 vol 1st cycle Zr-F waste*	0.7	0.32	45 sec	0.52

*Blended with 1 vol 2nd cycle waste.

It can be seen that the addition of a small amount of aluminum resulted in a substantial reduction of the amount of solids formed.

Calcining with a Minimum of Air

D.R. Ross; U.S. Patent 4,263,163; April 21, 1981 describes a process for making a shielded encapsulated radioactive material.

In conventional furnaces, it may be considered that two types of air are introduced. One type of air is for combustion purposes so that the fuel, such as a hydrocarbon gas, can be burned to give off heat energy. The second type of air can be considered to be an expansion air. The expansion air, along with the particles to be expanded is heated to be able to carry away the expanded solid particles. Because of the necessity of heating the expansion air, a considerable amount of heat energy is used. The expansion air, at ambient temperature, enters into the furnace, is heated and the temperature elevated to that temperature in the furnace, and this heated expansion air is used to carry away the expanded solid products and then the heated expansion air is exhausted to the atmosphere. In one manner of thinking, the heating of the expansion air is a waste of heat energy.

This process comprises a furnace having two opposed sets of refractories. The refractory may be furnace brick. These refractories are arranged in two circular paths. There is a lower refractory in a circular path and an upper refractory in a circular path with the upper refractory being positioned above the lower refractory. There is a means for rotating one of these refractories. Generally, the lower refractory is rotated. Also, there is means for introducing solid particles onto the lower refractory so the solid particles can be heat treated and, in certain instances, expanded. Also, there is a means to remove the expanded solid particles from the lower refractory and from the furnace.

The refractory can be porous so that a gaseous fuel can pass through the refractory and burn near the surface of the refractory. The furnace requires, essentially, only combustible air for burning the combustible fuel. The furnace does not need expansion air as the expanded solid particle or the heat treated particle is not removed from the furnace by means of expansion air. The expanded solid particle or the heat treated particle is removed from the furnace, mainly by force of gravity.

In certain instances, it is possible to use a solid fuel, such as coal, or to use a liquid fuel, such as fuel oil and to vaporize the liquid fuel prior to introducing it into the furnace.

With the furnace requiring, essentially, only combustible air for burning the combustible fuel, there is a saving in heat energy and fuel as it is not necessary to heat expansion air and fuel is not wasted in heating the expansion air.

Conversion of Nitrogenous Waste into Molecular Nitrogen

J.M. Dofson and T.E. Peters; U.S. Patent 3,862,296; January 21, 1975; assigned to General Electric Company have found that nitrogenous waste materials containing at least a portion of the nitrogen in a positive valence can be rapidly converted to a releasable gaseous molecular nitrogen by calcining the nitrogenous material in a fluidized bed in the presence of a reducing agent containing ammonium ions. The residual metallic and nonmetallic ions from the nitrogenous waste materials are disposed of as solid oxide calcine or treated for recovery if desirable. This process attains exceptionally high conversion of the waste materials to gaseous nitrogen since the ammonium ions impart a reducing nature to the calcination. The ammonium ions can be introduced in any convenient, readily heat decomposable form such as ammonia gas and ammonium compounds including ammonium hydroxide and ammonium salts.

The process is broadly applicable to the conversion of liquid nitrogenous waste materials, either containing ammonium ions or free of ammonium ions, including nitrogen-containing acids such as nitric acid and nitrous acid, nitrite salts, and nitrate salts such as uranyl nitrate, plutonium nitrate, ammonium nitrate, silver nitrate, aluminum nitrate, sodium nitrate, calcium nitrate, ferric nitrate and fission product nitrates such as ruthenium nitrate, lanthanium nitrate, cesium nitrate, strontium nitrate, etc.

The calcination process for the conversion of nitrogen-containing compounds to gaseous nitrogen can be represented by the following general reaction for a monovalent nitrate compound:

$$5NH_3 + 3MNO_3 \rightarrow 4N_2 + 1.5M_2O + 7.5H_2O$$

where M is any monovalent cation with representative cations being H^+, K^+ and Cs^+. Similar stoichiometric reactions are equally applicable to di-, tri- or other multiple valence cations with representative cations being UO_2^{+2}, Ca^{+2}, PuO_2^{+2} and Al^{+3}.

In a preferred application of this process, liquid plutonium nitrate wastes are converted to releasable gaseous materials, largely gaseous nitrogen and water vapor, and the waste liquid condensate streams have plutonium contamination removed to a concentration less than 10^{+5} grams of plutonium per liter. This enables release after scrubbing of both the gaseous product and the liquid product from this process without atmospheric contamination. This process involves a fluidized bed calcination of the plutonium-bearing nitrate wastes in the presence of a reducing agent containing the ammonium ion which converts the nitrates substantially to molecular nitrogen. In this process residual plutonium and other trace metallic ions would be deposited as oxide calcines in the fluidized bed for recovery by standard recovery processes or for disposal as a solid waste. Further protection against any release of plutonium to the environment is assured by fractional distillation of aqueous condensates in which nonvolatile contaminates would be retained from the distillation step for recycle to the calciner and the clean distillate water would be discharged.

Combustion in Molten Alkali Metal Carbonate Bath

L.F. Grantham; U.S. Patent 4,145,396; March 20, 1979; assigned to Rockwell International Corp. has found that when an organic waste contaminated with at least one element selected from the group consisting of strontium, cesium, iodine, and ruthenium is treated in a molten salt at elevated temperatures, it is possible to achieve a substantial reduction in the volume of the organic waste, and further, the selected element is retained in the molten salt. Broadly, the method comprises introducing an organic waste containing the selected element and a source of gaseous oxygen such as air into a molten salt comprising an alkali metal carbonate. The bath is maintained at a temperature of from about 400° to 1000°C and a pressure within the range of from about 0.5 to 10 atm to at least partially oxidize the organic waste. Complete combustion is generally preferred. Under such conditions, the volume of the organic waste is substantially reduced, and solid and gaseous combustion products are formed. The gaseous combustion products consist essentially of carbon dioxide and water vapor, and the solid products comprise the inorganic ash constituents of the waste; the selected element being retained in the molten salt.

The molten salt may be either a single alkali metal carbonate or a mixture of two or more alkali metal carbonates and may include at least one and up to about 25 weight percent of an alkali metal sulfate. The advantage of the inclusion of the sulfate is that it enhances the combustion of the organic material (for example see U.S. Patent 3,567,412). The preferred alkali metal carbonates are the lithium, sodium, and potassium carbonates.

Where it is desired to perform the combustion of the organic waste at a relatively low temperature, a low melting binary or ternary mixture of alkali metal carbonates may be utilized. For example, the ternary alkali metal carbonate eutectic consisting of 43.5, 31.5, and 25.0 mol percent of the carbonates of lithium, sodium, and potassium, respectively, melts at about 397°C. A preferred binary mixture is the sodium carbonate-potassium carbonate eutectic which melts at about 710°C. When the principal consideration is the cost of the molten salt, which ultimately is disposed of, a particularly preferred salt comprises sodium carbonate and optionally contains from 1 to 25 wt % sodium sulfate, which may be used at a temperature between about 750° and 1000°C.

The exact pressure and temperature utilized are not critical, provided, of course, that they are so selected as to be above the melting point of the salt and below its decomposition temperature. Generally, the temperature and pressure will be within the range of from about 700° to 1000°C and from 0.5 to 10 atm. A temperature of from about 800° to 900°C and a pressure of from about 0.8 to 1.0 atm are generally preferred, particularly with sodium carbonate.

Typical organic waste materials generally have a sufficiently high heating value to maintain the molten salt at a desired temperature. However, where the heating value of the waste material is insufficient to maintain the desired temperature, any carbonaceous material such as coal, tar, petroleum residuals and the like, may be added to the feed to increase its heating value.

The organic waste and a source of oxygen are introduced into the molten salt. Generally, the source of oxygen will be air in the interest of economy. Thus, the effluent gases will also include nitrogen when air is used and may also include unreacted oxygen. However, when it is desired to reduce the volume of gaseous products, pure oxygen can be used. Alternatively, of course, oxygen-enriched air also may be utilized.

In view of the economic importance of reducing the volume of organic waste contaminated with radioactive fission products and the obvious need to attain such radioactive products in a substantially stable solid form, this process will be described with reference to the treatment of such waste. In accordance with this method, such typical fission products as strontium, cesium, iodine and ruthenium are retained in the molten salt. In addition, it also has been found that numerous other radioactive products of fission also are retained in the molten salt, for example, the rare earths (La, Ce, Pr, Nd, Pm, Sm, and Eu) and yttrium would remain in the ash fraction as insoluble oxides. Further, the noble metals such as palladium and gold also would be retained in the molten salt. If radioactive amphoteric-like elements such as zirconium, niobium, molybdenum, technetium, and tellurium are present in the melt, they also will be retained in the molten salt as either an oxide or sodium salt compound. Thus, except for the inert gases, it has been found that most of the fission products are retained in the molten salt.

When the molten salt subsequently is processed (the molten salt is mixed with an aqueous medium and filtered to remove the ash, and the filtrate is subsequently cooled to precipitate sodium chloride crystals), the disposition of the various fission products will be as set forth below.

		. . . . Fission Product Disposition	
Molten Salt Process Stream	Elemental Class	Elements	Ultimate Chemical Form
Off-gas	Inert gases	Kr, Xe	Elements
Ash	Alkaline earths	Sr, Ba	Sulfate or carbonate
	Rare earths	La, Ce, Pr, Nd, Pm, Sm, Eu	Oxides
	Noble	Rh, Pd, Ag, Ru	Oxides or metals
	Amphoteric	Zr, Nb, Mo, Tc, Te	Oxides or sodium salt
	Miscellaneous	Cd, Y	Carbonate, oxide
NaCl crystals	Alkali metals	Rb, Cs	Chloride
	Halides	Br, I	Sodium salt

From the above table it is seen that substantially all of the fission products, even those of minor significance, are retained as stable solid disposable products (i.e., in the ash and sodium chloride) with the exception of the noble gases krypton and xenon. However, these noble gases are of minor importance, since they generally are removed in the uranium reprocessing plant and usually are not present in fission product contaminated organic waste.

Example: The following example demonstrates the combustion of a ruthenium-containing organic waste and retention of the ruthenium in a molten salt. Samples of ruthenium-contaminated waste were combusted in a molten salt bed comprising 85 wt % sodium carbonate, 5 wt % sodium sulfate, 5 wt % NaCl, and 5 wt % ash. The apparatus utilized was a laboratory scale combustor containing a 9 inch deep bed of salt. The off-gas was monitored continuously for carbon monoxide, carbon dioxide, hydrocarbons, oxygen, nitric oxide, nitrogen dioxide and ammonia. Particulate samples were taken during each test to determine the average particulate loading in the off-gas. Downstream of the particulate filter, dual aqueous scrubbers were used to trap ruthenium. Particulates and scrubber solutions were analyzed for ruthenium.

At the beginning of each test about one-half pound of the waste to be burned was placed in a feed hopper which was attached to the feed system. The combustion air feed rate was adjusted to about 2.5 scfm (about 1.0 ft/sec superficial velocity). The feed was then started and adjusted until the desired off-gas composition was attained (about 4 to 6% O_2, and 12 to 16% CO_2). When steady state conditions were attained, the particulate sampler was started. A mixture of paper, plastic, rubber mix and ruthenium (about 400 mg of ruthenium per 200 g of the mix) were introduced into the combustor at a rate of 7.1 g/min. The average combustion temperature was about 910°C. From analysis of the particulates collected in the off-gas filters it was determined that greater than 99.9% of the ruthenium was retained in the melt. The foregoing procedure was repeated using a cation exchange resin (sulfonic acid) containing ruthenium as the organic waste. Analysis showed that greater than 99.9% of the ruthenium was retained in the salt.

Waste Incinerating Furnace

E. Bregulla, A. Chrubasik and H. Vietzke; U.S. Patent 4,276,834; July 7, 1981; assigned to Nukem GmbH, Germany describe a heatable furnace for incinerating nuclear fission and/or fertile material waste, particularly plutonium- and/or uranium-containing organic waste by pyrohydrolysis with steam or burning with air oxygen in safe geometry.

The furnace comprises, in combination, a stationary cylindrical outer jacket terminating at its lower end in a funnel or conical shape, a rotatable inner cylinder likewise terminating at its lower end in a funnel or conical shape whose diameter is so regulated that the distance between the outer jacket and the inner cylinder guarantees a safe layer thickness and scrapers which are disposed on the inner surface of the outer jacket and the outer surface of the inner cylinder.

The inner cylinder is preferably installed in a manner that it is exchangeable so that according to the waste operated with, there can be used the corresponding inner cylinder diameter having the necessary safe layer thickness.

In order to avoid a neutron interaction at high plutonium concentrations the inner cylinder can be coated with neutron-absorbing material (e.g., B_4C). Besides there can be used as construction material of the furnace at least partially neutron-absorbing industrial materials. This furnace concept permits great variations in the layer thickness. According to the provided material inserted the layer thickness can be adjusted from 3.5 to 15 cm, if the inner cylinder is correspondingly changed. Therewith the furnace is suited for both highly enriched grades of U-235 or U-233 and also for high plutonium concentrations.

Example: For the pyrohydrolytic incineration of plutonium-containing organic wastes having a Pu content of 120 g/m³ (= 0.6 g Pu/kg waste) and a composition of:

Polyvinyl chloride (PVC)	50%
Rubber	20%
Cellulose	15%
Other synthetic resins	15%
Density	200 kg/m³

there was employed an annular gap furnace according to the process as described for a throughput of 17.5 kg waste per hour. The velocity of flow was limited to 0.2 m/sec in order to prevent discharge of dust. In order to completely gasify the organic portion with steam at 800° to 1000°C, a maximum residence time of 4 hr was adapted. As shown in the test series the Pu content in the ashes formed was less than or at 1.2%. For the maximum in case of accident it is established that the Pu content in the ashes should not exceed 10%. Under these conditions the annular gas furnace has the following dimensions.

	Millimeters
Inside diameter of the outer jacket	1,000
Outer diameter of the inner cylinder	780
Annular gap	110
Inner diameter of the ash cylinder	100
Length of the reaction zone	1,200

The ashes were removed intermittently (in cans) at the bottom. A neutron absorber under the above conditions is not necessary, for reasons of disturbing conditions a layer of boron carbide powder is arranged as an intermediate layer which here is limited to 80 to 90 mm by the double wall.

For larger diameters of the outer jacket and inner cylinder a throughput of above 35 kg/hr can be reached.

TREATMENT OF METALLIC FUEL SHELLS

Rupture of Protective Sheath

Prior to the processing of irradiated ceramic nuclear fuel to separate fission products from the fuel material, it is generally preferred to separate the fuel material from its sheath. Mechanical means are usually selected for this purpose because storage or disposal of the contaminated sheath material is easier if it is retained in solid form but with oxide fuel, for example, difficulties can arise because of the frangible nature of the fuel and the small diameter and thinness of the sheaths used. It is therefore common practice to shear such fuel elements into short lengths mechanically and to dissolve out the fuel material preferentially.

R.E. Strong; U.S. Patent 3,929,961; December 30, 1975; assigned to U.K. Atomic Energy Authority, England describes the treatment of an elongate irradiated nuclear reactor fuel element comprising ceramic nuclear fuel within a metal sheath by which the sheath is ruptured by subjecting the fuel element to local induction heating.

Rupture is observed to occur when a portion of a fuel element disposed lengthwise within an induction coil is exposed to a radio-frequency magnetic field whereby the sheath of the fuel element is heated to its melting temperature, and the fuel element is moved axially relative to the induction coil. The mechanism appears to be that a crack is first generated as a result of local melting and that then the crack is extended as heating continues by the relative axial movement of the element and the surrounding induction coil.

Rupture of a sheath makes it possible to dissolve out the fuel material enclosed by the sheath withou shearing the element into short lengths. Stainless steel sheaths enclosing oxide fuel material have been ruptured by suspending them within an energized coil fed with radio frequency power. A frequency between about 0.1 to 1.0 MHz has been found suitable, heating being confined to short lengths of sheath (about 5 cm long) by the use of current concentrators. A feed rate of 2.5 cm/sec through the coil produces satisfactory results. It is possible to rupture fuel elements arranged in clusters.

Conditioning of Zirconium Waste Shells

A particular problem in the handling of metallic shell wastes is the capability of zirconium, and its alloys, to react with oxygen and, at increased temperature, also with the nitrogen of the air, with the development of fire. Fires of radioactive zirconium wastes have been known to occur on several occasions. In addition to the usual undesirable effects of a metal fire, the fire may result in uncontrolled release of radioactivity. In reprocessing systems in which such shell wastes are routinely handled, special measures must be taken to accommodate this potential source of danger. One of these measures which has been used is the treating and storing of shell waste under water. With this method, continuous monitoring of the storage facility and of the water level is necessary and unavoidable. Further, there exists the latent danger of escape of contaminated water into uncontrolled regions. This process is therefore not suitable for permanent storage.

Another known process for conditioning the zirconium-containing waste shells is to fix them in a special cement slurry so that a solid block of metal concrete is obtained. Objections have been raised against this process, inasmuch as, depending on the type of shells involved, "gassing" of the concrete has sometimes been noted, which may possibly adversely influence the stability of the bond.

K.-L. Huppert and D. Fang; U.S. Patent 4,129,518; December 12, 1978; assigned to Gesellschaft zur Wiederaufarbeitung von Kernbrennstoffen mbH, Germany describe a method which permits handling and conditioning of the waste shells without danger and results in a product (waste product) which is suitable for permanent storage.

In the process, metallic waste shells of zirconium or zirconium alloys are subjected to a controlled oxidation at high temperature in a reaction furnace. The oxidation of the waste shells is effected at temperatures between 800° and 1500°C and an oxygen-containing gas, for example, air, pure oxygen or a mixture of air and oxygen, is introduced into the reaction furnace for this purpose. The reaction can be controlled without special measures so that the formation of the oxide takes place promptly and without the development of smoke.

By "controlled oxidation" is meant an increase by steps of the furnace temperature of 100°C per 15 minutes. The range for the amount of oxygen-containing gas per weight of waste shells is 50 to 100 fold of the stoichiometric amount. The range for the time period of the oxidation is 6 to 12 hours. The reaction takes place at normal pressure.

The oxidation is advantageously effected in a well known electrically-heated shaft furnace or, likewise, in a well known electrically-heated rotary drum furnace. The furnaces in which the oxidation reaction takes place preferably are provided with inserts which permit further transporting through the furnace of only the oxidized material.

The end waste product of the process is a whitish, granular mass of highly annealed zirconium oxide in mixture with other oxides, such as iron(III) oxide and tin(IV) oxide which originate from the alloy components. It can be an inert, particulate or powdery material. The end product is insoluble in water, diluted acids and liquors as well as organic solvents. The amphoteric or acid character of the zirconium oxide that is formed has the result that the more easily volatile oxides, such as lithium or cesium oxide, will not escape during the reaction, but are bound in the waste product in the form of zirconates or mixed oxides, respectively.

In the practice of this process, there is a realizable reduction in volume to about 50% of the metallic starting material, that is, the volume of the waste product generally is about 50% of the volume of the metallic starting material. Due to its granular consistency, the reaction product can be removed very easily from the reaction furnace by a well known air lifter which works similarly as an installation for transporting corn into a silo. With the air lifter the granular powder is sucked off and sent to the solidification plant. This is of particular advantage when shaft or crucible furnaces are employed.

Electrically-heated rotary drum furnaces which permit slow moving of the material are particularly well suited for effecting the abovedescribed reaction. By means of disc-shaped inserts, the movement of the material can be controlled so that only the smaller particles travel to the discharge end of the furnace and are thus completely oxidized through. The always-present fill of part of a finer-grained, already-oxidized, powdery material, which remains in the reaction furnace and moderates the conversion of the newly filled-in metal pieces, protects the newly-introduced shells from too strong a reaction.

The waste product, which substantially and essentially consists of zirconium dioxide, is already suitable for final storage in the resulting annealed form, that is, it can be placed in permanent storage directly as it comes from the reaction furnace without any further processing. For this purpose, the waste product is preferably filled into steel drums which are tightly sealed and stored. For example, it is possible to accommodate the processed waste shells from 2 to 14 reactor fuel elements in a 200 liter standard ton Straimer drum. Alternatively, the waste product can also be solidified into a concrete-like substance by means of a binder, such as, for example, a cement adhesive with a quantity of cement which corresponds to a quantity between one-fourth and one-seventh of the volume of the waste product or by mixing the waste product with molten bitumen. The resulting solidified mass is then suitable for final and permanent storage. Compared to the conventional methods for conditioning the waste shells, the process provides a stable, decay-free form for storage for which safety monitoring during storage is not required.

Example: In this example, 400 ml, corresponding to 405 g of piled zirconium-containing shell material were heated in an electrical furnace to 1000°C. A small quantity of air was blown in, for instance a fifty-fold surplus according to the stoichiometric amount needed. The oxidation of the metal proceeded slowly and was complete after about 6 hours. Glowing occurred only at the burrs of the chopped shell material. Smoke or gaseous reaction products were not observed. During the reaction, the metallic shells expanded and disintegrated into a whitish reaction product. Upon completion of the process, the reaction product had only a volume of 200 ml or 50% of the starting volume. Its weight was about 540 g. During the oxidation process, there was a slight advantageous moving of the mass which resulted in peeling off of the already oxidized layers of the shells. Through the moving of the shell material in a rotating furnace, for instance a furnace usual in the cement production, the already oxidized layers of the shells are peeled off and ground. The inserts have small holes through which the granular powder can flow.

OTHER THERMAL PROCESSES

Gasification of Graphite

A process for removal of graphite by combustion with oxygen or air has been conventionally employed as a pretreatment process in the reprocessing of spent fuel of graphite-moderated reactor such as high-temperature gas-cooled reactor or spent contaminated moderator or reflector of the graphite-moderated reactor. However, this conventional process involves a problem to be solved in lowering the amount of radioactivity which is liberated and minimization of the amount of radioactive waste, that is, in connection with the fixation or separation of radioactive C-14 in combustion off-gas. That is to say, this conventional process is unfortunately an extreme exothermic reaction and requires the fixation of the carbon dioxide gas containing radioactive C-14 by, for example, lime and, therefore, produces a large amount of waste.

S. Sugikawa, M. Maeda and T. Tsujino; U.S. Patent 4,228,141; October 14, 1980; assigned to Japan Atomic Energy Research Institute, Japan describe an alternative to the conventional process wherein the graphite is gasified without the generation of heat by using carbon dioxide gas as the gasifying agent and the carbon monoxide formed is pyrolyzed to recover the carbon and the carbon dioxide gas is recycled.

Since the rate of reaction of the gasification of the graphite by carbon dioxide gas is very slow, the gasification of the graphite by carbon dioxide gas absolutely requires a catalyst.

Iron group elements such as iron, cobalt, and nickel, alkaline metal such as sodium, etc., vanadium and halogen are conventionally used as the gasification catalyst; above all iron, cobalt and nickel are known to be high in activity.

One embodiment of this process wherein the gasification of the graphite is carried out at temperatures above 1000°C is described below by referring to iron, a commonly used gasification catalyst.

In accordance with this process a graphite block or compact is soaked in a catalyst solution of 1 to 14 mols per liter or nitric acid and 0.22 mol per liter of iron(III) nitrate [$Fe(NO_3)_3$] at temperatures of from 20° to 90°C. After the graphite is impregnated up to its center with 400 to 2,000 ppm of iron, the catalyst-impregnated graphite is dried at temperatures of from about 20° to 100°C for one hour and 100% of carbon monoxide (CO) or nitrogen (N_2) is introduced at temperatures of from 850° to 1000°C for 0.5 hour to activate the catalyst. When the CO_2 gas is introduced at a temperature above 1000°C, the graphite block or compact is gasified to carbon monoxide (CO) seven times or more as quickly as when no catalyst is used. That is, when the catalyst of this process is used, the rate of reaction is about seven or more times that when no catalyst is used.

One example of the gasification of the graphite carried out at a temperature below 900°C is described as follows. In accordance with this process, a fuel compact comprising coated-particle fuel and graphite matrix or a graphite block used for sleeve material or moderator or reflector is soaked in the solution of 1 to 18 mols of nitric acid (HNO_3), 0.6 to 1.5 mols of cobaltous nitrate $Co(NO_3)_2$ and 0.5 to 1.5 mols of sodium nitrate [$Na(NO_3)$] at temperatures of from 20° to 80°C for 1 to 5 hours to impregnate the compact or block with the solution. Then the thus solution-impregnated compact or block is heated at a temperature above 100°C for 1 to

2 hours in an atmosphere of air or nitrogen to remove water and nitric acid from the compact or graphite and mixed catalyst of cobalt and sodium is added to the fuel compact or graphite block uniformly, and then when the fuel compact or graphite block is contacted with 100% of carbon dioxide at a temperature below 900°C for 6 to 7.5 hours, 99% or more of the fuel compact or graphite block is completely gasified and removed and coated-particles having silicon carbide (SiC) outer layer can be recovered in the case of fuel compact.

Example: A fuel compact (25 mm o.d. x 8 mm i.d. x 40 mm length) comprising coated particles and graphite matrix was soaked in the solution of 0.66 M of $Co(NO_3)_2$, 0.5 M of $NaNO_3$ and 14 M of HNO_3 at 20°C for 5 hours to impregnate the fuel compact with the solution and, then the fuel compact was heated at 100°C for 2 hours to remove the water and nitric acid from the fuel compact; thus the mixed catalyst of cobalt and sodium was added to the graphite matrix. Thereafter, the fuel compact was allowed to contact with 100% of carbon dioxide at 850°C for 7.5 hours to gasify 99% or more of the graphite matrix and lowering of the reaction rate was not observed.

Treatment of Spent Ion Exchange Resins

K. Knotik, P. Leichter and H. Jakusch; U.S. Patent 4,235,738; November 25, 1980; assigned to Vereinigte Edlsthalwerke Aktiengesellschaft (VEW) and Oesterreichische Studiengegesellschaft fuer Atomenergie GmbH, Austria describe a method of and apparatus for the pretreatment of spent radioactive ion exchange resins prior to mixing with a binding agent and subsequent storage, in such a way as to prevent further radiolytic decomposition of the resin after such heat treatment and further to prevent the generation of flue ash and radioactive substances via a combustion of the resin.

For example, the resin heat-treatment step is carried out by heating a container containing the resin in a gas-tight furnace, in the presence of an inert or reducing atmosphere, to a relatively low temperature (e.g., ~500°C) which is sufficient to effect a carbonization of the resin but which is insufficient to cause evaporation or sublimation of the radioactive inorganic compounds which are carried by the resin.

Preferably, the spent resin is loaded into an open container, which is admitted into a vertically disposed furnace through a removable top thereof, and is seated on a perforated plate support at the bottom open end of the furnace. A clearance is provided between the wall of the container and the surrounding wall of the furnace, so that gases generated during the carbonization of the resin can flow downwardly through the furnace and the porous plate and into a vertically disposed condenser that is mounted below and in communication with the bottom end of the furnace. The resulting condensate, which is substantially free of radioactivity, can be removed via a closable condensate port disposed in a bottom end plate of the condenser.

In order to permit the evacuation of the furnace and the admission of the required nonoxidizing atmosphere, closable gas fittings are also provided in the bottom end plate of the condenser and a top end plate of the furnace, which end plate is employed to seal off the furnace after the resin-filled container is placed therein.

It is not necessary that the nonoxidizing atmosphere be represented by a gas separately introduced into the furnace. If desired, for example, the required atmosphere may be provided by a light, such as a high-temperature oil or melted paraffin, which is disposed in surrounding relation to the resin inside the container. Alternatively, the medium may be composed of the gases generated by the resin during carbonization.

In the event that the abovementioned liquid is used as the nonoxidizing medium, it is desired, where practicable, to distill off the liquid after the carbonization operation.

RECOVERY AND RECYCLING PROCESSES

URANIUM RECOVERY

Reuse of Wastewater Generated During Production of Uranium Dioxide

In preparing fuel elements for nuclear reactors it is desirable to employ UO_2 for this purpose. A widely used method is to process UF_6 to prepare ADU (ammonium diuranate) which is then calcined and reduced to UO_2. In preparing ADU, enriched UF_6 gas is reacted with water to produce an aqueous uranyl fluoride (UO_2F_2) solution which is then treated with an excess of ammonium hydroxide to cause a precipitate of ADU which is recovered as an aqueous slurry. The precipitate can be filtered out or centrifuged from the aqueous phase of the slurry. The resulting wastewater contains a high concentration of fluorides, primarily as NH_4F, with excess ammonia, and small quantities of dissolved uranium.

The disposal of this wastewater entails considerable problems and difficulty because of the toxicity of the fluorides as well as the excess ammonia and the small but significant quantities of radioactive uranium. On both ecological and health accounts it would be highly desirable to eliminate or greatly reduce this disposal problem, as well as to recover from the wastewater substantially entirely the traces of enriched uranium, which is a very costly material as well as being a hazard to health.

T.J. Crossley; U.S. Patent 3,961,027; June 1, 1976; assigned to Westinghouse Electric Corporation has found that the foregoing problems may be overcome, and at a low cost, by treating the wastewater resulting when ADU precipitate is separated from the aqueous component of the slurry in which it is present, as follows:

(1) Admixing the wastewater with sufficient lime to cause substantially all of the fluoride to form a calcium fluoride precipitate.

(2) Distilling off nearly all of the ammonia from this treated water containing CaF_2 precipitate, condensing the ammonia and recycling it to the UO_2F_2 treatment vessel.

150

(3) Centrifuging the calcium fluoride precipitate slurried in the substantially ammonia-free water to separate it from substantially all of the water to produce a small quantity of a nontoxic, essentially uranium-free, pasty CaF_2 product that can be employed in glass manufacturing or for making hydrofluoric acid, or can be safely disposed of by burial.

(4) After passing the water from the centrifuge through a polishing filter to remove the final traces of suspended solids, the water is passed through cationic ion-exchange beds to remove substantially all of the calcium and all of the impurities except for small amounts of ammonia and fluorides whereby to produce water of required purity that is recycled to either the UF_6 gas-water reaction area or to the ammonium hydroxide preparation and storage areas.

Utilizing the wastewater treatment procedures of this process wherein treated wastewater is recycled for use in converting more UF_6 into ADU, no water need be added or at most only small make-up additions need be made to the system. The only water losses from the system are in the ADU sludge going to the calciners and in the CaF_2 sludge, both taking relatively small quantities of water. This last amount of water loss can be very nearly compensated for by the water in the additional aqueous ammonia brought in to replace the ammonia present in the ADU. Consequently, no wastewater need be disposed of or discarded.

Nitrate Decomposition at Reduced Temperature and Pressure

Various industrial processes generate aqueous waste solutions of ammonium nitrate which pose problems with respect to release to the environment. For example, various chemical processes for the treatment or production of nuclear reactor fuels generate such waste solutions, these solutions often containing radioactive solutes such as impurities or incompletely recovered compounds of nuclear fuel metals (uranium, thorium, plutonium).

Preferably, the ammonium nitrate in such solutions would be decomposed to form environmentally acceptable waste products, any radioactive impurities or valuable fuel-metal compounds being recycled for recovery. However, practical rates of decomposition of ammonium nitrate are very difficult to achieve at atmospheric pressure. Furthermore, the decomposition of ammonium nitrate solutions under pressure has not been attractive hitherto because of the high temperatures and pressures required (e.g., 240°C and 500 psig) and because of the high corrosivity of the solutions at such temperatures.

P.A. Haas; U.S. Patent 4,225,455; September 30, 1980; assigned to the U.S. Department of Energy has found that the decomposition of ammonium nitrate and/or selected metal nitrates in pressurized, heated aqueous solution is accelerated when an effective proportion of both nitromethane (CH_3NO_2) and nitric acid (HNO_3) is incorporated in the solution.

An advantage deriving from this process is that the decomposition of nitrate in pressurized, aqueous solution can be effected by reduced temperatures and pressures. For instance, the inclusion of an effective proportion of nitromethane provides useful rates of decomposition at temperatures at least 70°C below those otherwise required. In other words, the nitromethane and nitric acid act in combination to lower the threshold temperature for the decomposition reaction. In some instances, the reduction in operating temperature so achieved makes it possible to reduce the decomposition zone pressure by as much as 400 psig. Such reductions in pressure and temperature are highly advantageous because, for example, corrosion rates are reduced to the extent that the system for carrying out the decomposition can consist of off-the-shelf stainless steel components rather than special materials which are most costly and less readily available. Furthermore, the use of moderate temperatures and pressures for the decomposition of NH_4NO_3 in the liquid avoids the formation of particulates in the waste gases; such particulates can be a problem in dry, high-temperature decompositions.

Figure 6.1 is a schematic diagram for continuously converting uranyl nitrate to uranium oxide in accordance with the process.

Figure 6.1: Process to Convert Uranyl Nitrate to Uranium Oxide at Reduced Temperature and Pressure

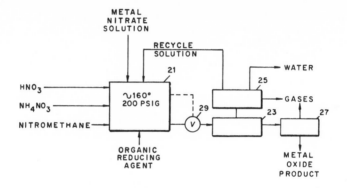

Source: U.S. Patent 4,225,455

Recovery of Uranium from Stripping Solutions

J.G. Cleary and G.E. Zymboly; U.S. Patent 4,265,861; May 5, 1981; assigned to Wyoming Mineral Corporation describe a process which substantially reduces the quantity of radioactive waste produced by the solution mining of uranium. In addition, the process recovers most of the uranium which precipitates with the calcium carbonate.

In the solution mining of uranium a stripping solution is prepared which is pumped into the underground uranium deposit through a number of injection wells. The stripping solution commonly consists of an aqueous solution of an oxidant and a bicarbonate. The oxidant is usually hydrogen peroxide because it is less expensive but potassium permanganate, sodium hypochlorite, or other suitable oxidant could also be used. The bicarbonate ion is usually obtained by adding ammonium bicarbonate but sodium bicarbonate or soluble carbonates could also be used.

The recovery leach containing the dissolved uranium is pumped to the surface for processing. A commercial recovery leach typically contains about 0.05 to 0.5 g/ℓ of dissolved uranium as ammonium uranyl carbonate, $(NH_4)_2UO_2(CO_3)_3$, if ammonium bicarbonate was used as the source of bicarbonate ion.

The recovery leach also contains small concentrations of highly radioactive radium. Typically, the precipitated calcium carbonate would contain about 500 to 1,000 piCi of radium per g of $CaCO_3$.

This process is useful with carbonate recovery leaches. The recovery leach is typically supersaturated with calcium carbonate, containing about 0.3 to 1.0 g/ℓ of calcium carbonate. Because the large concentrations of calcium carbonate in the recovery leach can result in the precipitation of calcium carbonate throughout the processing equipment, which would render it inoperable, it is first necessary to precipitate this calcium carbonate. Precipitation is preferably induced by the addition of ammonia to a pH of about 8.2. Carbon dioxide is also added slightly in excess of the calcium present (about 10%). The amount of ammonia can be about 1 to 2 g/ℓ and the amount of carbon dioxide 1.0 to 2.0 g/ℓ. Precipitation of the calcium carbonate can also be accomplished using carbon dioxide in combination with Na_2CO_3, $MgOH$, or $Ca(OH)_2$.

The calcium carbonate precipitate typically contains 20 to 30 lb of uranium per ton of calcium carbonate and about 9×10^8 piCi of radium per ton of calcium carbonate. About 15% of the uranium in the recovery leach is precipitated with the calcium carbonate. This precipitation can be accomplished in a reactor-clarifier. The precipitate can be removed as a slurry containing, for example, about 30% solids. The slurry is preferably sent to a settling pond to further separate the solids from the solution. The solids are then removed by suction pump, screw feeder, or other means and are sent to a dissolution reactor.

In the dissolution reactor, an acid is added which will dissolve the calcium carbonate. Hydrochloric acid is preferred as it is the least expensive, but nitric acid or other acids which do not form insoluble compounds with calcium (e.g., sulfuric acid) could also be used. Sufficient acid is used to effect the dissolution of all of the calcium carbonate. The carbon dioxide which is evolved can be collected if desired. If hydrochloric acid is used, the uranium forms soluble uranyl chloride at this stage.

The solution is then sent to a uranium reclamation system where uranium is removed from the solution. Uranium removal can be accomplished by solvent extraction, peroxide precipitation, or other suitable process. Solvent extraction gives a higher percentage yield and a cleaner product, but it is not preferred to a peroxide precipitation.

In solvent extraction, the aqueous solution is mixed with a counterflowing immiscible organic liquid containing a uranium extractant. The commercially used organic fluid is kerosene because it is inexpensive, and the commercial extractant is a mixture of diethylhexyl phosphoric acid (DEHPA) and trioctyl phosphene oxide (TOPO). Other organic fluids and other extractants such as amines or tributyl phosphate can be used if desired.

Peroxide precipitation can be accomplished by the addition of any peroxide to the solution to precipitate uranyl peroxide. Hydrogen peroxide is preferred as it is inexpensive, but Na_2O_2 or K_2O_2 could also be used. The amount of peroxide used should be about 0.12 lb/lb of U_3O_8 (i.e., stoichiometric) up to about a 10% excess. The pH of the solution should be adjusted to between 3.5 and 5.5 because below a pH of about 3 the uranium does not precipitate quantitatively and above a pH of about 5.5 the uranium precipitates as compounds other than uranyl peroxide. Less peroxide can be used at higher pH and at higher temperatures (i.e., up to about 50°C).

When the uranium is removed the solution is sent to a precipitator where the radium is precipitated out. This is accomplished by adding sulfate ions and barium or strontium ions which precipitates $BaSO_4$, $RaSO_4$ or $SrSO_4 \cdot RaSO_4$, respectively. The barium or strontium ions are preferably obtained by the addition of barium or strontium chloride, but other soluble barium or strontium compounds such as BaO or SrO, could also be used. The sulfate ions may be obtained by the addition of any inexpensive, soluble sulfate. Ammonium sulfate, sulfuric acid, sodium sulfate, or other suitable sulfates can be used. Radium sulfate is very insoluble, but is present in very small amounts. The amount of sulfate and barium or strontium ions should be about stoichiometric up to a 5% excess of stoichiometry of the amount needed to form $MSO_4 \cdot RaSO_4$ where M is Ba or Sr.

The solid $MSO_4 \cdot RaSO_4$ is radioactive and must be stored as radioactive waste. This process reduces the amount of this radioactive waste from about 18.2 ft^3/ton of calcium carbonate to only about 2.0 ft^3/ton of calcium carbonate. The effluent, a solution of calcium chloride, is not radioactive. It can be added to ground water and deep-well disposed or placed in ponds to crystallize and recover the calcium chloride.

Removal of Uranium from Aqueous HF Solutions

H. Pulley and S.F. Seltzer; U.S. Patent 4,234,555; November 18, 1980; assigned to the U.S. Department of Energy describe a simple and effective method for removing uranium from aqueous HF solutions containing trace quantities of the same. The method comprises contacting the solution with particulate calcium fluoride to form uranium-bearing particulates, permitting the particulates to settle, and separating the solution from the settled particulates. The CaF_2 is selected to have a nitrogen surface area in a selected range and is employed in an amount providing a calcium fluoride/uranium weight ratio in a selected range. As applied to dilute HF solutions containing 120 ppm uranium, the method removes at least 92% of the uranium, without introducing contaminants to the product solution.

Example: The process was tested by mixing selected quantities of particulate CaF_2 in six 0.200 ℓ samples of an aqueous HF solution containing 120 ppm uranium. The solution comprised 20 wt % HF. The CaF_2 powder was acid-grade Fluorspar. This powder had a nitrogen surface area of 1.6 m^2/g (based on the well-known BET measurement). Tyler-sieve data for the powder were as follows: 45% of the powder passed through a 325 mesh screen and 15% was retained by the screen; 27% was retained on a 200 mesh screen; 9% was retained on a 100 mesh screen; and 4% was retained on a 65 mesh screen. Each of the samples containing particulate CaF_2 was stirred at room temperature for 1 hour. Following stirring the resulting slurries were either filtered promptly or the supernate was separated by decanting. The resulting solutions were analyzed for uranium by gamma-spectrometry and for calcium by atomic absorption.

The accompanying table shows the CaF_2-to-uranium weight ratios employed in each of the six tests, together with the results obtained. It will be noted that removal of 50% of the uranium was accomplished at a CaF_2/U ratio of 8 and that removal of 92% of the uranium was accomplished at ratios exceeding 37. As shown, the product solutions contained very little calcium—only 9 ppm if the solution was not filtered and less than 0.2 ppm if it was filtered. Thus, the process was found to remove uranium effectively while avoiding contamination of the product solution.

Note that in the table Sample No. 4 was allowed to stand overnight before decanting; all other samples were filtered before analysis.

Sample No.	Lbs CaF$_2$ per 13,500 Gal. Solution	CaF$_2$/U by Weight	Initial U Conc., ppm	Final U Conc., ppm	U Removal, %	Final Ca^{++} Conc., ppm
1	56	4	120	72	40	<0.2
2	84	6	120	71	41	<0.2
3	112	8	120	60	50	<0.2
4*	112	8	120	59	51	9
5	500	37	120	10	92	<0.2
6	1000	74	120	10	92	<0.2

Fuel Reprocessing in Molten Nitrate Salts

G. Brambilla, G. Caporali and M. Zambianchi; U.S. Patent 3,981,960; Sept. 21, 1976; assigned to AGIP Nucleare SpA, Italy describe a process in which ceramic nuclear fuel is reprocessed through a method wherein the fuel is dispersed in a molten eutectic mixture of at least two alkali metal nitrates and heated to a temperature in the range between 200° and 300°C. That heated mixture is then subjected to the action of a gaseous stream containing nitric acid vapors, preferably in the presence of a catalyst such as sodium fluoride. Dissolved fuel can then be precipitated out of solution in crystalline form by cooling the solution to a temperature only slightly above the melting point of the bath.

The compositions and melting points of eutectic mixtures of alkali metal nitrates which may be used with advantage are:

	Mol Percent	Temperature (°C)
LiNO$_3$-NaNO$_3$-KNO$_3$	30-17-53	120
LiNO$_3$-KNO$_3$	41-59	133.5
LiNO$_3$-NaNO$_3$	54-46	193
KNO$_3$-NaNO$_3$	50-50	220

Example: 10 g of U$_3$O$_8$ coming from a sintered tablet of UO$_2$, broken up by oxidation at 450°C in the air stream during 3 hours, was dispersed in 100 g of an eutectic mixture of molten salts KNO$_3$-LiNO$_3$ (67-33% by wt, melting point 133.5°C) prepared by melting the pure reagents in suitable proportions at 350°C.

The reaction was effected in a 100 or 150 cc quartz glass placed in an electric oven having a thermo-regulated well, and the temperature was kept at 250°C.

In the same bath a stream of argon gas has been let to bubble this stream carrying HNO$_3$ vapors, produced by reaction between NaNO$_3$ and H$_2$SO$_4$ in molar proportions 1:1, in a 250 cc glass flask, kept by a waterbath at the constant temperature of 85°C.

After 3 hours and with an argon stream of 20 ℓ/hr conveying 3 g/hr of HNO$_3$, the uranium oxide was entirely dissolved with the formation of a quite clear yellow solution.

The density of the nitrates at 160°C being equal to 1.942 g/cc, the solution resulted in a concentration in UO_2 equal to 180 g/ℓ.

In another test was dissolved 15 g of U_3O_8 in 85 g of the same molten nitrates at 250°C; after 3.5 hours of reaction and with a consumption of 12 g of HNO_3, there was obtained a solution at about 300 g/ℓ of UO_2.

Dissolution Process for ZrO_2-UO_2-CaO Fuels

B.E. Paige; U.S. Patent 3,965,237; June 22, 1976; assigned to the U.S. Energy Research and Development Administration describes a process whereby ZrO_2-UO_2-CaO fuel is dissolved by immersing the fuel within a dissolver vessel in zirconium-dissolver product. As used herein, zirconium-dissolver product is the solution which results from the dissolution of the zirconium or Zircaloy cladding from the reactor fuel with hydrofluoric acid. The zirconium-dissolver product is a fairly well-defined solution whose composition does not vary substantially and which is essentially 6.8 M total fluoride with from 1.0 to 1.3 M zirconium.

Although other possible steps can be followed to immerse the oxide fuel wafers in the zirconium-dissolver product, from a practical standpoint the oxide fuel will become immersed as a result of the dissolution in the dissolver vessel of a complete fuel assembly. In accordance with standard techniques, a fuel assembly of the zirconium-clad ZrO_2-UO_2-CaO-type fuel is charged to the dissolver vessel. Hydrofluoric acid of about 7 M and preferably 6.8 M is added to the dissolver vessel to dissolve the zirconium cladding. Upon dissolution of the zirconium cladding by the hydrofluoric acid, the ZrO_2-UO_2-CaO fuel wafers will be exposed and immersed in the resulting zirconium-dissolver product.

Nitric acid is added to the zirconium-dissolver product in the dissolver vessel to facilitate the dissolution of the fuel wafers which have been exposed to the solution by the dissolution of the cladding material. The nitric acid added serves as an oxidant which must be added to oxidize and solubilize UF_4 as is necessary in any hydrofluoric acid dissolution of high uranium content fuel. The addition of the nitric acid results in the successful complete dissolution of the oxide fuel wafers at an acceptable dissolution rate and without a calcium fluozirconate residual.

This dissolution process offers the advantage of reduced corrosion in comparison with previously used processes.

It is preferred that the fluoride-to-zirconium ratio be maintained at 4.5 to 5.2. The ratio is maintained in this range to minimize corrosion and yet prevent precipitation of calcium fluozirconate, $CaZrF_6$, which can form at too low a ratio.

A preferred embodiment of the method includes the step of adding additional hydrofluoric acid to the dissolver solution after a period of time subsequent to the addition of the nitric acid to facilitate dissolution of the ZrO_2, since, as the ZrO_2 dissolves, the fluoride-to-zirconium mol ratio decreases, which also decreases the rate of dissolution and can cause solution instability. The addition rate of the additional hydrofluoric acid is controlled to prevent excessive corrosion and to maintain the ratio at 4.5 to 5.2. Consequently, the preferred concentration is approximately 0.1 M nitric acid with a fluoride-to-zirconium ratio of about 5.1.

URANIUM-THORIUM RECOVERY

Working Up of Uranium-Thorium Wastes in Kernel-Casting Processes

P. Börner and H.-J. Isensee; U.S. Patent 4,124,525; November 7, 1978; assigned to Nukem GmbH, Germany describe a process for working up and the return without loss of uranium-thorium wastes into the kernel casting process which does not require an extractive separation step and in which there is no precipitation of the metal hydroxide in the neutralization of a nitric acid uranium-thorium solution.

This problem was solved by evaporating a solution of uranium-thorium oxide waste dissolved in a nitric acid-hydrofluoric acid mixture up to the appearance of nitrous gases, the residue diluted in the hot condition with water and this solution brought to a pH between 2.5 and 3.5 with ammonia at a temperature below 40°C.

The solution recovered by the process can be adjusted to any heavy metal (i.e., uranium-thorium) concentration up to about 300 g/ℓ at a pH of 3 to 3.5, is stable for over 1 month and is miscible with PVA (polyvinyl alcohol). This solution can be adjusted through further addition of uranium or thorium to the desired uranium-thorium ratio for the casting solution.

The nitric acid-hydrofluoric acid solution can contain, for example, HNO_3 and HF in a molar ratio of from 100 to 1 to 400 to 1. The exact concentration of the nitric acid is not critical so long as it is strong enough that the mixture has a pH of not over 1.

According to the process, all wastes can be returned without loss during the production process. This is likewise true for wastes which have a uranium-thorium ratio deviating from the current production.

The (U-Th)O_2 waste is, in a given case after suitable pretreatment for removal of carbon and coating, dissolved with double the volume amount of HNO_3 (65%) and 0.06 N hydrofluoric acid (based on the HNO_3) in a flask equipped with a reflux condenser. The time up to complete solution depends on the degree of fineness and the U-Th ratio of the scrap. The solution is then freed of excess HNO_3/HF by distilling twice. Thereby, air or another gas is introduced for stirring up and delaying boiling in the heavy syruplike solution. Distillation is continued to the appearance of NO_2 vapors. The thickened residue must still be diluted in the hot state portionwise with water. Generally, it is diluted with water to a heavy metal concentration of 500 to 650 g/ℓ. Since the pH is around 1 it is still not possible to form a U-Th sol in this condition. In the subsequent neutralization to pH 2.5 to 3.5 with ammonia the solution must be unconditionally held below 40°C, in order to prevent a sol formation at a pH above 2. A dark red color indicates such sol.

The thus-recovered, mostly yellow U-Th solution containing about 250 g/ℓ of heavy metal can be added in any amount to the casting solution and yields after customary casting and further treatment particles which correspond to the known production quality, as well as after coating.

Example: From assorted uncoated nuclear waste there were dissolved 3 kg of uranium-thorium [corresponding to 3.410 kg of (U-Th)O_2] in 6.8 ℓ of HNO_3 (65%) and 17 ml of HF (40%) with boiling, using a reflux condenser. Depending on the order of fineness, the time for dissolving was 30 to 70 hours.

By replacing the reflux condenser with a distillation bridge having a condenser connected thereto as well as a definite introduction of gases or air, there was next carefully distilled off 4.4 ℓ of nitric acid. The syrupy residue was diluted with 3 ℓ of water and neutralized cold by the addition of about 1.5 ℓ of NH_4OH (25%) to a pH of 2.5 to 2.8. By filling up to a final volume of 12 ℓ at 20°C there was formed a solution containing about 250 g/ℓ of heavy metal.

The typical analysis of such a solution, for example, yields the following concentrations and impurities:

Thorium	220.0 g/ℓ
Uranium	27.6 g/ℓ
Ammonium nitrate	133.5 g/ℓ
Boron	15 ppm
Fluorine	2,400 ppm
Silicon	<30 ppm

This solution which is maintainable for weeks was mixed with two different casting formulations with final concentrations of 120 g/ℓ. The nuclei cast and sintered therefrom corresponded to the customary quality requirements in regard to chemical and physical properties.

Thus, the following analytical values were found, for example, for sintered kernels with an average atomic ratio of U to Th of 1 to 10, produced with an addition of 10 to 20% scrap.

. . **Kernels with 10% Addition of Scrap** . .

Thorium	79.80%
Uranium	8.23%
U:Th, atomic wt ratio	1:9.7
Boron	≤0.5 ppm
Fluorine	<3.0 ppm
Silicon	50.0 ppm

. . **Kernels with 20% Addition of Scrap** . .

Thorium	79.87%
Uranium	8.02%
U:Th, atomic wt ratio	1:9.6
Boron	≤0.5 ppm
Fluorine	5.0 ppm
Silicon	38.0 ppm

Concentric Spherical Fuel Elements

M. Hrovat, H.-J. Becker and H. Huschka; U.S. Patent 4,134,941; January 16, 1979; assigned to HOBEG Hochtemperaturreaktor-Brennelement GmbH, Germany describe a type of pressed spherical fuel element made of graphite for high-temperature reactors consisting of a graphite nucleus (or core) containing only fertile particles (breeder), a graphite shell containing only fuel particles and a further outer shell of pure graphite and an especially advantageous process for reprocessing this fuel element after the irradiation in the reactor. The three layers of the fuel element are concentric.

The particular advantage of this separate arrangement of fuel and fertile material according to the process is that a simple separation of the uranium-containing particles from the thorium-containing particles is offered in the Head-End stage of the reprocessing of the fuel elements after the irradiation in the reactor. In the Head-End stage the burned-down fuel elements are subjected to a burning process in oxygen at about 1000° to 1200°C. In this process the graphite structural material is burned off, whereby the fertile and fuel particles become exposed. In order to make possible a uniform burning off of the surface of the sphere it is advantageous to burn the spheres in a rotating cylindrical furnace. According to the process, the burning of the spherically shaped fuel elements takes place in two steps. In the first step, the spheres of, for example, 60 to 40 mm diameter are burned off and thereby only the uranium-containing fuel particles are exposed. In the second step, there takes place the burning of the nucleus of the sphere and the recovery of the thorium-containing fertile material particles.

Example: As fuel particles there were employed spherically shaped kernels of UO_2 having a diameter of 210 μm. These particles were twice provided with pyrolytically deposited carbon layers having a total thickness of 160 μm. The coated particles with a diameter of 560 μm and a density of 2.2 g/cc contained 23 wt % uranium.

The fertile material particles (ThO_2) having a kernel diameter of 617 μm were likewise double-coated with pyrolytically deposited carbon layers having a total thickness of 160 μm. The coated particles having a diameter of 905 μm and a density of 3.99 g/cc contained 63 wt % thorium.

As graphite molding powder there was employed a mixture consisting of 64 wt % natural graphite, 16% of graphitized petroleum coke and 20% novolak (phenol-formaldehyde) resin binder.

The fuel and fertile material particles were encased with the graphite molding powder in separate operations with addition of methanol in a rotating drum. The set amounts were so chosen that there was formed on the fertile material particles an encasing layer of 160 μm and on the fuel material particles an encasing layer having a thickness of 240 μm.

For the production of the spherical nuclei 48 g of the encased coated fertile material particles together with 30 g of graphite molding powder of the type set forth above were transferred into a rubber mold, mixed thoroughly and preliminarily pressed into a spherical nucleus at a pressure of 50 kg/cm^2.

In a second operation this nucleus was arranged in a second rubber mold with the help of three interval spacers in the center of the mold and the rest of the volume of the rubber mold filled with a mixture consisting of 41 g of encased coated fuel particles and 20 g of graphite molding powder. After that, the compression took place at a pressure of 80 kg/cm^2.

Subsequently these preliminarily pressed spheres were provided, according to a process known itself and described in Hrovat German Offenlegungsschrift 1,646,783, with a shell of the same graphite molding powder and finally molded under high pressure (3 metric tons/cm^2). The spheres were heated to 800°C for 18 hours to carbonize the binder resin, and after the cooling roasted in a further operation at 1800°C. After the final temperature treatment the spheres were turned to the predetermined diameter (6 cm). The finished element contained 18 g of thorium in the 40 mm diameter nucleus of the sphere and 2 g of uranium in the 5 mm thick fuel zone. The measured breaking load through crushing between two parallel steel plates was 2,300 kp.

PLUTONIUM RECOVERY

Plutonium Dioxide Dissolution in Hydriodic Acid

B.L. Vondra, O.K. Tallent and J.C. Mailen; U.S. Patent 4,134,960; January 16, 1979; assigned to the U.S. Department of Energy describe a method for dissolving solid material containing PuO_2, the method comprising contacting the solid material with an aqueous solution at least 3 M in HI to form an aqueous solution containing Pu values. This aqueous solution can be evaporated to dryness leaving a dry residue containing Pu values which is readily dissolvable in HNO_3 of above about 2 M and amenable to conventional nuclear fuel reprocessing such as Purex.

Alternatively, Pu values can be separately recovered from other metal values present in the aqueous solution by cation exchange chromatography. Pu values load a cation exchange resin from 0.5 to 2.0 M HI and are selectively eluted by HI above 4.0 M.

The dissolution rate is an increasing function of the HI concentration and the temperature. Below 3 M HI and 75°C the dissolution rate is impractically low. The preferred HI concentration is, therefore, the maximum tolerable by equipment and other constraints and the preferred temperature is the boiling point of the HI solution. Concentrated HI (about 6.5 M) boils at about 128°C. The HI dissolver solution should contain a sufficient amount of reductant such as H_3PO_2 or red phosphorus to prevent evolution of I_2 from the oxidation of I ions. H_3PO_2 is typically supplied for use with HI solution as a reductant stabilizer; however, other reductants compatible with process constraints can be used as a stabilizer.

The amount of reductant needed is related to process conditions and the amount of Pu which goes into solution. Since Pu(IV) is reduced to Pu(III) during the process, at least an amount stoichiometric to Pu(IV) $\rightarrow$ Pu(III) reduction is needed. Generally, about 0.5 wt % H_3PO_2 is sufficient to prevent I_2 formation; however, the amount may vary due to the presence in the solution of impurities or the tendency of HI to oxidize from exposure to light. Additionally, PuO_2 dissolution rates in HI are substantially lower if sufficient reductant to prevent I_2 formation is not present. Of course, it is a matter of routine testing to determine the minimum amount of reductant needed to prevent I_2 evolution and provide high dissolution rates in a particular system.

The following examples demonstrate the dissolution of PuO_2 in HI. The PuO_2 used for Examples 1 and 2 was refractory PuO_2 microspheres having a surface area of 0.012 m^2/g and a bulk density of 11.0 g/cc (96% theoretical), which stimulates the most difficultly soluble PuO_2 residues from spent reactor fuel.

Example 1: A series of dissolution tests were conducted to determine the dissolution rate of PuO_2 microspheres by digesting 0.4 to 0.8 g of PuO_2 in 6 ml volumes of stirred aqueous HI solution in glass equipment at 42°, 75° and 100°C. The tests were made using 2.48, 3.69, 6.73 and 7.53 M HI solutions which were stabilized with 1.2 wt % H_3PO_2. The microspheres had been calcined at 1150°C in an argon-4 wt % hydrogen mixture to produce dense PuO_2 containing little or no excess oxygen.

During the dissolution tests, liquid samples were withdrawn at temperature and analyzed for plutonium by gross alpha and alpha-pulse-height techniques. A conventional spectrophotometer was used for valence analysis of plutonium in dissolver solutions. The fraction of PuO_2 dissolved was less than 0.25 in each test used to obtain the data presented in Tables 1 and 2. Table 1 depicts the amount of plutonium dissolved versus time for 0.5 g of PuO_2 microspheres digested in 6 ml of aqueous HI solution at the several acid concentrations and temperatures.

It is apparent that higher temperatures result in increased dissolution rates. The dissolution rate is shown to increase with additional amounts of PuO_2 present. Table 2 shows the dissolution rates of PuO_2 at the several acid concentrations at various temperatures for 0.5 g of PuO_2 microspheres digested in 6 ml of dissolvent. The dissolution rate is shown to be an increasing function of both temperature and HI concentration. The rates were determined from statistical analysis of the data sets of Table 1. The net surface area of undissolved PuO_2 was assumed to have remained approximately constant or to have increased only slightly due to surface roughening.

Table 3 shows the dissolution rate and normalized dissolution rate of PuO_2 microspheres in 6 ml of 6.73 M HI with a mean HI activity of 9.6 at 100°C. The normalized dissolution rate is defined as the mmol of plutonium dissolved per hr per m^2 of PuO_2 surface area in the sample.

Table 1: Concentration of Plutonium Dissolved as a Function of Dissolution Time and Temperature

Time (hr)	Concentration of HI (M)	Temperature (°C)	Dissolved Plutonium (M)
0.67	2.48	100	0.0006
1.67	2.48	100	0.0013
2.75	2.48	100	0.0021
3.08	2.48	100	0.0032
0.67	3.68	100	0.0018
1.67	3.68	100	0.0041
2.75	3.68	100	0.0068
3.08	3.68	100	0.0082
0.67	7.35	100	0.0188
1.67	7.35	100	0.0448
2.75	7.35	100	0.0536
3.08	7.35	100	0.0678
0.50	2.48	75	0.00007
1.50	2.48	75	0.0002
2.50	2.48	75	0.0004
0.50	3.68	75	0.0003
1.50	3.68	75	0.0008
2.50	3.68	75	0.0014
0.50	7.35	75	0.0044
1.50	7.35	75	0.0109
2.50	7.35	75	0.0160
3.50	7.35	75	0.0183
0.50	2.48	42	0.00009
1.50	2.48	42	0.00016
2.50	2.48	42	0.00019
3.50	2.48	42	0.00022
0.50	3.68	42	0.0009
2.50	3.68	42	0.0011
3.50	3.68	42	0.0012
0.50	7.35	42	0.0002
1.50	7.35	42	0.0004
2.50	7.35	42	0.0012
3.50	7.35	42	0.0016

Table 2: Dissolution Rates of PuO_2 in Aqueous Solutions

Concentration of HI (M)	Dissolution Temperature (°C)	Plutonium Dissolution Rate (mmol/hr/g)
7.35	100	0.2196
3.68	100	0.0312
2.48	100	0.0120
7.35	75	0.0564
3.68	75	0.0072
2.48	75	0.0024
7.35	42	0.0060
3.68	42	0.0012
2.48	42	0.00048

Table 3: Effect of PuO_2 Sample Size on Dissolution Rate

Weight PuO_2 (g)	Dissolution Rate (mmol/hr)	Normalized Dissolution Rate (mmol/hr/m^2)
0.40	0.0556	11.75
0.50	—	12.75
0.60	0.0846	11.75
0.80	0.1212	12.65

The data obtained from the dissolution tests of this example, when combined with mean HI activity coefficients, corresponding HI concentrations and degrees of dissociation used to calculate the activities, permitted the determination of the order of the dissolution reaction with respect to the HI activity. The dissolution reaction at the three temperatures of 42°, 75° and 100°C was shown to be second order with respect to the HI activity. The rate constants for these three temperatures were 5.7×10^{-5}, 3.6×10^{-4} and 1.55×10^{-3}, respectively. Based on the essentially constant normalized dissolution rates of Table 3, it appears that the dissolution rate is first order with respect to PuO_2 surface area.

The following example illustrates HI dissolution followed by evaporation to dryness and dissolution of the dry residue in HNO_3.

Example 2: A 1 g quantity of PuO_2 microspheres was dissolved in 25 ml of 6.35 M HI at about 128°C in less than 8.0 hours. The HI solution was stabilized with 1.2 wt % H_3PO_2. The resulting dissolver solution had a dark blue color indicating the presence of Pu(III). The dissolver solution was distilled to dryness. The dried cake was divided into three approximately equal parts. One part of the dried solids was dissolved in 10 ml of 8.0 M HNO_3 in 0.25 hour to yield a 0.123 M Pu(IV) solution with less than 1 mg/ml iodide concentration.

Stripping of Plutonium from Organic Solvents

G. Cousinou and M. Ganivet; U.S. Patent 3,980,750, September 14, 1976; assigned to Commissariat a l'Energie Atomique, France describe a method of selective stripping of plutonium from organic solvents containing plutonium and in some cases uranium by reduction of the plutonium. The method applies either to organic solvents loaded with plutonium and uranium such as tributylphosphate or to organic solvents loaded only with plutonium such as trilaurylamine.

The method is primarily characterized in that the initial loaded organic solvent is contacted in countercurrent flow with a solution containing a reducing agent constituted by an aromatic organic compound selected from the group comprising the substituted hydrazines, the polyamines, the aminophenols, and an agent whose intended function is to remove the nitrites which are formed. Suitable agents which are open to selection for the removal of nitrites are hydrazine, phenylhydrazine, sulfamic acid as employed either alone or in a mixture.

In accordance with a characteristic feature of the process, the aromatic organic compound which constitutes the reducing agent can also perform the function of agent for the removal of nitrites. To this end, phenylhydrazine is particularly advantageous.

A. Bathellier and M. Germain; U.S. Patent 3,981,961; September 21, 1976; assigned to Commissariat a l'Energie Atomique, France describe a related process distinguished by the fact that the stripping process is carried out by countercurrent contacting of the organic solvent with an aqueous solution of a salt of hydroxylamine and formic acid, the concentration of the formic acid in the reaction medium being within the range of 1 and 5 M. The stripping solution is usually nitric acid. The hydroxylamine salt is advantageously selected from the group constituted by nitrate, formate, acetate and hydroxylamine propionate. Use of hydroxylamine in the form of a nitrate or of a formate has the advantage of avoiding the introduction of an anion which is foreign to the system.

Example: It is desired to strip plutonium from a solvent constituted by 20% trilaurylamine in Solgil 54 B which has previously been loaded with 16 g/ℓ of plutonium by contacting with a nitric acid solution of this element. Stripping is carried out in the bank (1) of mixer-settlers as illustrated diagrammatically in the following figure. The bank is made up of 16 stages.

Figure 6.2: Bank of Mixer-Settlers for Selective Stripping of Plutonium

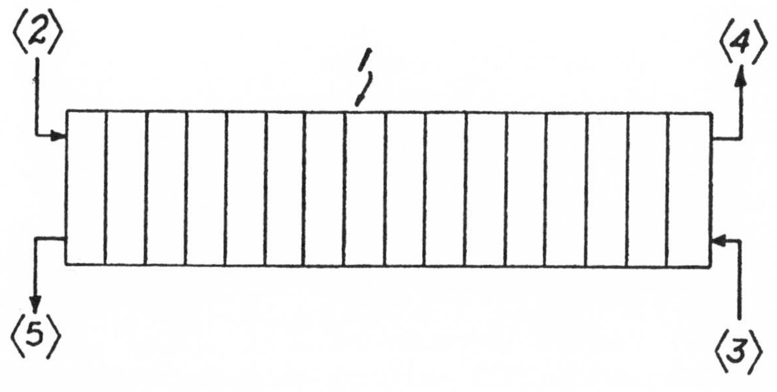

Source: U.S. Patent 3,981,961

The plutonium-loaded solvent has the following composition: 0.32 M trilauryl-amine in Solgil 54 B; 0.066 M $Pu(NO_3)_4$; 0.07 M HNO_3. The solvent is introduced into the bank of mixer-settlers at **2** at a volume rate of flow of 2.5 V (V being a reference volume which has been established at the outset).

A solution having the following composition is introduced at **3** at a volume rate of flow of 1 V: 2 M HCOOH; 0.35 M $NH_2OH \cdot HNO_3$; 0.1 N HNO_3.

The ratio of throughputs of the aqueous phase to the organic phase is 0.4 and the residence time of each phase within each stage of the bank of mixer-settlers is 10 minutes. At the end of 100 hours of operation, a constant composition is obtained in each stage. At this point, the solvent withdrawn from the bank at **4** contains only 0.3 mg/ℓ of plutonium. The stripping efficiency is, therefore, 99.998%. The aqueous solution withdrawn from the bank at **5** contains 40 g/ℓ of plutonium with the following distribution: 15% Pu(IV) and 85% Pu(III).

Pyrochemical Separation of Plutonium

G. Brambilla and G. Caporali; U.S. Patent 4,092,397; May 30, 1978; assigned to Agip Nucleare, SpA, Italy describe a method for the pyrochemical separation of plutonium from an irradiated nuclear fuel element discharged from a fast reactor and containing uranium oxide and plutonium oxide, or uranium carbide and plutonium carbide mixtures thereof, which comprises the following steps:

- (a) severing the irradiated nuclear fuel element;
- (b) disaggregating the fuel element by placing it in a bath of a molten alkali metal nitrate and/or alkaline earth metal nitrate heated to 480° to 500°C;
- (c) effecting dissolution of the oxides and/or carbides in the bath produced through step (b) by passing nitric vapors through the bath at a temperature of 250°C;
- (d) filtering insoluble fission products out of the bath produced by step (c);
- (e) decomposing the plutonium compound in the filtrate from step (d) by raising the temperature of the filtrate to 300°C and holding it in an atmosphere of nitric vapors so that plutonium is precipitated; and,
- (f) filtering out the plutonium precipitated in step (e).

If the separated plutonium should not fulfill the requested specification, it must be further purified, either with the dry method in a multistage run or by the aqueous way prior to being sent to the refabrication of the fuel element to be admixed with other natural or depleted uranium.

Impoverished uranium, the fissile value of which is nil, is, conversely, discarded from the cycle and stored in situ in an unalterable and insoluble form. It can also be recovered, if justified by economical reasons, in the form of an alkali metal uranate by subsequent decomposition with heat at a higher temperature. The molten nitrates are conversely regenerated from the fission products which were left in solution by thermal decomposition at temperatures which are still higher and by flowing over inorganic exchangers, and then they are recycled.

Example: A sintered pellet of UO_2-PuO_2 containing 18.3% of PuO_2, weighing 1.225 g was disaggregated at 480°C in a molten salt bath composed by 100 g of an eutectic mixture $NaNO_3$-KNO_3 (45.7-54.3% on a weight basis), then dissolved at a temperature of 250°C by having a stream of vapors of nitric acid as entrained by an inert gas flowing and bubbling through the same bath.

Upon dissolution, the reactor was closed so as to maintain a nitric atmosphere over the bath and the temperature abruptly raised to 360°C and then maintained at 300°C during 18 hours.

After that time period, more than 99% of the initially present plutonium was precipitated and 5.4% of uranium.

Dissolution of Plutonium Dioxide

O.K. Tallent; U.S. Patent 3,976,775; August 24, 1976; assigned to the U.S. Energy Research and Development Administration describes a method for bringing PuO_2 into aqueous solution comprising adding a silver compound to an aqueous dissolving mixture comprising PuO_2, HNO_3 and an effective catalytic amount of fluoride, the silver compound being soluble in the dissolving solution and supplying silver ions of oxidation state greater than 1.

According to the process, HF is no longer required in stoichiometric amounts with respect to Pu. While the rate of dissolution may be increased by adding greater amounts of fluoride, all that is required for dissolution is an effective catalytic amount, since the fluoride ion is now returned to the solution rather than being tied up in the PuF^{+3} complex ion. An effective catalytic amount is that amount, independent of Pu stoichiometry, which will effectively catalyze the PuO_2 dissolution for the particular HNO_3 concentration and may be routinely determined for the desired application. According to the process, the necessary fluoride ion concentration may now be reduced to less than 0.01 M to suit the corrosion resistance of the dissolving vessel.

Example: One-half gram samples of refractory PuO_2 microspheres were digested at 100°C in Teflon equipment in 40 ml volumes of 8 M HNO_3-0.02 M HF solution with and without Ag_2O_2 addition. The data in the table show that in 174 hours of dissolution there is 51.0% more total plutonium dissolved with Ag_2O_2 addition than without addition. The data in the table show also that the dissolution with Ag_2O_2 results in large fractions of the dissolved plutonium being oxidized to the plutonyl (PuO_2^{+2}) state.

Hours	Percentage Increase in PuO_2^{+2} Concentration	Percentage Increase in Total Pu Concentration
51	65.5	8.0
75	204.3	10.1
129	247.8	28.7
174	325.0	51.0

Note: 100 mg Ag_2O_2 added every 24 hours.

The increased plutonium dissolution rates resulting from the Ag_2O_2 additions can be seen in Figure 6.3 where plutonium concentrations in the dissolvents are plotted as a function of dissolution time for dissolutions with and without Ag_2O_2 additions. The curves in the figure show that the initial dissolution rates are the same with and without Ag_2O_2 additions, but that the rate without the addition decreases much faster than does the rate with the addition.

Figure 6.3: Plutonium Concentration vs. Time with and Without
Ag_2O_2 Addition

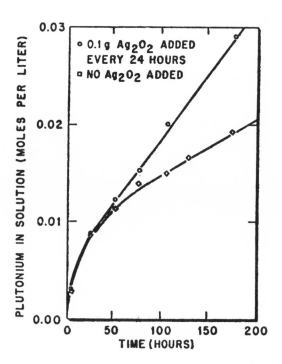

Source: U.S. Patent 3,976,775

Removal of Plutonium and Americium 241 from High Salt Content Waste

W.W. Schulz; U.S. Patent 4,156,646; May 29, 1979; assigned to the U.S. Depart-ment of Energy describes a combined precipitation and ion exchange process for efficiently removing plutonium and americium from plutonium reclamation facility salt wastes. As shown in Figure 6.4, on the following page, the process involves the addition of sodium hydroxide to adjust the plutonium reclamation facility salt waste to 0.5-2 M hydroxide ion concentration to precipitate metal hydroxides (such as iron hydroxide, calcium hydroxide and magnesium hydroxide) and remove 80 to 90% of the plutonium and greater than 99.9% of the americium 241. The solids removed from the metal precipitation step may thereafter be concentrated such as by filtration and drying to be converted to a small volume of nonleachable borosilicate glass or may be placed in suitable containers for retrievable storage.

The supernatant solution or liquid derived from the metal precipitation step is processed through a sodium titanate $[Na(Ti_2O_5H)]$ powder to reduce the concen-trations of both plutonium and americium 241 to below their maximum permissible concentration in water in an uncontrolled zone. The sodium titanate powder is pref-erably a size between 40 and 140 mesh (U.S. Standard Sieve Series). It was found that sodium titanate powder has a very high affinity and capacity for sorbing both plutonium and americium from alkaline solutions.

Figure 6.4: Process to Remove Plutonium and Americium 241 from
High Salt Content Waste

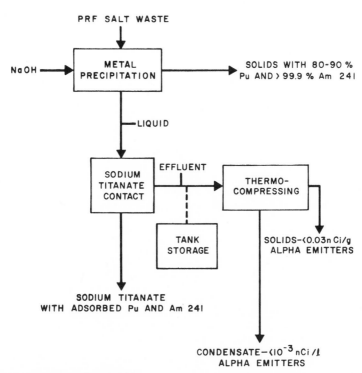

Source: U.S. Patent 4,156,646

The sodium titanate sorbent with the adsorbed plutonium and americium 241 may subsequently be stored as a retrievable alpha waste, or may be compacted under heat (i.e., hot pressed at about 1100°C) to a highly immobile, monolithic solid which can then be stored under various conditions. The effluent resulting from the sodium titanate contact step can either be routed to existing underground waste tanks or, alternatively, may be heated under pressure to evaporate the liquid to form a condensate which may be reused in the plant since it has less than 10^{-3} nCi/ℓ of alpha, while the solids may be heated in a dryer to achieve a dry solid which may subsequently be stored as chemical wastes or a very low level radioactive waste having less than 0.03 nCi/g alpha emitters.

The efficiency of freshly precipitated iron hydroxide in scavenging plutonium and other actinides from aqueous waste solutions is well known, having been used successfully at other sites for this scavenging purpose.

The plutonium reclamation facilities salt waste contains, typically, from 0.1 to 0.3 M aluminum nitrate. To eliminate a possibility of permanently precipitating gelatinous, hard to separate aluminum hydroxide, it is desirable in the precipitation step to adjust the plutonium reclamation facilities salt waste to 1 to 2 M hydroxide ion concentration. Even at these terminal hydroxide concentrations, which are much higher than those employed in the hydroxide precipitation steps in other sites, americium and plutonium are still effectively carried by the hydroxide precipitate. It has been found that metal hydroxide scavenging of americium 241 from plutonium reclamation facilities salt waste is extremely insensitive to change in precipitation temperature, agitation time, terminal hydroxide ion concentration (in the range of 0.5 to 2 M) and whether sodium hydroxide is added to the waste or the waste is added to the sodium hydroxide.

A single precipitation of iron and other metal hydroxides from plutonium reclamation facilities salt waste solution yields a solution still containing more than the desired 10 nCi/g total alpha activity. Significant additional decontamination of this clarified alkaline solution liquor can be obtained by its passage through one or more beds of titanate sorbent.

A sodium titanate bed measuring 0.04 m^3 (15 cm diameter) will adequately decontaminate the liquor derived from the metal precipitation step of about 120 m^3 of salt waste from the plutonium reclamation facility. In this respect, this volume of sodium titanate will reduce the concentration of both americium 241 and plutonium in 1,000 to as much as 4,000 column volumes (CV) of the liquor to less than the maximum permissible concentration of each radioisotope in water in an uncontrolled zone. The decontaminated salt waste effluent from the titanate bed containing less than 10 nCi/g total alpha activity may be routed to suitable underground waste tanks or can be dried such as by a spray dryer or a wiped film evaporator and stored as a chemical waste. It may also be mixed with sufficient kaolin or bentonite clay to form the silicate mineral cancrinite and the resulting solid containing entrapped sodium nitrate may be thereafter stored as a chemical waste.

The sodium titanate powder contact step reduces the americium 241 concentration to less than 0.0002 to 0.0016 μCi/ℓ and the plutonium concentration to less than 10^{-3} to 0.004 μCi/ℓ. For reference, the maximum permissible concentrations for plutonium 239 and americium 241 in water in an uncontrolled zone are 0.005 and 0.004 μCi/ℓ, respectively.

Ferric Ion Scavenging Agent

L.E. Bruns and E.C. Martin; U.S. Patent 3,987,145; October 19, 1976; assigned to the U.S. Energy Research and Development Administration describe a method to improve the recovery of plutonium in a plutonium scrap recovery process.

In accordance with the process, ferric ions are added into the aqueous feed solution of a solvent extraction process employing tributyl phosphate in an organic base for the recovery and partitioning of actinide values. The process includes an extraction unit or process portion in which substantially all of the actinide values are extracted from the aqueous feed into the organic, and a subsequent process unit or portion in which at least one species of the actinide values is stripped into an aqueous strip solution. Precipitates formed of iron and degradation products of tributyl phosphate are removed from the extraction portion of the process to enhance the stripping or partitioning of values in the subsequent portion as well as the extraction characteristics of the tributyl phosphate.

It is of importance in the process that the ferric ion concentration be controlled within 0.01 to 0.2 M, preferably 0.05 to 0.1 M. It is also of importance to note that the ferric ions must be added into the feed solution as opposed to a later stream.

This permits the dibutyl phosphate to be precipitated as formed without opportunity to complex or precipitate with plutonium values.

Example: A typical feed solution containing the following composition is fed into a solvent extraction process: 2.5 M HNO_3; 0.8 M Al^{+3}; 0.4 M AlF^{+2}; 0.4 M Mg^{+2}; 0.2 M Ca^{+2}, 0.1 M Na^+; 0.2 M other cations; 5.0 g/ℓ $UO_2^-(NO_3)_2$; 10 g/ℓ $Pu(NO_3)_4$ [some $PuO_2(NO_3)_2$] 8.5 M total NO_3^- in aqueous solution. Various of these cations represent radioisotopes that can produce radiolytic degradation of TBP. To this feed is added sufficient ferric nitrate, $Fe(NO_3)_3$, to form 0.1 M ferric ion concentration. The feed solution is contacted within a sieve plate air-pulsed column with a countercurrent flow of 20% by wt TBP in CCl_4. Ferric dibutyl phosphate precipitate is withdrawn from the aqueous to organic interface at the column top. The organic effluent discharged from the column bottom containing the extracted actinide values is expected to have no more than about 0.0001 M dibutyl phosphate.

System Producing Contaminated Plutonium

There is serious concern throughout the world that the increasing deployment of reprocessing capacity for nuclear reactor fuel will increase the likelihood of further proliferation of nuclear weapons. This concern is due to the fact that essentially all processing methods used to date are derived from the processes originally developed during or shortly after World War II for producing plutonium for nuclear weapons. This concern has been led by the United States Government, which since 1977 has taken the position that such processes are undesirable for civilian power use because they potentially make purified fissionable materials available and, therefore, susceptible to diversion by terrorist groups. In addition, there is concern because such reprocessing plants could readily be converted to the extraction of weapons material by a change of intention by a government which had previously pledged by treaty to forego the production of nuclear weapons.

M. Levenson and E.L. Zebroski; U.S. Patent 4,278,559; July 14, 1981; assigned to Electric Power Research Institute describe a system whereby plutonium can be made highly resistant to diversion through a combination of changes in the method of reprocessing spent fuel and changes in the design of a reprocessing plant. The concept is that plutonium in weapons-usable form is not produced at any point in the process, that mixtures containing plutonium are always kept sufficiently radioactive, that attempted diversion is easy to detect, and that no readily available change in the process operation can be made to yield weapons-usable material.

The design of the plant is such that even if the plant were occupied by hostile forces, the length of time, the skills and resources required, and the complexity of converting it over to produce separated plutonium are comparable to the efforts required to assemble a crude military-style reprocessing plant from scratch. Furthermore attempts to modify the process involve overt changes in equipment and in flow patterns. Such attempts are readily detected so that there is timely warning available that diversion is being attempted.

Nowhere in the process is plutonium completely separated from uranium and fission products. Plutonium is always diluted with an excess of uranium such that it cannot be directly made into a nuclear weapon. The residual penetrating radiation from long-lived fission products facilitates detection of attempted theft or diversion of even small amounts of material. The residual radiation also presents a biological hazard to a diverter if larger amounts are handled.

In this process, the plant and the method of reprocessing are such that anyone who diverts a significant amount of plutonium from the plant must take with it sufficient radioactive fission products to require the use of another chemical separation process in order to obtain usable material. It is the intent of the process to make the materials produced in the process comparable in resistance to diversion (for weapons purposes) to an equivalent amount of spent fuel from conventional light water reactors.

The process comprises a diversion resistant method such that substantial amounts of additional capital, equipment, time, organization, force, and sophisticated skills are required to obtain material of weapons-grade purity. In addition, the plant is so designed that the modifications required to obtain material of the weapons grade are extensive and easily detectable by a wide variety of surveillance and inspection methods and by either local or international inspectors.

One feature of the system is that approximately three-fourths of the equipment currently used in a conventional reprocessing plant is eliminated. A conventional plant typically has several solvent extraction or other auxiliary purification cycles and their associated equipment to achieve sufficient purity in the product streams. The system uses less than one complete cycle. The absence of the associated pumps, tanks, and plumbing of the subsequent cycles provides for a simpler operation as well as precludes the production of weapons-usable material.

A further benefit of the system is that the cost of reprocessing spent fuel is substantially reduced. The recycle of off-specification product streams is avoided by design, so that high operating factors for the plant are practical. Although the cost of the subsequent remote fabrication of fuel is increased, there is a potential net reduction in total cost due to the reduction in the size of the facility, the reduction in the complexity of the process, and the increase in assurance of effective safeguards which can reduce institutional obstacles and physical security costs.

The foregoing and other objects are achieved by a method and apparatus for processing spent nuclear reactor fuel. Spent fuel is first dissolved to produce a spent fuel stream containing plutonium, uranium and a varying mixture of short-lived and long-lived fission products in solution. Next, the plutonium and uranium (together with a substantial quantity of radioactive fission products) are extracted using an organic solvent and discharged together into a product stream. This product stream is thereafter partitioned into a uranium product stream and a plutonium-containing product stream. The plutonium-containing product stream contains some fission products and a quantity of uranium equal to or greater than the quantity of plutonium. In all of these steps and in every piece of apparatus plutonium continuously is both diluted with uranium and contaminated with substantial levels of radioactive fission products.

URANIUM-PLUTONIUM RECOVERY

Separation of Zirconium from Uranium and Plutonium by Hydriding

As is generally known, the spent fuel elements of nuclear reactors and compositions or mixtures thereof contain residual fissionable components that may be further utilized as a nuclear fuel. Besides these fissionable components, there are also present fuel elements and mixtures thereof which contain construction metallic components having a low coefficient of absorption for thermal neutrons. Such construction metallic components are zirconium alloys containing tin, niobium, titanium, and other elements. The fissionable components are, in general, hard compounds of uranium and plutonium produced by ceramic methods. The fuel component is usually enclosed in thin-wall tubes made of zirconium alloys forming elements that are connected into or associated with larger assemblies enclosed in a casette tube. In addition, the assemblies contain fastening construction parts from nonfissionable metals and alloys. The single parts of the fuel assembly, when removed from the reactor, are deformed and are highly radioactive.

When nuclear fuels are reprocessed, difficulties arise during separation of the fissionable and nonfissionable components required for biological protection. The difficulty in separating fissionable and nonfissionable elements is enhanced by the different mechanical and chemical properties of these components. Uranium and plutonium compounds are extremely hard and brittle; zirconium alloys, on the other hand, are firm and tough.

B. Cech, E. Kaderabek and T. Hanslik; U.S. Patent 4,024,068; May 17, 1977; assigned to Ceskoslovenska akademie ved, Czechoslovakia describe a process utilizing hydrogen in the separation of uranium, plutonium, and their compounds from composite substances and assemblies or mixtures comprising metallic and ceramic components where zirconium or zirconium alloys are the metallic components, and uranium and/or plutonium compounds represent the ceramic component.

The process is based on a working procedure in which the original system is reacted with hydrogen and the formed hydrides are separated from the mixture physically and/or chemically. In a typical procedure, the system is treated with hydrogen, under pressure and at an elevated temperature. The pressure can be varied in the range from 20 to 50 atmospheres while the working atmosphere may be varied and suitably be from 200° to 700°C, most suitably in the range from 250° to 460°C.

The process is based on the fact that fuel elements and assemblies or mixtures containing nonfissionable construction parts of zirconium, niobium, titanium, and their alloys may be separated, before processing, by being preheated in a hydrogen atmosphere at an elevated pressure. Under these suitable conditions, the aforesaid components are rapidly converted to hydrides of high-melting metals that are brittle and easily separable. The cladding and construction parts of the fuel elements are then disintegrated either by virtue of their weight or mechanically. Very little energy is required. Steel parts that do not react with hydrogen to form hydrides may also be separated in this manner. When the parts from the high-melting metals are disintegrated, the ceramic fuel is liberated.

Example: A heterogeneous assembly of ceramic pellets of cylindrical shape having a diameter of approximately 8 mm and a height of 5 to 10 mm, composed mainly of UO_2 and enclosed in a zirconium alloy tube with a 1% by wt of niobium and a wall thickness of approximately 1 mm, is placed in a heated autoclave. After filling the autoclave with electrolytic hydrogen, the hydrogen pressure is elevated to 30 atmospheres and the pressure vessel is closed. The autoclave is heated with simultaneous rotation up to a temperature of 450° to 500°C. Before this temperature is reached, hydrogen pressure in the autoclave decreases as a result of the rapid exothermic reaction of hydrogen with the zirconium alloy. Under these conditions, the duration of the reaction does not exceed 5 minutes. The vessel is then cooled, washed at standard pressure with nitrogen and the contents placed in a separation column where uranium oxide on one side and zirconium and niobium hydrides on the other side are mechanically separated.

Pyrochemical Reprocessing of Liquid Metal Cooled Nuclear Fuels

W.E. Miller, J.F. Lenc and I.O. Winsch; U.S. Patent 3,867,510; February 18, 1975; assigned to the U.S. Atomic Energy Commission describe a pyrochemical fuel reprocessing system for spent, short-cooled LMFBR fuel assemblies in which the fuel assemblies are declad by dissolution in a liquid zinc melt with the plutonium and uranium oxides contained therein being separated and then reduced by subsequent contact with a molten zinc-magnesium-calcium reduction solvent. To recover the reduced uranium and plutonium metal dissolved in the molten solvent, a bed of particulate calcium nitride or magnesium nitride is immersed in the 750° to 800°C solvent, the solvent being continually agitated so that it passes through the particle bed. The nitride bed remains immersed in the solvent for an effective period of time to react the uranium and plutonium with the nitride particles to form insoluble particles of uranium nitride, plutonium nitride and uranium-plutonium nitride mixtures, the calcium or magnesium being retained in the liquid-metal solvent.

After the effective contact period, the bed of reacted particles is removed from the liquid-metal solvent and then retorted at approximately 925°C in order to remove any remaining solvent, as well as to volatilize any unconverted calcium nitride or magnesium nitride. These recovered particles of UN, PuN and UN-PuN are then available for further reprocessing.

In addition, the process can be utilized to form uranium nitride, plutonium nitride and uranium-plutonium nitride nuclear reactor fuels. In this particular embodiment, uncontaminated uranium, plutonium, or mixtures thereof are dissolved in a suitable liquid-metal solvent such as zinc, zinc-magnesium or zinc-magnesium-calcium. Insoluble nitrides of uranium, plutonium, or uranium-plutonium mixtures are formed by reacting the dissolved uranium, plutonium or uranium-plutonium mixtures with a bed of calcium nitride or magnesium nitride particles, either contained in a perforated basket immersed in the liquid-metal solvent or added to the liquid-metal solvent in bulk form.

Recovery of the insoluble nitride product is obtained by either retracting the perforated basket containing the product, vacuum-distilling off residual solvent metal and volatilizing unconverted calcium or magnesium nitrides; or if the perforated basket is not utilized, by vacuum-distilling off all of the solvent metal to recover the nonvolatile nitride product. As a result, substantially pure particles of uranium mononitride, plutonium mononitride, or uranium-plutonium mononitride mixtures are obtained and can be directly utilized as nuclear reactor fuels.

Plutonium-Enriched Nuclear Fuel

K.H. Puechl; U.S. Patent 4,182,652; January 8, 1980 describes a process whereby plutonium-containing nuclear fuel material is provided whose plutonium content is as small as practicable considering the plutonium available and the demand of nuclear reactors for fuel. To achieve a desired fissile content this nuclear fuel material includes less than 1% of plutonium and the remainder uranium of U-235 enrichment which together with the plutonium is sufficient to attain the required fissile content. Alternatively, the material includes less than 1% plutonium, uranium of modest enrichment (less than 1%) and enriched uranium sufficient to attain the desired fissile content. There are also provided a fuel rod or element or assembly containing the abovedescribed nuclear fuel material and a nuclear reactor in which such fuel rods are distributed throughout all regions of the core. In such a reactor local power peaking is suppressed; an equilibrium fuel cycle is attainable.

The nuclear fuel material in accordance with this process is produced during the processing of the spent nuclear fuel of a reactor to derive plutonium. During this processing, a solution of spent uranium and plutonium is separated from the fission products. The fuel material can be produced at this stage of the process by adding a solution of appropriately enriched uranium to the solution of spent uranium and plutonium in sufficient quantity to achieve the required proportions and fissile content. The fuel material may also be produced at a later stage of the process by separating the plutonium from the uranium and adding to the plutonium solution a solution of enriched uranium of lower enrichment than for the above solution of uranium and plutonium to achieve the required proportion and fissile content. In each case, the solute is precipitated from the solution and typically reduced to a mixed oxide powder of uranium and plutonium from which the fuel is formed.

The concentration of plutonium in the fuel material depends on the plutonium available and the fuel demands. During the period 1976 to 1980 the projected recovery of plutonium from spent fuel, when distributed over all fuel required for projected light-water reactors, would yield a fissile plutonium content of about 0.16% and a total plutonium content of 0.21%. The remainder 99.79% would be all enriched uranium or partly enriched uranium and partly spent uranium. Over the time span 1981 to 1995, similar distribution of plutonium (making allowances for plutonium required for contemplated fast breeder fueling) results in the lowest realistic concentration being about 0.42% total plutonium content or 0.29% fissile plutonium content. Discharged nuclear fuel after normal service contains approximately 0.6% fissile plutonium or 0.9% plutonium. This can be considered as an upper level of practicability for the process.

Recovery of Nuclear Fuel from Scrap Material

Described by *L.A. Divins and L.E. Short; U.S. Patents 4,230,672; October 28, 1980 and 4,177,241; December 4, 1979; both assigned to General Electric Company* is a process for recovering in solution form enriched nuclear fuel compounds from scrap materials, including spent filter media. The process comprises the steps of:

(a) calcining the precursor material, other than the filter media, to yield an oxidized material or calcine;

(b) comminuting the oxidized material to form a particulate material of particle size less than about 250 microns;

(c) contacting the particulate material with an acid containing some recycled acid in a mechanically agitated leaching zone of a nuclear-safe configuration to dissolve the nuclear fuel compounds and yield an acid solution of the nuclear fuel compounds;

(d) mulching the filter media so that it is in the form of a particulate material;

(e) contacting the mulched media with the acid solution in a leaching zone;

(f) filtering the insoluble solids from the acid solution in a filtering zone;

(g) separating suspended solids from the acid solution in a clarification zone;

(h) recycling at least a portion of the acid solution from the clarification zone to the leaching zone; and,

(i) collecting the remainder of the acid solution for subsequent treatment such as with an organic solvent in a liquid-liquid extraction zone to recover the nuclear fuel from the solution.

The process can also include an optional preliminary screening step to insure that only the oxidized material being treated in the comminution step will be above the desired size range for contacting with the acid.

The process is particularly useful for recovering nuclear fuel values from ventilation filters and incinerator ash and is particularly useful when the nuclear fuel is an oxide of uranium or mixture of uranium and plutonium oxides. The process also includes an apparatus for recovering nuclear fuel values from scrap materials, including spent filter media. The apparatus includes calcining means for calcining the scrap material, other than filter media, to form an oxidized material which is introduced into comminuting means for reducing the oxidized material into a particulate material of given particle size range. The comminuting means discharges the particulate material into a slab-shaped, nuclear-safe leaching means for contacting the particulate material with an acid solution, a portion of which can be a recycled acid solution. The slab-shaped leaching means provides substantially complete dissolution of the nuclear fuel portion of the particulate material with the aid of mechanical agitation. This yields an acid solution containing undissolved solids.

The apparatus further includes mulching means for mulching the nuclear-fuel-containing filter media to yield mulched filter media in particulate form. Filter leaching and repulping means receives the mulched filter media containing nuclear fuel material and also receives the acid solution from the leaching means for contacting the filter media with the acid. The filter leaching and repulping means discharges the acid solution containing undissolved or insoluble solids into a separator means. The separator means removes the major portion of the undissolved or insoluble solids from the acid solution and discharges the acid solution to a surge and reagent heat tank means. In this tank, an adjustment can be made such as adding additional acid, if desired. The tank provides a constant flow output to a clarification means that removes suspended residual solids from the acid solution. A portion of the acid solution is recycled in a recycle line to the leaching means for further dissolution of the nuclear fuel, and the remainder of the acid solution is pumped to a storage tank or treated to recover the nuclear fuel.

Figure 6.5 is a flow diagram showing the sequential steps of the process.

Low Tributyl Phosphate Concentration

A.L. Mills, E. Lillyman and P.G. Bell; U.S. Patent 3,959,435; May 25, 1976; assigned to United Kingdom Atomic Energy Authority, England describe a process for the treatment of irradiated nuclear fuel in which fissile material in solution in nitric acid is contacted with tributyl phosphate diluted with an inert organic diluent to transfer the fissile material to the organic phase which is then separated from the aqueous phase and backwashed with aqueous sulfuric acid, the concentration of the tributyl phosphate in the inert diluent being less than 7 vol %.

The effect of lowering the tributyl phosphate concentration in this way is to decrease the density of the organic phase and reduce its viscosity. These changes in density and viscosity cause the separation of the organic and aqueous phases after contacting, allow more organic liquor to be fed through a given plant, and hence the processing of more fuel. It has been found also that a higher decontamination

Figure 6.5: Sequence for Recovery of Nuclear Fuel from Scrap Materials

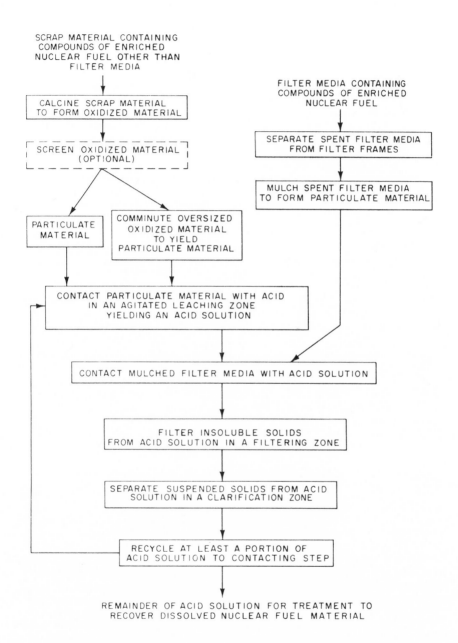

Source: U.S. Patent 4,230,672

factor is achieved with respect to the major active fission products found in irradiated nuclear fuel. The full advantages are only obtained if aqueous sulfuric acid is used for backwashing the organic phase after contacting with the diluted tributyl phosphate. Using the conventional very weak nitric acid backwash gives rise to large volumes of liquid which are difficult to handle in subsequent extraction cycles and may reduce the decontamination factors achieved.

Separation of Carbon from Radioactive Materials

In the reprocessing of some nuclear fuels of HTGR (high-temperature gas-cooled reactors) types it is necessary to separate large amounts of carbon from the active material (U, Pu, Th) which is present in the form of little spheres. This carbon constitutes the coating of the spheres, the matrix wherein they are dispersed and the structural graphite which constitute a real fuel element. The proportion by weight between active material and carbon varies from a minimum of one-tenth to one-one-hundredth and more, according to the possibility and convenience of eliminating part of structural graphite before the effective reprocessing.

G. Dolci and R. Renzoni; U.S. Patent 3,984,519; October 5, 1976; assigned to Snamprogetti, SpA, Italy describe a method for eliminating carbon from nuclear fuel elements in the reprocessing thereof in a closed system wherein such carbon is caused to combine with hydrogen to form methane in an attack zone, and the methane is then caused to flow from the attack zone to a regeneration zone where it is cracked into carbon and hydrogen.

The process for the transport of the carbon may be illustrated as follows: Consider a closed system. In a zone of this system, which is called the "attack zone," the temperature and pressure conditions are such as to produce a reaction between the carbon and the fluid therein present. The so-formed gaseous product, by simple diffusion or by circulation, however accomplished, goes to another zone of the same system, which is called the "regeneration zone."

In this zone, the temperature and pressure conditions are such as to produce a reaction which is the reverse of that which took place in the attack zone. Consequently, carbon will be deposited and the attack fluid will be regenerated. In other words, the system will go back to the initial state with the only difference that the transport of carbon took place (and the subsequent inevitable losses of energy).

Various reversible reactions, long well known, can be differently used in the system. Therefore, in order to illustrate more particularly the process some examples, i.e., a certain number of specific embodiments, are shown by way of unrestrictive example.

A first practical example of what is abovementioned is given by the equilibrium reactions of the system: $C + CO_2 \rightleftharpoons 2CO$. The proportion of the two gases in equilibrium (in the presence of carbon) depends only on the pressure and on the temperature. The pressure increase shifts the equilibrium to the left (carbon is deposited) and the temperature increase shifts the equilibrium to the right (carbon is absorbed).

Therefore, it is possible to adjust these two parameters in such a way that in the attack zone the reaction may be shifted widely to the left and so that in the regeneration zone the contrary takes place.

As pointed out above, the only change due to the operation of the system is represented by the transport of carbon from one zone to another and by the inherent energy losses.

A second example is given by the reversible equilibrium reactions of the system: $C + 2H_2 \rightleftharpoons CH_4$. Also in this case, the proportion of the two gases in the presence of carbon depends on the pressure and on the temperature.

The preceding considerations are also valid, except that in this case the pressure increase shifts the reaction to the right (absorption of the carbon) and the temperature increase shifts the reaction to the left (cracking).

As a matter of fact, a result similar to that of the two abovementioned examples may be obtained by means of more than two reactions which may be convenient for particular reasons, provided that at the end of the cycle the system returns to the initial state with the exception of the occurred transport of carbon, which remains always the useful object to be reached.

A practical example of the third possibility is given by the following reactions:

$$C + H_2O \rightarrow CO + H_2$$
$$CO + 3H_2 \rightarrow H_2O + CH_4$$
$$CH_4 \rightarrow C + 2H_2$$

After separation of water by condensation and cracking of pure methane, the state of the system has gone back to the initial one and only the carbon has been transported.

The process eliminates entirely the problem represented by the discharge of large amounts of contaminated gases to the atmosphere and, moreover, reduces or eliminates the need of filters, as the particles of fission products which possibly pass in the zone wherein carbon is deposited will go to waste storage.

Fuel Rod Reprocessing Plant

A pamphlet published by the Energy Research and Development Administration in May 1976 entitled "Radioactive Waste Management at Hanford" describes one type of traditional reprocessing plant which happens to be designed for recovery of neptunium as well as plutonium and uranium. The Purex Extraction portion of the Hanford Reprocessing Plant has processing equipment for sequentially dissolving depleted fuel rods, treating such solutions with extractants for the recovery of components such as plutonium and uranium and for the isolation of fission products, actinides, and other high-level radioactive waste, which are further processed in buildings different from the Purex plant.

The depleted fuel rods are dissolved to provide an aqueous solution which is subjected to solvent extraction to recover the plutonium, uranium, and neptunium. Each of the neptunium, uranium and plutonium streams is purified by a series of processing steps. The fission products are separated and processed in an appropriate manner.

By reason of the varying demands relating to the ultimate disposition of the fission product and the contemplated end use of the uranium fraction and the plutonium fraction, different processing schemes for fuel rods are appropriate. It is highly desirable that the process be sufficiently flexible to permit some modification of the processing subsequent to start-up. Substantially all operations within a reprocessing plant are conducted by remote control, and accordingly are quite costly. Some of the reprocessing plants built in the last two decades have been abandoned because of the excessive cost of retrofitting to install equipment needed for modifications of the process. There has been a long-standing demand for a reprocessing plant having the flexibility to permit alterations of the process without excessive downtime after start-up of the plant.

In fuel rod reprocessing plants such as the Purex facility at Hanford, it has generally been the custom to provide a canyon in which the vehicle access area was a system of overhead cranes and the various items of processing equipment were positioned linearly along the length of the canyon. Such an arrangement has numerous advantages from the standpoint of minimum cost for initial construction. However, such linear canyon has numerous disadvantages when dealing with retrofitting apparatus for modifying the process. Moreover, the reliance upon overhead cranes in the vehicle access area has disadvantages when dealing with emergencies requiring the overhead crane to shift from one spot to another along the length of the path of the crane. If there was leakage of liquid from a processing tank and/or pipe, there was a hazard of a plurality of processing vessels being contaminated because the canyon partook of the nature of a single compartment.

Certain processing units in a reprocessing plant have a lifetime significantly less than the contemplated total lifetime of the reprocessing plant. In a Hanford canyon-type of structure, the plant is shut down and a processing unit is dismantled and replaced by a substitute unit in order to deal with the replacement problem. Thus, the downtime for such necessary maintenance has been a significant burden in the operation of reprocessing plants. Once in operation, a reprocessing plant represents an investment of such large size that there are financial incentives for minimizing the downtime of a reprocessing plant.

M.J. Szulinski; U.S. Patent 4,261,952; April 14, 1981; assigned to Atlantic Richfield Company describes a fuel rod reprocessing plant which includes: the combination of an array of a plurality of deep compartments having horizontal cross sections which are generally rectangular such array desirably being such that many of the compartments are in a rectangular grid arrangement; a vehicle access zone which is generally above the array of deep compartments; rectangular covers for the compartments, each cover being liftable by a vehicle in the vehicle access zone; an equipment support floor in each compartment; processing equipment on support floors on each of several deep compartments, there being communication lines extending from the processing equipment to processing equipment in other compartments, a significant portion of such communication lines between compartments being near the top of the compartments and accordingly immediately beneath the vehicle access zone; a control room shielded biologically from the compartments and the vehicle access zone; monitoring and control means adapted to permit control room operation of the process equipment in the deep compartments.

Also included are: a sump pit beneath the support floor in each deep compartment, adapted for monitoring and detecting liquid spilled in such compartment; drainage systems from sump pits to annular storage tanks adapted to minimize accumulation of large masses of plutonium-containing liquid, appropriate for some compartments. The number of compartments in the array of compartments should be at least one-fourth greater than the number of compartments required for normal operation, whereby processing equipment scheduled for substitution can be constructed in a vacant compartment prior to disconnection of corresponding processing equipment scheduled to be withdrawn from service. Operations can be resumed after making the piping connections for the processing equipment scheduled for substitution, and the downtime for the reprocessing operations can be lessened when scheduling replacement of equipment.

Fuel Cycle Management

A.E. Smith; U.S. Patent 4,018,697; April 19, 1977; assigned to Atlantic Richfield Company describes a process whereby nuclear fuel is employed for a first cycle in a light water reactor and thereafter subjected to reprocessing for the removal of fission products, for the removal of higher actinides, for the recovery of plutonium, and for the recovery of uranium. Such recovered uranium includes about 0.1 to 0.7% uranium 236. Higher concentrations of U-236 in subsequent cycles are plausible. The difficulties attributable to the U-236 isotope are managed by separately enriching the recovered uranium to provide an upgraded fraction containing from about 1.0 to 3% U-235. From 30 to 60% of U-236 initially present can be shifted to the tails fraction, with only about 40 to 70% of the U-236 in the enriched fraction. The U-236 tends to follow the U-235 in the enrichment step so that the U-236 concentration (as distinguished from distribution of total U-236) in the upgraded fraction is greater than initially.

Such enriched upgraded recovered uranium is mixed with plutonium to provide a mixed oxide fuel containing from about 1 to 7% plutonium oxide. No enriched uranium derived from virgin uranium is mixed with or diluted with recovered uranium because of the significant penalty attributable to the presence of the U-236 isotope, particularly in the dilute parasite range.

Example: A nuclear power plant requires for each reloading about 30 tons of nuclear fuel equivalent to 3.3% U-235 in purified virgin uranium oxide. A batch of spent rods is reprocessed, the mechanical losses being about 1%. The spent fuel contains 0.79% U-235 and 0.45% U-236, but based upon UO_2 content, the recovered uranium contains 0.827% U-235 and 0.47% U-236. Such uranium is enriched by 1.62 separative work units per kilogram of product at a feed-to-product ratio of 3.2252 to provide an upgraded fraction of about 8.8 tons (29.3% of reload requirements) containing 2.01% U-235 and 0.823% U-236.

The spent fuel contains about 0.91% plutonium of which about 72% is fissile, providing a fuel equivalency (based upon 1.25% fissile plutonium equalling 1% U-235) of 0.515%. A batch of fuel weighing about 9.09 tons is prepared by coprecipitating in atomic intimacy the recovered plutonium and the upgraded uranium, such fuel containing 1.953% U-235, 2.846% plutonium (equivalent to 1.623% U-235) and about 0.823% U-236. At this concentration, the amount of extra U-235 needed to compensate for the U-236 is only about one-third of the U-236 concentration even though about three-fifths (60%) of the U-236 concentration is required at concentrations such as 0.05% described as the dilute parasite effect. The avoidance of such dilute parasite effect by working with concentrations of U-236 permitting compensation with U-235 at about 33% instead of about 60% represents a significant advantage of the process.

Approximately 0.276% U-235 (or equivalent) is necessary to compensate for such 0.823% U-236, and it is for this reason that the recycled fuel is designed to be 3.576% equivalent U-235 to provide fuel value corresponding to 3.3% virgin U-235.

Because such adjustment for the presence of U-236 is achieved by including about 0.346% fissile plutonium (125% of 0.276% contained in about 0.480% plutonium, 72% fissile) instead of U-235 and because prolonged exposure of U-236 to the nuclear reactor leads to transmutation to higher actinides, additional U-236 is not generated as part of the measures for its presence. Moreover, because only about 55% of the fissile equivalency is U-235, the marginal production of U-236 during a second cycle is lessened. The mixed oxide fuel tends to produce less U-236 for a given U-235 content permitting patterns of transmutation of more U-236 than generated when the fissile plutonium contributes more than about 66% of the fuel value.

The 9.09 tons of mixed oxide fuel can be employed in making fuel rods for about 30.3% of the reloading and 69.7% of the rods can be made from virgin uranium.

On recycle of such 30 tons of fuel, the depleted rods would contain about 0.78% U-235, about 0.65% U-236 and about 1.34% plutonium. After enrichment of the recovered uranium at a feed-to-product ratio of 2.79, employing 1.0924 separative work units per kilogram of product, there are 10.34 tons of upgraded uranium containing 1.5% U-235 for use in the coprecipitation zone, from which are recovered 10.743 tons of mixed oxide fuel containing about 3.75% plutonium.

Nuclear fuel materials are so costly that the return on investment on materials being processed can be a significant factor in the marginal cost of nuclear fuel. Fuel cycle management systems must be evaluated in part upon the time required to convert spent rods into nuclear fuel rods suitable for recycling. The speed with which the spent rods could be subjected to reprocessing, the recovered uranium enriched to provide the upgraded fraction, the coprecipitation to provide mixed oxide fuel, and the manageability of following similar procedures through a series of recycling steps represent the advantages of the process.

OTHER ACTINIDE RECOVERY PROCESSES

Recovery of Actinides Using Di-Hexoxyethylphosphoric Acid

One major problem with recovering the actinides from large volumes of acidic high-level radioactive waste solutions is to find a method which will do so effectively and economically, since the actinides are present in several valence states and difficult to recover together. The extractant, di-hexoxyethylphosphoric acid (HDHoEP) is known to extract tetra- and hexavalent actinides from an acidic solution, along with some fission products and rare earths. However, because the extractant has a strong affinity for the actinide elements, there has heretofore been no effective means for recovering or backextracting the actinides from the HDHoEP extractant and for separating the actinides from the coextracted elements.

E.P. Horwitz, W.H. Delphin and G.W. Mason; U.S. Patent 4,162,230; July 24, 1979; assigned to The U.S. Department of Energy have developed a process for partitioning or separating actinide values from an acidic radioactive waste solution using HDHoEP by which one is able to recover the actinide values from the extractant, separate the actinides from the coextracted fission products and rare earth, and separate some of the actinide values from each other.

The process comprises the steps which are given on the following page.

(a) adding hydrazine and hydroxylammonium nitrate in equimolar amounts to the waste solution to adjust the oxidation state of neptunium and plutonium in the waste solution to +4, thereby forming a feed solution;

(b) contacting the feed solution with an organic extractant of 0.1 to 1.5 M di-hexoxyethylphosphoric acid in a water-immiscible aromatic or aliphatic hydrocarbon diluent whereby the actinide values, the rare earths, some fission product values and iron are selectively extracted from the feed solution thereby loading the extractant;

(c) contacting the loaded extractant with a first aqueous strip selected from the group consisting of a nitric acid solution and an oxalate solution, the oxalate solution selected from the group consisting of oxalic acid and tetra- or trimethylammonium hydrogen oxalate, the nitric acid soltuion selectively stripping americium, curium and rare earth values and the oxalate solution selectively stripping neptunium, plutonium, fission product zirconium, niobium and molybdenum and iron values, the uranium values remaining in the extractant, whereby some of the actinide values and some of the other values are stripped from the extractant, forming a partially loaded extractant; and,

(d) contacting the partially loaded extractant with the other of the strip solutions of step (c), whereby the remaining actinide values and some of the other values are stripped from the extractant, thereby recovering the actinide values from the waste solution.

The process also includes the following steps:

(e) adding nitric acid to the oxalate strip solution containing the neptunium, plutonium and fission product zirconium, niobium, and molybdenum values in an amount sufficient to destroy the complexing ability of the oxalate, thereby forming a second feed solution;

(f) contacting the second feed solution with a second organic extractant of about 0.1 to 0.5 M tricaprylmethyl ammonium nitrate in an inert water-immiscible aromatic or aliphatic hydrocarbon diluent whereby the neptunium and plutonium values are selectively extracted from the feed solution; and,

(g) contacting the extractant with a formic acid strip solution about 0.5 to 1.5 M in formic acid, whereby the neptunium and plutonium values are stripped from the extractant, thereby recovering the neptunium and plutonium values.

Actinides from Na_2CO_3 Scrub Solutions

Efficient removal of actinides and fission products from sodium carbonate scrub waste solutions presents problems, e.g., neutralization of the carbonate solutions with HNO_3 followed by cation exchange results in poor metal ion absorption on the resin, precipitate formation of some of the metal ions as complexes of the degradation products and extensive column plugging. Acidification of the carbonate solutions with excess HNO_3 followed by extraction with TBP (tri-n-butyl phosphate) or preferably DHDECMP (dihexyl-N,N-diethyl carbamylmethylene phosphonate), results in rapid buildup of high concentrations of acid degradation products which prevent efficient back-extraction. Thus, there is no effective recovery of actinide values from the scrub waste solutions for utilization or long-term storage.

E.P. Horwitz and G.W. Mason; U.S. Patent 4,208,377; June 17, 1980; assigned to the U.S. Department of Energy eliminate many of the above problems in rendering actinide values recoverable from sodium carbonate scrub waste solutions.

It was found that by extracting the radiolytic and hydrolytic degradation products away from the scrub solutions, it becomes relatively easy to recover the actinide values from the waste solution for further processing or storage. Thus, for rendering the actinide values recoverable, the sodium carbonate scrub waste solution containing the actinide and other values and radiolytic and hydrolytic degradation products from neutral organo-phosphorus extractants is made acidic with mineral acid to form a feed solution. The feed solution is then contacted with a water-immiscible highly polar organic extractant which selectively extracts the radiolytic and hydrolytic degradation products away from the feed solution while the actinide values remain in the feed solution, and the feed solution is separated from the organic extractant. The actinide values are readily recoverable by evaporating the water from the feed solution for processing or storage or the solution may be recycled back into the high-level waste process stream.

The process is advantageous in that the water-immiscible highly polar extractant may be any of a number of readily available relatively inexpensive alcohols, carboxylic acids or ketones easily recovered, purified and recycled. For example, the degradation products are readily stripped by contacting the extractant with a sodium carbonate solution. The extractant may then be recycled while the carbonate solution containing the degradation products is concentrated and stored, for example, by incorporation into concrete. The recovered actinide and fission product values are preferably recycled back into the normal high-level liquid waste process stream for handling and ultimate recovery or disposal with the actinide values already present in the stream eliminating the need for a separate process stream.

Example: The molar composition of a waste scrub solution was: Na_2CO_3 0.21, $NaHCO_3$ 0.038, $NaNO_3$ 0.38, HDBP (dibutyl phosphoric acid) 0.015, H_2MBP (monobutyl phosphoric acid) 0.005, U(VI) 0.0075, Pu(IV) trace, Am(III) trace.

To 2.5 ml of the above waste scrub was added 2.65 ml of 8 M HNO_3 to prepare 5.31 ml of a feed solution 4.0 M in HNO_3. The feed solution was then contacted with 3.2 ml of 2-ethyl-1-hexanol extractant which had been equlibrated with HNO_3 for several minutes, allowed to settle, and separated. The loaded extractant was contacted with 1.1 ml of 1.5 M HNO_3-0.05 M oxalic acid to scrub any actinides which may have been coextracted. The nitric acid scrub was added to the carbonate solution forming about 6.5 ml of aqueous raffinate.

The extractant was first contacted with 1.6 ml water to wash the nitric acid from the extractant and then with 6.4 ml 0.05 M Na_2CO_3 to strip away the HDBP and H_2MBP degradation products. Analysis of the raffinate showed that six extractant and six scrub stages would give a decontamination factor from all actinides of about 10^5, from the degradation product H_2MBP of 10^3 and from HDBP of $>10^6$.

Oxalate Precipitation of Actinide Elements

D.O. Campbell and S.R. Buxton; U.S. Patent 4,025,602; May 24, 1977; assigned to the U.S. Energy Research and Development Administration describe a two-step process for separating actinide values from a nitric acid nuclear fuel reprocessing waste stream containing actinide values, lanthanide values, and other metal values.

It comprises a first partitioning step to provide a trivalent fraction enriched in actinide and lanthanide values and a first waste product substantially free of actinide values, and an actinide/lanthanide partitioning step in which the trivalent fraction is partitioned to provide an actinide waste product and a second waste product substantially free of actinides.

In this improved process, the first partitioning step comprises: adjusting the nitric acid concentration of the reprocessing waste to about 0.1 to 1.0 M; in a precipitation zone, contacting the reprocessing waste with an excess of a source of oxalate ions to cause a major portion of the actinide and lanthanide values to precipitate as solid oxalates providing a supernate solution of the remaining actinide and lanthanide values and the major portion of the other metal values in the presence of dissolved oxalate; and separating oxalate precipitate from the precipitation zone to provide a trivalent fraction for the actinide/lanthanide partitioning step. The supernate solution is contacted with a sufficient quantity of strong acid ion exchange resin to cause the loading of actinide and lanthanide values onto the resin, providing a raffinate substantially free of actinide. Then the actinide and lanthanide values are eluted from the loaded resin with 3 to 6 M HNO_3 to provide an eluate solution enriched in actinide and lanthanide values.

The eluate stream may be concentrated and actinide and lanthanide values therein may be combined with the trivalent fraction prior to the actinide/lanthanide partitioning step. Alternatively, actinide and lanthanide values from the eluate can be combined with additional reprocessing waste for recycle. One embodiment of the method involves the addition of a complexant such as nitrogen oxides, soluble nitrite, hydroxylamine, etc. to the reprocessing waste to cause the complexation of ruthenium values prior to oxalate precipitation.

Example: Reprocessing waste from a Purex plant comprises 5,900 liters 2.4 M HNO_3 containing americium and curium values (Ac); rare earth values (RE), barium and strontium values (Ba, Sr) and other fission product values (FP). This reprocessing waste is combined with sugar to decompose HNO_3 and is evaporated to provide a precipitation feed of 2,000 liters of 2 M HNO_3. Off-gases comprising HNO_3, nitrogen oxides and CO_2 are condensed and passed to a HNO_3 recovery system where they are condensed to recover HNO_3 and water.

To the precipitation feed in a precipitation zone is added the precipitant, 750 liters of oxalic acid, 0.8 M, and N_2O_2 is bubbled through to complex ruthenium. In addition, 3,250 liters H_2O is added as a wash and diluent to provide a HNO_3 concentration of 0.67 M for the precipitation. Under these conditions, about 95% of the Ac, RE content precipitates as an Ac-RE solid along with about 48% of the Ba, Sr content. The precipitation zone should be cooled to enhance solids recovery. This Ac-RE solid is a trivalent fraction enriched in actinide and lanthanide values. The Ac-RE solid is contacted with strong refluxing nitric acid, about 12 M, to destroy oxalate, evaporated and dissolved in 1,800 liters 1.0 M glycolic acid to provide a feed for subsequent Ac-Re partitioning.

Figure 6.6: Flow Diagram of Process to Recover Actinide Values from Nuclear Fuel Waste Stream

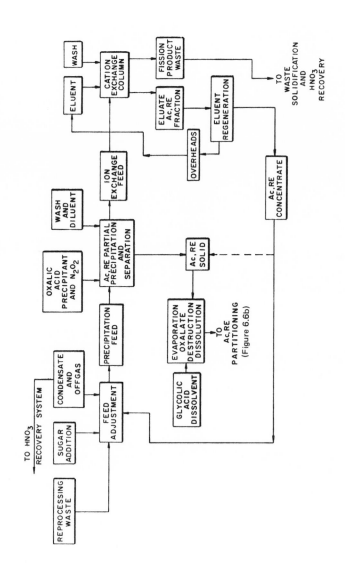

(a) First partitioning step.

(continued)

Figure 6.6: (continued)

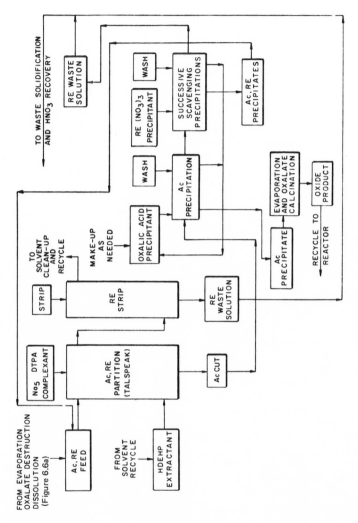

(b) Partitioning of trivalent fraction.

Source: U.S. Patent 4,025,602

The supernate from the partial precipitation is 6,000 liters 0.67 M HNO_3-0.1 M in oxalate ion and contains about 5% of the original Ac-RE content, about 48% of the original Ba-Sr content and about 100% of FP content and makes up the ion exchange feed. The ion exchange feed is passed through an ion exchange column containing 50 liters of Dowex-50 ion exchange resin followed by a 50-liter wash with 0.01 M HNO_3. It is preferred that the ion exchange step on an industrial scale be carried out at about 70° to 80°C to enhance flow rates through the column and reduce pressure. The solution passing through the column is the fission product waste and constitutes a first waste product substantially free of actinides. This fission product waste consists of 6,050 liters of 0.67 M HNO_3, containing about 0.02% of original Ac-RE content, about 48% of original Ba-Sr content and about 100% of the original FP content, and is suitable for subsequent waste solidification and HNO_3 recovery by conventional means.

The column is then eluted with 500 liters 4 M HNO_3 and washed with 50 liters of 0.01 M HNO_3 to provide the Ac-RE fraction consisting of 550 liters 3.64 M HNO_3 and containing about 5% of the original Ac-RE content and about 4% of the original Ba-Sr content. The Ac-RE fraction is then evaporated partially to provide about 530 liters HNO_3 for solvent recycle and about 20 liters of about 14 M HNO_3 which may either be combined with the reprocessing waste to provide precipitation feed or combined with the recovered Ac-RE solid (indicated by the dotted line in Figure 6.6a) to aid in oxalate destruction. If recycled to make up the precipitation feed, the Ac-RE contents of successive fractions will be proportionately increased and if added to Ac-RE solid precipitate, the Ac-RE feed for the actinide/lanthanide partitioning will then be practically 100% of original Ac-RE content of the waste.

As mentioned, the product of the glycolic acid dissolution is the Ac-RE feed for the actinide/lanthanide partitioning step shown in Figure 6.6b and comprises a 1,800-liter 1.0 M glycolic acid solution containing Ac-RE values and about 48% of the original Ba-Sr content. The Ac-RE feed is contacted with a complexant of 1,800 liters of 1.0 M glycolic acid (aqueous) 0.05 M in sodium diethylenetriaminepenta-acetate (Na_5DTPA) and the resulting solution contacted with an extractant consisting of 2,700 liters di(2-ethylhexyl)phosphoric acid (HDEHP) 0.8 M in diethyl-benzene. The aqueous phase is the Ac cut, and the RE content is stripped from the organic phase with about 1,200 liters 6 M HNO_3 to provide a RE waste solution of 1,200 liters 6 M HNO_3 containing about 99.8% of the RE content of the Ac-RE feed and less than 0.01% of the actinide content. The RE waste fraction provides a second waste fraction substantially free of actinides.

The Ac cut is 3,600 liters of 1.0 M glycolic acid and contains about 99.99% of Ac content in Ac-RE feed, and is 0.125 M in Na^+. The Ac fraction is then contacted with 900 liters 0.93 M oxalic acid to precipitate about 99% of the Ac content to provide an Ac precipitate which is evaporated and calcined to provide an actinide waste product which is an oxide and may then be fabricated into fuel and recycled to a reactor.

The supernate from the Ac precipitation is then contacted with excess RE nitrate solution in successive scavenging precipitations to precipitate residual Ac values which are recovered and recycled to the Ac-RE feed. After at least two scavenging precipitations, the precipitation supernates are reduced in Ac content to less than about 0.001% of the Ac content of the feed and are suitable for waste solidification and HNO_3 recovery.

As shown in the example and experimental demonstration, the column capacity required per ton of reprocessed fuel is greatly decreased. The reduced loading of fission products on the column substantially decreases the quantity of interfering ions entering the actinide/lanthanide partitioning step thereby providing an ultimate actinide product reduced in metal impurities.

RECOVERY OF FISSION PRODUCT METALS

Ruthenium Recovery by Solvent Extraction

Conventional processes for the reprocessing of irradiated nuclear fuel elements generally involve a first stage of dissolving the nuclear fuel by means of nitric acid, which leads to the obtainment of a nitric solution containing not only uranium and plutonium but also certain fission products, particularly ruthenium contained in by no means negligible quantities in irradiated fuels. Thus, 1 ton of irradiated uranium in a light water nuclear reactor contains approximately 2,300 g of ruthenium.

Thus considerable interest is attached to the recovery of ruthenium not only in order to ensure a satisfactory purification of the uranium and plutonium, but also due to the catalytic and photochemical metallurgical properties of ruthenium which make it an element sought in numerous applications. However, in irradiated fuel processing processes the recovery of ruthenium causes certain problems because during the first uranium and plutonium extraction cycle in an organic solvent such as tributyl phosphate the ruthenium fraction which is also extracted in this solvent is only reextractable with difficulty and the solvents used consequently have a high residual activity.

R. Fitoussi, S. Lours and C. Musikas; U.S. Patent 4,282,112; August 4, 1981; assigned to Commissariat a l'Energie Atomique, France describe a process for the recovery of ruthenium present in an aqueous nitric solution characterized in that the ruthenium is extracted in an organic solvent by bringing the nitric solution into contact with an organic phase comprising an organo-phosphorus compound having at least one electron donor sulfur atom in the presence of a compound able to displace the NO^+ ions of the ruthenium complexes present in the nitric solution.

According to the process, the organo-phosphorus compound is advantageously a dialkyl-dithiophosphoric acid such as di-(2-ethyl-hexyl)dithiophosphoric acid.

The process has the advantage of leading to a quantitative recovery of the ruthenium present in a nitric solution obtained, for example, by dissolving irradiated nuclear fuel elements. Thus, in this case, the uranium(VI) and the plutonium are not in practice extracted in the organic solvent, so that it is possible to recover the ruthenium with satisfactory yields.

The compound which is able to displace the NO^+ ions of the ruthenium complexes present in the nitric solution is advantageously sulfamic acid or hydrazine. When sulfamic acid is used, the latter is preferably added to the aqueous phase in a quantity such that the sulfamic acid concentration of the aqueous phase is between 0.05 and 1 M/ℓ. Advantageously, extraction takes place at a temperature between 20° and 80°C, preferably at 70°C.

Example: This example relates to the recovery of ruthenium from a 3 N nitric acid solution containing $2.6 \cdot 10^{-3}$ M/ℓ of radioactive ruthenium in the form of mixed nitratoruthenium and nitrosoruthenium complexes at valences II, III and IV. In the example, an organic solvent constituted by di(2-ethyl-hexyl)dithiophosphoric acid (DEHDTP) diluted in dodecane is used for extracting the ruthenium, the DEHDTP concentration of the organic phase beng 0.5 M/ℓ.

Extraction is carried out in an apparatus which is thermostatically controlled by water circulation by bringing into contact within the apparatus 20 cc of aqueous phase and 20 cc of organic phase and by stirring the two phases in the presence of at least one rotary bar magnet.

During extraction, the phases present are sampled accompanied by stirring and after separating them by centrifuging, each of them is analyzed by gamma spectrometry in order to determine their respective ruthenium concentrations. This makes it possible to calculate the distribution or partition coefficient D of the ruthenium which is equal to the ratio of the ruthenium concentration of the organic phase to the ruthenium concentration of the aqueous phase.

In a first series of experiements, this extraction is carried out by using a solvent after adding to the aqueous phase 0.25 M/ℓ of sulfamic acid and extraction is performed at 20°, 50° and 70°C, while determining the ruthenium partition coefficient as a function of time for each temperature.

The results obtained are first given in Figure 6.7a, which shows the variations in the partition coefficient D of ruthenium as a function of time in minutes for extractions carried out, respectively, at temperatures of 70°C (curve I), 50°C (curve II) and 20°C (curve III). It can be seen that the partition coefficient increases with temperature, the best results being obtained when extraction is performed at 70°C.

In a second series of experiments, ruthenium extraction is carried out either at 70° or at 50°C from aqueous phases having different sulfamic acid concentrations, while once again determining in each case the partition coefficient D of the ruthenium as a function of time.

The results are given in Figure 6.7b, which illustrates the variations in the partition coefficient D as a function of time for extraction processes carried out at 70°C with sulfamic acid concentrations of 0.5 M/ℓ (curve I) and 0.01 M/ℓ (curve III) and for extraction processes carried out at 50°C with sulfamic acid concentrations of 0.4 M/ℓ (curve II) and in the absence of sulfamic acid (curve IV). It can be seen that the partition coefficient D is low in the absence of sulfamic acid (curve IV) and that the addition of sulfamic acid makes it possible to improve this partition coefficient. Moreover, it has been found that on the basis of a sulfamic acid concentration of 0.4 M/ℓ, the ruthenium extraction kinetics substantially do not change.

In a third series of experiments, ruthenium extraction is carried out by adding hydrazine to the aqueous phase extraction being performed at 70°C. The results obtained are given in Figure 6.7c, which shows the variations in the partition coefficient D of ruthenium as a function of time for extractions carried out on the basis of aqueous phases having, respectively, hydrazine concentrations of 0.1 M/ℓ (curve I) and 0.01 M/ℓ (curve II). It can be seen that hydrazine also leads to an improvement in the partition coefficient D, but that it is less effective than sulfamic acid.

Figure 6.7: Variation of Partition Coefficient D of Ruthenium as Function of Time

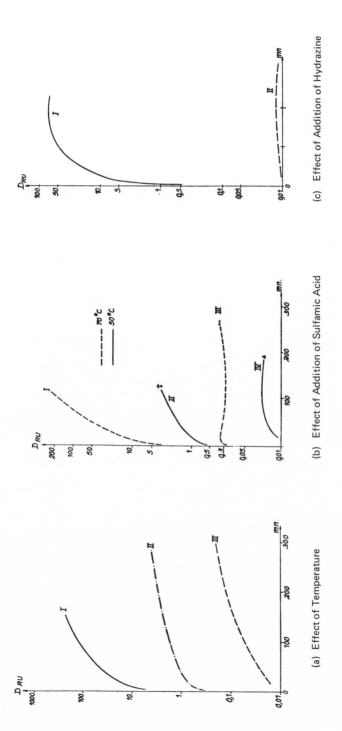

(a) Effect of Temperature

(b) Effect of Addition of Sulfamic Acid

(c) Effect of Addition of Hydrazine

Source: U.S. Patent 4,282,112

Removal of Ruthenium by Reduction with Hydrazine

Y. Berton and P. Chauvet; U.S. Patent 4,116,863; September 26, 1978; assigned to Commissariat a l'Energie Atomique, France describe a method of decontamination of a radioactive effluent containing at least ruthenium which essentially consists in forming a precipitate of cobalt sulfide in the effluent to be treated.

Thus, when the method is carried into effect for the decontamination of an effluent which results from reprocessing of fuels, that is to say, which contains especially strontium, cesium, ruthenium and antimony, precipitates of nickel ferrocyanide, of cobalt sulfide and of barium sulfate are formed in the effluent and the sludges thus obtained are then separated. In the event that the treatment is carried out in a tank made of material which does not afford resistance to corrosion by the sulfide ions in an acid medium, the abovementioned precipitates are formed by successively adding to the effluent sulfuric acid, nickel ferrocyanide, sodium hydroxide, ammonium sulfide, cobalt sulfate or nitrate and barium nitrate.

A preferential mode of execution of the method for decontamination of an effluent of the type mentioned above consists in adding the following constituents to the effluent: sulfuric acid in a proportion of 3,000 to 18,000 mg/ℓ of sulfate ions; nickel ferrocyanide in the form of a suspension of preformed colloidal precipitate corresponding to doses of 300 mg/ℓ of ferrocyanide ions and 100 mg/ℓ of nickel ions; a quantity of sodium hydroxide which makes it possible to adjust the pH of the effluent to a value of the order of 8.5, ammonium sulfide in a proportion of 200 mg/ℓ of sulfide ions; cobalt nitrate or sulfate in a proportion of 250 mg/ℓ of cobalt ions; and barium nitrate in a proportion of 1,500 to 2,000 mg/ℓ of barium ions.

In accordance with a first preferential arrangement of the process, it is possible to improve the entrainment of the ruthenium contained in the effluent even further in order to achieve a greater reduction in residual activity of the effluent after treatment, this residual activity being mostly due to the ruthenium in a proportion of 80 to 90%.

In accordance with this preferential arrangement, the method essentially consists in subjecting the effluent to be treated to a reduction process prior to formation of the precipitate of cobalt sulfide, the reduction being preferably carried out by addition of hydrazine or of a hydrazine salt to the effluent.

It can be noted that the preliminary reduction of the effluent by hydrazine or a hydrazine salt makes it possible to increase the entrainment of the ruthenium to an even greater extent and a further advantage of the reducing agent lies in the fact that it does not produce sludges.

It is preferable in this mode of operation to add 50 to 500 mg/ℓ of hydrazine or hydrazine salt to the effluent.

In accordance with a second preferential arrangement of the process which is more particularly directed to the decontamination of radioactive effluents containing at least ruthenium and antimony, it is possible to ensure satisfactory removal of the antimony contained in the effluent and also to achieve a further improvement in ruthenium decontamination of the effluent.

In accordance with this second arrangement which may or may not be associated with the first arrangement relating to the reduction pretreatment, the method essentially consists in additionally forming in the effluent to be treated a precipitate of a hydroxide of an element of column IV-a of the periodic table which is preferably selected from the group comprising titanium, thorium and zirconium and in separating out the sludges thus obtained.

Recovery of Palladium and Technetium Using Tricaprylmethylammonium Nitrate

E.P. Horwitz and W.H. Delphin; U.S. Patent 4,162,231; July 24, 1979; assigned to the U.S. Department of Energy describe a method for recovering palladium and technetium values from a nitric acid nuclear fuel reprocessing waste solution containing these and actinide, rare earth and fission product values. The method comprises: (a) adjusting the nitric acid concentration of the solution to 0.5 to 3.0 M; (b) contacting the solution with an extractant of 0.05 to 0.5 M tricaprylmethylammonium nitrate ($TCMA \cdot NO_3$) in an inert water-immiscible aromatic or aliphatic hydrocarbon diluent whereby the palladium and technetium values are selectively extracted from the solution; and, (c) contacting the extractant containing these values with an aqueous solution 4 to 8 M in nitric acid whereby the palladium and technetium values are selectively stripped from the extractant, thereby recovering the palladium and technetium values.

Example 1: A synthetic waste solution was prepared by mixing nitric acid solutions of salts of nonradioactive isotopes of fission products and rare earths. The quantities of fission product elements used were for liquid light water reactor fuel irradiated to 33,000 Mwd/metric ton of heavy metal. The products from 1 metric ton of the fuel are assumed to be present in 5,600 ℓ of a 2.9 M HNO_3 (HAW waste stream) or 5,900 ℓ of a 2.4 M HNO_3 (EEW waste stream from exhaustive tributyl phosphate extraction of the HAW waste stream). Separate portions of the synthetic waste were spiked with palladium and technetium for testing.

A countercurrent extraction process was set up using 0.1 M $TCMA \cdot NO_3$ in DEB (diethylbenzene), synthetic EEW waste solution which was 2.4 M in nitric acid and a scrub of 0.44 M HNO_3. The temperature was 25°C. The phase ratio of feed:organic scrub was 1.0:1.43:0.43. After four extractions and two scrub stages the extractant contained 98.0% of the palladium and 99.9% of the technetium present in the feed.

Example 2: The extractant from Example 1 containing the palladium and technetium values was contacted with 8 M HNO_3 strip solution at 25°C with an organic: aqueous phase ratio of 1:1. After four stages of contact, about 4.0% of the palladium remained in the extractant, as did about 7% of the technetium. This gave an overall palladium recovery from the feed solution of about 94%.

Recovery of Technetium, Rhodium and Palladium from Acid Waste

W.W. Carlin, W.B. Darlington and D.W. Dubois; U.S. Patent 3,891,741; June 24, 1975; assigned to PPG Industries, Inc. describe a process whereby fission products, e.g., technetium, rhodium and palladium, are recovered by treatment of an aqueous acidic waste stream produced by nuclear fuel processors. More specifically, the highly acidic aqueous waste stream, Purex acid waste (PAW), which generally has an acid concentration of about 8 molar, is filtered to recover undissolved rhodium and the filtrate treated with alkaline reagent until its acid concentration is reduced to about 1 molar or less, preferably about 0.25 molar.

Alternatively, the waste stream can be partially neutralized first and then filtered to recover the rhodium metal. The partially neutralized liquid (filtrate) is electrolyzed in a first electrolytic cell under controlled cathodic potential conditions and at a potential at which technetium is deposited upon the cathode. At such a potential, palladium and rhodium, as well as ruthenium, will also be deposited on the cathode. Following electrolysis, the electrolytic cell liquor, depleted of the aforementioned metals, is returned to the nuclear fuel processor.

The metals deposited on the cathode are removed, either mechanically or chemically, e.g., by dissolution with acid. Rhodium, which is substantially insoluble in acid, is removed as a solid from the resulting acid solution by filtration and purified. Purification can be accomplished by dissolving the rhodium by alkali metal, e.g., sodium or potassium, bisulfate fusion techniques and electrolysis of an acidic, aqueous solution prepared from the melt.

In one embodiment of the process (Figure 6.8), the acid solution of metals removed from the cathode of the first electrolytic cell is treated with a strong oxidizing agent such as perchloric acid and distilled to thereby separate ruthenium and technetium as volatile oxides from the palladium. The remaining aqueous, acidic solution containing palladium is electrolyzed under controlled cathode potential conditions to recover palladium as a deposit on the cathode. This deposit can be removed from the cathode and consolidated.

The distillate from the aforementioned distillation, which contains technetium and ruthenium, is also electrolyzed under controlled cathodic conditions to deposit technetium and ruthenium on the cathode and to recover the strong oxidizing agent, e.g., perchloric acid, for reuse. Following removal of the technetium and ruthenium from the cathode, an alkaline solution of the metals is subjected to organic solvent extraction techniques to separate technetium from ruthenium. Ruthenium is returned to the nuclear fuel processor and technetium separated from the organic solvent by steam distillation and recovered as elemental technetium or as a metal salt product.

Example: *Recovery of fission product metals* — A simulated acidic waste solution of fission products (PAW) is prepared with 8 molar nitric acid. The waste solution contains 15 g/ℓ of ruthenium, 10 g/ℓ of palladium, 3 g/ℓ of rhodium and 5 g/ℓ of technetium. 100 ml of the simulated acid waste solution is filtered through a 0.8 μ Millipore filter to remove any rhodium solids that are present. The filtrate is partially neutralized with 50% sodium hydroxide so that the resulting solution is approximately 0.2 M in nitric acid (pH 1). This solution is poured into a 150 ml glass beaker cell equipped with a platinum-coated titanium anode and a titanium cathode installed on a magnetic stirrer.

Electrolysis of the 0.2 M nitric acid solution is conducted at a controlled cathode potential of –0.40 volt (versus a silver/silver chloride reference electrode) for about forty-eight hours. The current density on the cathode is initially 70 mamp per cm^2. At the end of electrolysis, the current density is 0.3 mamp/cm^2. About 99% of the palladium, rhodium and technetium and about 60% of the ruthenium in the simulated solution is plated onto the cathode. The spent electrolyte is discarded and the cathode washed with 10 ml of deionized water. The metal deposit on the cathode is anodically stripped from the cathode with 30 ml of 5 M nitric acid. The palladium, ruthenium and technetium in the metal deposit are dissolved in the acid, while the rhodium remains as small solid flakes in the resulting solution.

Figure 6.8: Recovery of Palladium, Rhodium and Technetium by Electrolysis

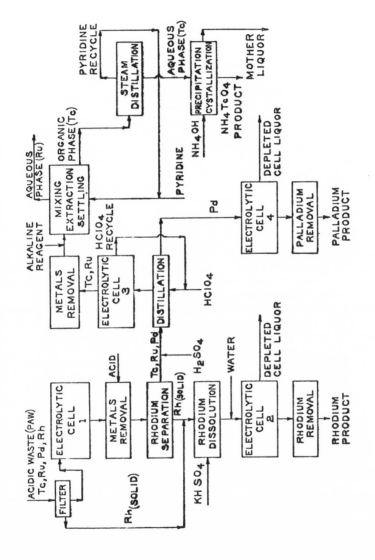

Source: U.S. Patent 3,891,741

Separation of rhodium — The acid solution of anodically stripped metals is filtered through an 0.8 μ Millipore filter to remove the rhodium solids. These solids are found to contain a trace amount of ruthenium contamination. 2.5 mg of the rhodium solids are mixed with 5 g of potassium bisulfate and heated for 30 minutes of red heat (500°C). On cooling, 50 ml of deionized water is added to the solid fusion product to dissolve it.

The resulting aqueous solution is charged to an electrolytic beaker cell having a platinum-coated titanium anode and a titanium cathode and being similarly constructed to the first-mentioned beaker cell and electrolyzed at a controlled cathode potential of –0.31 volt (versus a silver/silver chloride reference electrode) for 24 hours. The maximum current density is 2 mamp/cm^2. About 96% of the rhodium in the aqueous solution is deposited on the cathode. The rhodium deposit is recovered by draining the electrolyte from the cell, washing the cathode with 5 ml of deionized water and mechanically stripping the rhodium metal deposit from the cathode.

Separation of palladium — The filtrate obtained from filtering the acid solution of anodically stripped metals is mixed with 10 ml of concentrated sulfuric acid and the mixture distilled in a 250-ml, 3-necked still pot equipped with a Friedrich condenser and a 150-ml receiver. The still is operated at a head temperature of 120°C until no more nitric acid is carried over into the receiver. The nitric acid distillate is put aside for reuse. The still pot is cooled and 50 ml of deionized water and 25 ml of 72% perchloric acid is added to the pot. The perchloric acid mixture in the pot is distilled until the appearance of SO$_3$ fumes, which indicates that all of the technetium and ruthenium, as well as the perchloric acid, have been distilled.

The pot residue, which contains palladium, is cooled, diluted with deionized water to a volume of 80 ml and charged to an electrolytic beaker cell having a platinum-coated titanium anode and a titanium cathode. The cell is constructed similarly to that of the first-mentioned beaker cell. The aqueous palladium-containing solution is electrolyzed in the cell at a cathode potential of –0.04 volt (versus a silver/silver chloride reference electrode) until the current decreases to 0.002 amp. Palladium metal is recovered from the cathode by draining the electrolyte from the cell, washing the cathode with 10 ml of ddeionized water and removing the palladium deposit mechanically from the cathode. Palladium recovery is 96.9%.

Separation of technetium — The perchloric acid distillate (70 ml) is charged to an electrolytic beaker cell of the same type as described above and the distillate electrolyzed at a cathode potential of –0.4 volt (versus a silver/silver chloride reference electrode). Electrolysis is continued for 24 hours at an initial current density of 70 mamp/cm^2. The current density at the end of the electrolysis is 0.3 mamp/cm^2. The perchloric acid electrolyte is drained from the cell and saved for reuse. The cathode is washed with 10 ml of deionized water and then anodically stripped into 30 ml of 1 M nitric acid. The technetium recovery is 99% and the ruthenium recovery is 57%.

The acid solution containing technetium and ruthenium is made alkaline with solid sodium hydroxide and diluted to 50 ml. The alkaline solution is extracted with two 50-ml portions of 2,4-dimethylpyridine in a separatory funnel and 96% of the technetium recovered in the organic phase. The organic phase is steam distilled in a 250-ml 3-necked flask equipped with a Friedrich condenser.

Figure 6.9: Recovery of Technetium Using Flocculant and Electrolysis

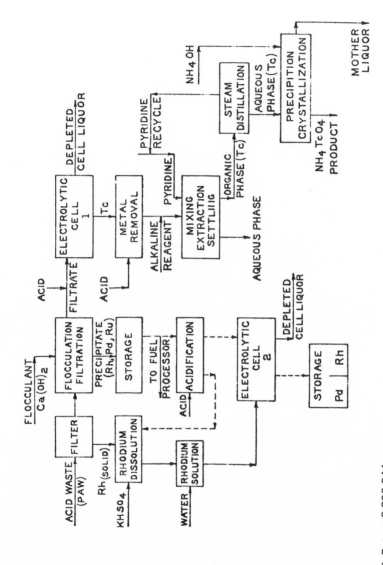

The organic phase distills over at 97°C leaving an aqueous phase containing technetium free of ruthenium, palladium and rhodium. The technetium is recovered as a salt by mixing ammoniacal solution with the aqueous phase and crystallizing out ammonium pertechnetate at 0°C.

In a similar process, *W.W. Carlin; U.S. Patent 3,890,244; June 17, 1975; assigned to PPG Industries, Inc.* describes the recover of technetium from aqueous, acidic waste solutions thereof. The acidic waste solution is mixed with a flocculant, e.g., an alkaline earth metal hydroxide or oxide, to precipitate certain fission products, e.g., palladium, rhodium and ruthenium, contained in the acidic waste solution. Technetium remains in solution and in the resulting supernatant alkaline aqueous phase.

The supernatant alkaline aqueous phase is made acidic and electrolyzed in an electrolytic cell under controlled cathodic potential conditions to deposit technetium on the cathode. Elemental technetium is removed from the cathode. Technetium is separated from other plated fission product metals by extraction from an alkaline solution with an organic extractant such as pyridine having affinity for technetium. Technetium is separated from the organic extractant by steam distillation and the resulting aqueous phase treated with ammoniacal reagent to precipitate technetium as ammonium pertechnetate.

The precipitate from the flocculation step is removed to an interim storage vessel from whence it can be returned to the nuclear fuel processor for disposal. In a further embodiment, the precipitate is acidified to form an aqueous acidic solution of fission product metal values and the solution electrolyzed in an electrolytic cell under controlled cathodic potential conditions and at a potential sufficiently negative to plate out from the solution those fission product metals desired. The metal deposit is stripped from the cathode and stored until its radioactivity has diminished. The metal-depleted cell liquor is returned to the nuclear fuel processor.

Figure 6.9 is a flow diagram of the above process.

Recovery of Palladium, Rhodium and Technetium from Alkaline Waste

W.W. Carlin and W.B. Darlington; U.S. Patent 3,922,231; November 25, 1975; assigned to PPG Industries, Inc. describe a process for separating palladium, rhodium and technetium values from an alkaline aqueous solution containing the metal values in the form of anions as well as other fission product metal values, the aqueous solution resulting from processing irradiated nuclear fuel, which comprises, in combination, the steps of:

    (a) contacting the alkaline aqueous solution with an anion exchange resin to adsorb palladium, rhodium and technetium values thereon;

    (b) eluting adsorbed palladium, rhodium and technetium values from the anion exchange resin to obtain an eluate containing the eluted metal values;

    (c) separating technetium as its volatile oxide from an aqueous solution containing the eluted metal values while retaining eluted palladium and rhodium values in the solution and collecting the thus-separated technetium oxide;

Figure 6.10: Process to Recover Palladium, Rhodium and Technetium from Alkaline Waste

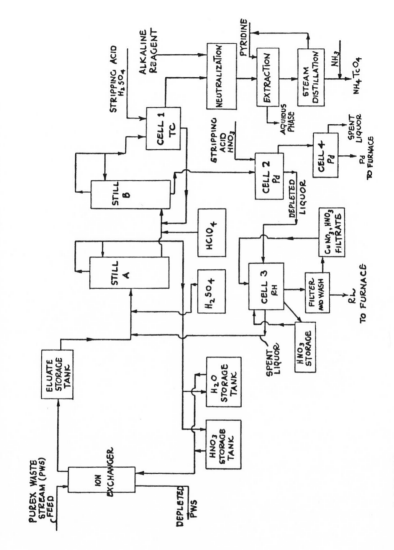

Source: U.S. Patent 3,922,231

(d) subjecting an acidic solution of the collected technetium to an electrolyzing current under controlled cathodic potential conditions sufficient to deposit technetium on the cathode;

(e) subjecting a substantially technetium-free acidic solution of the eluted palladium and rhodium values to an electrolyzing current under controlled cathodic potential conditions sufficient to deposit palladium on the cathode but insufficient to deposit thereon any substantial quantity of rhodium; and,

(f) subjecting the palladium-depleted solution of step (e) to an electrolyzing current under controlled cathodic potential conditions sufficient to deposit rhodium on the cathode.

Figure 6.10 is a flow diagram illustrating the above process.

Example: 1,800 ml of aged alkaline Purex waste stream from the U.S. Atomic Energy Commission installation at Hanford, Washington was passed sequentially through two 12-ml anion exchange columns packed with Amberlite IRA-938, a macroreticular anion exchange resin, at a flow rate of 3 to 5 ml/min. The feed stream contained 32 mg of palladium, 29 mg of rhodium, 22 mg of ruthenium and 52 mg of technetium. After the 1,800-ml charge of Purex feed had passed through both columns, the two columns were washed separately with two bed volumes (BV) of 0.1 M nitric acid and one BV of 0.25 M nitric acid. The columns were then each eluted sequentially with two BV of 3.0 M HNO_3, 9 BV of 6.0 M HNO_3, 1 BV of 8.0 M HNO_3 and 1 BV of distilled water. The eluants from the two columns were combined and 10 ml of concentrated sulfuric acid added. The resulting solution was distilled to remove water and nitric acid sequentially.

The pot residue was transferred to a second distillation flask and 15 ml of 70% perchloric acid was added. The perchloric acid was distilled from the distilling flask until sulfur trioxide fumes began to appear. The distillation was stopped and the pot centers were allowed to cool. An additional 15 ml of 70% perchloric acid was added and the distillation repeated.

The contents of the distilling flask were washed into a beaker and diluted to about 100 ml. Palladium was deposited from this solution on a platinum cathode by electrolysis at a cathodic potential of -0.04 volt versus a standard silver/silver chloride reference electrode. An anode of the laboratory electrolytic cell was platinum.

Upon completion of the palladium deposition, the platinum cathode was removed and a copper-plated platinum cathode substituted. The cathode voltage was then changed to -0.31 volt versus the silver/silver chloride reference electrode and the rhodium was deposited on the cathode. The combined perchloric acid distillates were charged to an electrolytic cell wherein technetium and ruthenium are deposited upon a titanium cathode at a controlled cathodic potential of -0.4 volt versus the silver/silver chloride reference electrode. The anode of this laboratory cell was also platinum.

Recovery of Palladium and Cesium Using Chloride Ions

D.O. Campbell; U.S. Patent 3,979,498; September 7, 1976; assigned to the U.S. Energy Research and Development Administration describes a method for recovering Cs and Pd values from fission product waste solution comprising oxidizing Pd ions present in the solution to Pd^{+4}, contacting the solution with a source of chloride ions to precipitate the Cs and Pd values as Cs_2PdCl_6 and recovering the precipitate.

As a preferred embodiment to minimize handling and the addition of material to the waste, the waste solution which leaves the first extraction column in the fuel processing facility is held for several days to allow very highly active short-lived wastes to decay. The solution is then electrolytically oxidized by passing sufficient current through the solution to oxidize the Pd^{+2} present to Pd^{+4}. Sufficient HCl is added at this time to precipitate Cs fission product as Cs_2PdCl_6. This compound contains the valuable ^{137}Cs isotope as well as ^{133}Cs, ^{134}Cs, and ^{135}Cs. Because chloride is associated with other ionic species in the solution, an excess of about 50% HCl with respect to $PdCl_6^{-2}$ ion formation is sufficient. Because $CsPdCl_6$ is slightly soluble, the precipitate recovery can be improved by adding additional (or recycle) Pd to the solution, lowering the temperature, or other well-known methods of depressing the solubility of a slightly soluble precipitate.

The precipitate, which now contains most of the Cs and Pd typically present, is separated centrifugally, dried and ready for use as a gamma source. Alternately, the Pd may be separated from the Cs by reducing the Pd^{+4} with conventional reducing agents and precipitation with hydroxide.

Because Pd is normally present in fission product waste in stoichiometric excess (with respect to Cs_2PdCl_6 formation), Pd recovery may be increased by adding additional stable Cs and (if needed) Cl ions to the waste solution remaining after recovery of the initial precipitate. This will cause the precipitation of additional Cs_2PdCl_6 which will contain some of the ^{137}Cs which had previously remained in solution due to the slight solubility of the initial precipitate. Therefore, by contacting the fission product waste solution with a source of stable Cs ions, additional Cs_2PdCl_6 will precipitate and may be recovered, thereby increasing both the Pd recovery and the ^{137}Cs removal efficiency.

MISCELLANEOUS RECOVERY PROCESSES

Cobalt Recovery Through Ion Exchange

K. Fujita, S. Takeuchi, H. Yamashita and F. Nakajima; U.S. Patent 4,178,270; December 11, 1979; assigned to Hitachi, Ltd., Japan describe a process characterized by supporting a hydrous metal oxide on a carrier under such a condition that the hydrous metal oxide and the carrier have zeta potentials of opposite polarities, and more particularly by contacting a carrier with a solution containing a metal salt capable of forming a hydrous metal oxide by hydrolysis, adjusting a solution to a pH range of making a zeta potential of resulting anhydrous metal oxide and that of the carrier having different polarities from each other, and conducting hydrolysis of the metal salt, thereby depositing the hydrous metal oxide resulting from the hydrolysis as such on the surface of the carrier, while establishing a strong bondage between the deposited hydrous metal oxide and the carrier.

The iorganic ion exchanger thus prepared is used for recovery of useful metals, especially uranium in seawater, and recovery of cobalt from water in nuclear reactor, and removal of heavy metal ions from industrial effluent water.

Zeta potential can be readily measured in various manners. For example, a measurement based on electrophoresis is widely employed as the ordinary method.

In the process, the hydrous metal oxide is specifically a metal oxide to which water molecules are bonded by hydrolysis. For example, it is known that hydrous zirconium oxide is represented by the general formula, $ZrO_2 \cdot nH_2O$, where n is mainly in a range of 1 to 2, hydrous titanium oxide by $TiO_2 \cdot 0.5-2H_2O$, hydrous iron oxide by $\gamma \cdot Fe_2O_3$ or Fe_3O_4 to which 1–2.5 water molecules are bonded.

According to the process, the carrier supports about 10% by wt of the hydrous metal oxide in the hydrolyzed state, on the basis of the carrier, and at least 5% by wt thereof, even after shaking.

In carrying out the process, zeta potentials of hydrous metal oxide and carrier with changes in pH are measured individually in advance, and the carrier is dipped in an aqueous solution of metal salt capable of forming the hydrous metal oxide. Then pH is adjusted so that the zeta potential of resulting hydrous metal oxide and that of the carrier can have opposite polarities on the basis of the zeta potentials measured in advance, and then the metal salt is hydrolyzed.

The hydrous metal oxides applicable to the process include hydrous oxides of metals employed in the ordinary ion-exchangers such as titanium and zirconium, and the carriers applicable to the process include silica gel, alumina, activated carbon, etc. Hydrous oxide of such metal supported on alumina carrier as titanium, zirconium, or titanium and manganese, zinc, tin, zirconium, silicon, or rare earth element, is suitable as an adsorbent for recovering uranium from seawater.

An inorganic ion-exchanger for catching the impurities in water in nuclear reactor must be freed by force in advance from readily bleedable components to prevent the bleeding of the supported hydrous metal oxide. It is desirable to vigorously shake the carrier supporting the hydrous metal oxide in water in a pH range for effecting the hydrolysis of the metal salt.

An adsorbent for catching radioactive cobalt ions contained in the water in nuclear reactor at an elevated temperature is prepared by supporting a hydrous titanium oxide on an alumina carrier according to the present method, and then calcining the carrier at a temperature of 700° to 1100°C. The calcination is carried out to prevent bleeding of the hydrous titanium oxide into hot water. In the calcined catching agent, rutile-type titanium oxide is supported on gamma- or psi-type alumina, and 5 to 20% by wt of titanium oxide is supported on the alumina, on the basis of the alumina.

Example 1: 20 g of γ-alumina dried in air at 120°C for 2 hours was admixed with 50 ml of a commercially available aqueous hydrochloric acid-acidified titanium tetrachloride solution, tightly sealed, and left standing for 24 hours, thereby sufficiently impregnating the carrier with the titanium tetrachloride solution even to its inside. Then, the impregnated carrier was transferred into a beaker, brought in contact with 100 ml of water adjusted to pH 3.5 by hydrochloric acid, and then left standing for 24 hours, thereby hydrolyzing the titanium tetrachloride in the carrier. After completion of the reaction, the carrier was sufficiently washed with distilled water, and air dried.

Example 2: The γ-alumina impregnated with titanium tetrachloride in the same amounts and the same manner as in the comparative Example 1 was brought in contact with water adjusted to pH 7 with aqua ammonia, left standing for 24 hours thereby hydrolyzing titanium tetrachloride in the carrier, washed with distilled water, and air dried.

Example 3: Inorganic ion-exchangers of hydrous titanium oxide supported on alumina carriers prepared under the same conditions as in Example 2 were calcined individually at 600°, 740°, 840° and 1050°C in air for 5 hours to obtain calcined inorganic ion-exchangers. 1 g each of the calcined inorganic ion-exchangers was placed individually in cylindrical columns and pressurized water at 150°C passed through the cylindrical columns. Bleeding of titanium was not detected at all in the effluent water from the ion-exchangers calcined at 600°–1050°C.

Then, sample water being adjusted to pH 4.1 with hydrochloric acid and containing 3 ppm cobalt was passed through the cylindrical columns at a column inlet temperature of 150°C for 7 hours. After the passage, the inorganic ion-exchangers were melted with potassium pyrosulfate and cobalt (caught) contents were measured by atomic absorption analysis after dissolution of the melt in dilute sulfuric acid. Results are given below:

Calcination Temperature (°C)	Co (Caught) Content (mmol/g exchanger)	Co Catching (%)
600	0.10	61
740	0.15	92
840	0.13	80
1050	0.11	67

Alumina carrier supporting no hydrous titanium oxide was calcined at 740°C and the sample water containing cobalt was passed through the carrier layer where 0.02 mmol of cobalt was caught per g exchanger. It is seen from the foregoing that titanium takes part in catching the cobalt. The titanium oxide calcined at the elevated temperature can have a Co ions-catching capacity, because it seems that rehydration of titanium oxide takes place in water at the elevated temperature and the cobalt ions are caught by ion exchange with hydroxy group on the surface thereof.

Regeneration of Ammonium-Containing Wash Waters

It is known that during the production of nuclear fuels or breeder materials hydrated oxide particles are precipitated from salt solutions, especially nitrates of uranium, plutonium and thorium, by hydrolysis. This process is oftentimes triggered by the decomposition of hexamethylene tetramine by heat and sometimes involves the addition of stabilizing agents such as urea.

The chemicals used for the precipitation and the by-products of the reaction must be removed from the actinide hydrate oxide particles prior to further processing. It is known to wash the particles with water containing 2 to 3% of ammonia. After repeated use of wash water, it will contain small amounts of heavy metal salts in addition to larger amounts of ammonium nitrate and organic substances. The ammonium solutions contain also carbonate from the air and from the disintegration of the organic substances. Therefore, before recycling, the wash water also must be free from the carbonate because this will dissolve the actinide particle fuel.

H. Lahr and G. Krug; U.S. Patent 4,209,399; June 24, 1980; assigned to Deutsche Gesellschaft fur Wiederaufarbettung von Kernbrennstoffen GmbH, Germany have found that wash waters can be regenerated to be free of radioactive products and other inorganic and organic substances. If the loaded wash waters are evaporated to a concentration of more than 900 g/ℓ of salt, preferably more than 1,000 g/ℓ of salt, and the resulting CO_2-containing vapors from the evaporation are scrubbed with a hot alkali lye, a carbonate-free regenerate is obtained which, after optional replenishment of the ammonium content, can be used again for the scrubbing of hydrated oxide particles. The treatment with the hot alkali lye is preferably performed at a temperature between 95° and 115°C, more preferably about 105°C. During this hot alkali lye treatment a portion of the alkali is transformed into the corresponding carbonate by the CO_2-containing vapors.

Preferably, as the alkali lye, a sodium hydroxide solution is employed. This sodium hydroxide solution preferably has a strength of 25 to 35% by wt, more especially about 30% by wt. As indicated, during the scrubbing process it gradually changes to a sodium carbonate solution. It is the carbonate-containing wash waters which are effective to dissolve the oxide hydrate particles so as to insure that the resulting waters, when condensed, are virtually free of carbonate before recycling.

As a result of the scrubbing of the evaporate with the alkali lye there is obtained a soda solution having a concentration of about 40%. This resultant solution containing traces of alkali lye can be evaporated employing a wiper-blade evaporator to dryness. The resultant solid alkaline component, e.g., soda with traces of the corresponding solution, is therefore the only waste product from the process, in addition to filter materials mentioned below.

It has been discovered in accordance with another embodiment that in a second vessel the inorganic and organic components can be removed from this residue by introducing therein concentrated nitric acid especially nitric acid having a concentration of between 60 and 65% by wt, preferably about 65% by wt. It is thereafter heated to a temperature greater than 100°C, preferably more than 105°C, causing disintegration of ammonium nitrate and other impurities, especially organic impurities, into gases such as carbon dioxide, nitrogen, and nitrogen oxides. These off-gases are going through a filtering and washing unit to the atmosphere.

According to a further embodiment, the residue resulting from evaporation of the wash waters is mixed with nitric acid, repumped into the abovementioned second heating apparatus and additional concentrate resulting from evaporation of the wash waters is thus continuously added. Desirably, the reaction mixture flows in countercurrent to the off-gases resulting from decomposition of the inorganic and organic residues. When a maximum volume in the vessel is obtained, the feed of nitric acid and wash water concentrate is interrupted. With suitable construction for the heating apparatus, the resulting solution can then be concentrated therein. With sufficient decontamination effect of the distilling unit, no radioactive waste results except, of course, for the filtering material employed.

It has been found further that chlorides in amounts of less than 1.5%, preferably less than 1%, have a catalytic effect on the disintegration of the residues in the second step of the process and favorably influence the same. The salts remaining in solution following disintegration of the inorganic and organic impurities, especially the nitrates of the actinides, can be recycled as such or, following additional concentration, for production of heavy metal oxide hydrate particles.

Treatment of Phosphoric Acid Ester-Hydrocarbon Waste Liquids

S. Drobnik; U.S. Patent 4,022,708; May 10, 1977; assigned to Gesellschaft fur Kernforschung mbH, Germany describes a process whereby phosphoric acid ester-hydrocarbon waste liquids from extraction and/or reextraction cycles contaminated by at least one of the groups including radioactive materials, decomposition and/or hydrolysis products of phosphoric acid esters, nitric acid esters, nitrous acid esters, nitro compounds and complex compounds of at least one of these compounds with ruthenium, zirconium, niobium, uranium or plutonium are treated with aqueous solutions of inorganic compounds for separation of the hydrocarbons from the phosphoric acid esters and impurities.

In an explanation of the method and in order to highlight the advantages of this method as against previous methods for the removal of the organic radioactive liquids mentioned, the process steps needed for preparation of the waste liquids for storage will be summarized in two flowsheets, process flowsheet 1 starting with the admixture of the concentrated phosphoric acid; process flowsheet 2 starting with the saponification of the phosphoric acid esters and the saponifiable impurities in the waste or washing and cleaning liquids, respectively. Process flowsheet 1, in addition, includes one example mentioning the quantitative and radioactivity data of a waste liquid prepared for storage according to the process and containing TBP (tributyl phosphate) and kerosene.

Process Flowsheet 1:

(1) Waste liquid is mixed with concentrated phosphoric acid, the phases generated are separated from each other into one lighter phase (phase 1 L) and one heavier phase (1 S). For instance, 1 m^3 of a liquid consisting of 5 vol % TBP and 95 vol % kerosene and radioactive impurities of 1 Ci is mixed with 42.25 kg (corresponding to 24.7 ℓ) of 85% H$_3$PO$_4$. This generates approximately 950 ℓ of phase 1 L with a phosphorus content of less than 1 mg/ℓ and approximately 75 ℓ of phase 1 S with the adduct compound TBP·2H$_3$PO$_4$ and a radioactivity of approximately 1 Ci.

(2) For decomposition of the adduct compound phase 1 S is treated with an excess of water, the phases produced are separated from each other into a lighter phase (2 L) and a heavier phase (2 S). For instance, approximately 75 ℓ of phase 1 S and 150 ℓ of water generate 175 ℓ of phase 2 S consisting of diluted phosphoric acid with a radioactivity of approximately 0.85 Ci which is supplied to the wastewater decontamination plant and from there to solidification with bitumen; and approximately 50 ℓ of phase 2 L of TBP which contains approximately 0.05 mol H$_3$PO$_4$ and a radioactivity of approximately 0.15 Ci.

(3) Phase 2 L is saponified; a mixture which is azeotropic at the temperature of saponification is distilled off, collected in the receiver and cooled. Two phases occur, a lighter phase (phase 3 L) which is burst and a heavier phase (phase 3 S) which is returned to the saponification process. For instance, 44 kg of 50% caustic soda solution is added to approximately 50 ℓ of phase 2 L and saponified for 6 to 7 hours at 120°C. At this temperature, about 50 ℓ of azeotropic butanol/water with a radioactivity of 10^{-4} Ci/m^3 will be produced by distillation. The aqueous residue remaining in the saponfication vessel after the period of saponification is cooled to room temperature, diluted with approximately 50 ℓ of water to avoid phosphate precipitation, and solidified into a block in bitumen in the well-known manner. The solidification product of the quantities

mentioned has a volume of 53 ℓ and radioactivity of 0.03 Ci/ℓ. The solidification products introduced into drums in a molten condition are ready for nonpolluting storage. The aqueous distillate escaping in the solidification with bitumen is taken to the wastewater decontamination system where part of it is recycled for solidification.

(4) Phase 1 L is washed and cleaned and reused afterwards. For instance, the approximately 950 ℓ of phase 1 L are mixed with 50 ℓ of 0.1 molar sodium oxalate solution of pH of 10 to 11 and washed. The two phases generated are separated from each other into one heavier phase (phase 4 S) consisting of a sodium oxalate solution, which is taken to the wastewater decontamination system, and a lighter phase (phase 4 L) consisting of kerosene with a radioactivity of approximately 10^{-4} Ci/m^3, which is run through an active charcoal filter. Afterwards, about 950 ℓ kerosene are obtained with a radioactivity of 10^{-5} to 10^{-6} Ci/m^3 which is used to produce fresh TBP-kerosene solutions.

Process Flowsheet 2:

(1) Waste or washing and cleaning liquids, respectively, are treated with an aqueous solution of a base and phosphoric acid ester and saponifiable impurities are saponified at elevated temperatures. A mixture which is azeotropic at the saponification temperature is distilled off, collected and cooled. Two phases are generated, one lighter phase (phase 5 L) which is burnt, and a heavier phase (phase 5 S) which is recycled to the process of saponification.

For instance, a waste liquid containing 1 m^3 of TBP with a radioactivity of approximately 1 Ci is added to 8.8 kg of 50% caustic soda solution per vol % of TBP, referred to the TBP content of the liquid, and saponified for 7 hours at a temperature between 120° and 135°C. During that time 12 ℓ/vol % of TBP in the waste liquid before saponification is obtained as distillate. The distillate essentially consists of azeotropic butanol/water with minor quantities of hydrocarbons (<40% of the quantity distilled) and has a phosphorus content of approximately 0.01 mg/ℓ and a radioactivity of 10^{-5} to 10^{-4} Ci/m^3.

(2) After cooling of the reaction mixture, that is the liquid residue in the saponification vessel after the period of saponification, this is treated with an excess of water and the two phases are separated from each other into one organic phase containing hydrocarbons (phase 6 L) and one aqueous phase containing the impurities and phosphates (phase 6 S).

For instance, 1,100 ℓ of water is added to the reaction mixture remaining in the saponification vessel after step (1). This results in the bulk of hydrocarbons, which is the amount of hydrocarbons originally contained in the waste liquid minus the small quantity of hydrocarbons removed from the saponification product, with a phosphorus content of 0.1 mg/ℓ and a radioactivity of approximately 10^{-4} Ci/m^3 (phase 6 L) and a volume of approximately 1,300 ℓ of aqueous phase (phase 6 S) with a dry residue of 220 g/ℓ.

(3) Phase 6 S is solidified. For instance, the 1,300 ℓ of phase 6 S are introduced into molten bitumen at a temperature of 150°C and solidified. This generates 11 ℓ of solidification product per vol % of TBP in the waste liquid before saponification and an aqueous distillate which is taken to the wastewater decontamination system. The solidification product contains about 40 wt % of salts and has a density of approximately 1.5 g/cc. Almost all the radioactive materials originally contained in the quantity of waste liquid are solidified in it. After filling in drums of the liquid solidification products the products can be taken to nonpolluting storage.

(4) Phase 6 L is washed with concentrated phosphoric acid for purification and afterwards reused for the fabrication of fresh phosphoric acid ester-hydrocarbon solutions.

For instance, phase 6 L is washed with 10 ℓ of 85% phosphoric acid; afterward, the two phases are separated. The phosphoric acid is taken to the wastewater decontamination system and to solidification with bitumen, respectively. After purification, the hydrocarbon phase has a residual radioactivity of only approximately 10^{-6} Ci/m^3 and is reused in the cycle for reprocessing solutions.

According to process flowsheet 1, almost 100% (more than 99%) of the hydrocarbons are recovered with a decontamination factor 10^6. The volume reduction relative to the waste liquid prepared for storage is about 20%.

According to process flowsheet 2, some 70 to 80% of the hydrocarbons are recovered. The decontamination factor of the hydrocarbons is also 10^6. The volume reduction of waste is approximately the same as in flowsheet 1.

L.A. Humblet, L. Salomon and H.R. Eschrich; U.S. Patent 4,039,468; August 2, 1977; assigned to Societe Europeenne pour le Traitement Chimique des Combustibles Irradies (Eurochemic), Belgium describe a similar process for the treatment of organic wastes resulting from the extraction of metal elements by a solvent of the phosphoric acid ester type in the presence or in the absence of an organic diluent which comprises the following main steps.

(a) Separation of the phosphoric acid ester from its organic diluent in such a way that the ester phase contains the degradation products and also the elements to be finally insolubilized.

(b) Conversion of the phosphoric acid ester into phosphoric acids on one side and organic compounds mainly consisting of hydrocarbons on the other side.

(c) Solidification of the liquid radioactive phosphoric acids in the form of inorganic phosphates, which can eventually be incorporated into a suitable matrix material.

According to a feature of the process, the organic diluent is unpolar and the separation of the phosphate and the organic diluent is made by means of a polar agent which is selectively miscible with the phosphate. The conversion of the phosphoric acid ester is made by pyrolysis (and partially by hydrolysis and acidolysis).

In a preferred embodiment, the ester is tributyl phosphate (TBP) and its diluent is kerosene, whereas the polar agent is concentrated phosphoric acid.

Instead of an alkyl phosphate such as tributyl phosphate or diethylhexyl phosphate, other esters of phosphoric acid may be used such as aryl phosphates (for example, triphenyl phosphate). It is also possible to use alkylaryl phosphates or mono- or di-alkylphosphates such as monobutyl phosphate or dibutyl phosphate.

In this embodiment, the first step of the process takes place at room temperature and consists of mixing the solvent with concentrated phosphoric acid (preferably with the phosphoric acid formed in the second step of the process) in quantities sufficient to dissolve all the TBP in the phosphoric acid phase. The free kerosene (diluent of the TBP), obtained in such a way, will contain no TBP nor any trace of the metal elements, if the correct working conditions are used. These metal elements which can be fission products or actinides are, under these conditions, concentrated in the phosphoric solution of TBP. This separation of the TBP and kerosene can be effected in a system comprising two or three phases, the only condition required being that the phase containing finally the TBP is at least 4 M in phosphoric acid (H_3PO_4).

The separation of the two or three phases can be made batch-wise. However, one of the features of the process is a continuous separation: either in a column (pulsed or not), which is fed at the bottom with the solvent to be treated and at the top with phosphoric acid at convenient flow rates, the kerosene freed from its TBP and metal elements, leaving continuously the column by the upper outlets, or in a preferred embodiment, in a centrifugal extractor which is simultaneously fed with the solvent and the concentrated phosphoric acid in the convenient proportions, the centrifugation resulting in a clear-cut separation which contributes to an optimal decontamination of the kerosene.

In the same preferred embodiment, the second step of the process consists in the decomposition of the TBP in acid medium at a temperature of about 175°C or more, leading to the formation of a very concentrated phosphoric acid (heavy inorganic fraction) and simple organic compounds (essentially hydrocarbons), which are combustible and volatile at the reaction temperature. The composition of the organic fraction (distillate) can be directed by selecting the working conditions so as to form light hydrocarbons (butenes, volatile at room temperature) or compounds condensable at room temperature (carbon chains of 8 or more carbon atoms). The reaction is preferably carried out under inert atmosphere, e.g., by bubbling nitrogen, which causes at the same time some agitation during the reaction. According to the wanted result, the nitrogen stream can be humidified. In all cases, the organic fractions carry less than 0.005% of the total activity, which means a decontamination factor for the organic substrate of more than 10^4.

According to an additional feature of the process, decomposition of the TBP is performed by keeping the TBP acid phase in a thin film so as to cause a quick removal of the formed gases in order to prevent their polymerization. Although various devices can be used for that purpose, namely a column with a coil or a series of grooved inclined plane boxes, the pyrolysis reactor is preferably a wiped film pyrolyzer in which the liquid phase flows down by gravity, whereas the gases flow in countercurrent.

According to an important feature of the process, the energy needed for the conversion of the ester is obtained by the combustion of the hydrocarbons produced by this dealkylation and/or the organic diluent originating from the first step of the process (separation of the ester and its diluent). Moreover, catalysts can be added to the reaction medium in order to speed up the ester conversion or in order to prevent the formation of polymers or also to favor the conversion of the organic compounds of the solvent into simple (saturated or not) hydrocarbons, which are volatile at the reaction temperature. These catalysts are preferably selected among steam and metals of the platinum series.

In the same preferred embodiment, the third step of the process comprises the solidification of the liquid radioactive waste, which takes place by the reaction of the concentrated phosphoric acids of this waste on aluminum oxide at a temperature higher than 150°C, preferably between 150° and 600°C, in such a way that either a granular solid (e.g., in a fluidized bed reactor) or a monolithic product is obtained. The granular product formed can be incorporated into bitumen, into synthetic resins (e.g., resins of the epoxy- or phenoxy-type), glass or a metal matrix.

The aluminum oxide, which is used for the solidification of the liquid radioactive waste, may have previously been used as a filter agent for the kerosene in the first step of the process.

The process is very suitable for the treatment of spent solvents deriving from processes used for the recovery of radioactive elements, in particular uranium and plutonium, from irradiated fuel elements.

Compared with previously known processes, this process has the following advantages:

> it requires no pretreatment resulting in a transfer or dilution of radioactivity;
>
> it can be realized with a reagent (phosphoric acid) produced by the process itself;
>
> it is nonpolluting and no effluent is discharged, neither in the earth nor in the atmosphere, during the treatment of the solvent;
>
> it is an economical process in the sense that during the reactions involved in it enough fuel is produced for supplying the energy needed for the process;
>
> it allows a concentration of the radionuclides present in the original solvent waste by a factor of 20 to 100;
>
> the inorganic concentrate obtained can eventually be added to the aqueous medium-level or high-level liquid radioactive waste, in which it can play a favorable role for the insolubilization of the radionuclides during the solidification of the wastes, whatever be the selected solidification process.

Example: 1 ℓ of a spent 30 vol % TBP solution in kerosene containing, besides the solvent degradation products, fission products with a total β-γ activity of 10 mCi and α-emitters with a total of 1 mg plutonium and 0.1 g uranium is used. 1 ℓ of this solution is contacted in a separatory funnel with 250 ml of 14.7 M phosphoric acid during 2 minutes.

After settling of the liquids, three phases are separated, i.e.:

(1) a light organic phase having a volume of 672 ml and a density of 0.76 (the same as that of the used kerosene), which contains no acid nor water and less than 0.01 mCi β-γ activity, and in which Pu and U are not detectable (the analysis shows that the solution still contains about 1% TBP);

(2) a heavy organic phase (intermediate phase) having a volume of 400 ml and a density of 1.14, containing 4.1 M H_3PO_4 and 9.2 mCi β-γ activity, 0.1 mg of Pu and 10 mg of U (the analysis shows that this phase contains about 5% kerosene);

(3) a heavy aqueous phase having a volume of 151 ml and containing 13.2 M H_3PO_4, about 0.8 mCi β-γ activity, 0.9 mg of Pu and 0.09 g of U.

A second contact of the light organic phase with 14.7 M H_3PO_4 in the same volume ratio of 5 gives a colorless and inactive solution of 665 ml of kerosene. Likewise, three consecutive contacts of the intermediate phase (TBP-H_3PO_4 mixture) with 14.7 M H_3PO_4 result in the elimination of kerosene from the intermediate phase without any less β-γ activity therein.

PROCESSING OF RADIOACTIVE GASES

GENERAL PROCESSES

Perchloro- or Perfluoroalkene Gas Scrubbing Liquid

In the prior art various systems and techniques have been devised for reducing or eliminating the noxious contaminants produced as by-products of industrial processes. Of particular interest are the contaminants which are released to the atmosphere as a result of combustion. These contaminants are of the type which are normally released from a stack in the form of what is normally referred to as smoke. Smoke, depending upon its source, may contain suspended particulates, hydrocarbon vapors, soot, and aerosols.

Soot has been removed from effluent stream by filters. Hydrocarbon vapors including tars, however, are not affected by a filter and condense on the filter to clog the same or pass on through unremoved. Another method of soot removal involves burning the soot with an excess of air. However, this creates further emission for which some technique of abatement is needed.

In general, the use of water scrubbers has solved many of these problems. Water, however, still does not act as a solvent for many hydrocarbon vapors and thus either permits these vapors to pass on through to the atmosphere or to condense at some undesirable point within the system.

In the nuclear processing industry, the avoidance of hydrogen-bearing materials is of particular importance because of the neutron-moderating properties of hydrogen. Thus, it is unwise to use an aqueous gas scrubber on a system containing fissionable material. The possibility of the scrubbing liquid flooding the system and creating a critical mass condition is remote, but such possibility should be avoided if at all possible. Aqueous scrubbers are thus unsatisfactory for systems containing fissionable material.

W.J. Lackey, R.S. Lowrie and J.D. Sease; U.S. Patent 4,276,063; June 30, 1981; assigned to The U.S. Department of Energy have found that fully chlorinated and/or fluorinated hydrocarbons have remarkably effective properties for scrubbing noxious contaminants from a gas stream. In general, these properties are:

(1) high solubility for hydrocarbons;

(2) heat capacity of greater than 0.2 cal/g°C;

(3) low flammability;

(4) boiling point of greater than 75°C; and

(5) capacity to wet soot.

Thus, in accordance with this process, it has been found that fully chlorinated and/or fluorinated hydrocarbons possessing the above-listed properties are capable of scrubbing noxious effluents from combustion gases with a higher efficiency than water while at the same time avoiding the use of hydrogen-bearing material in nuclear material processing.

While the term fully chlorinated and/or fluorinated hydrocarbon is used in describing the scrubbing liquids of this process, the perchloro- and perfluoroalkenes are the liquids of principal interest. Most particularly, the scrubbing liquids include perchloroethylene, hexachloropropylene, perchlorobutadiene, tris perfluorobutyl-amine, perfluoroacetone, perfluoro ethers, and perfluoro thioethers.

The scrubbing liquids embodied by this process have been found to possess scrubbing properties superior to water in the following regards:

(1) hydrocarbon gases and liquid are readily dissolved within the scrubbing liquid in appreciable quantities;

(2) soot is wetted by the scrubbing liquid thus leading to the formation of a soot-solvent slurry;

(3) neutrons are not moderated by the scrubbing liquid since they are nonhydrogenous; and

(4) the liquids are noncorrosive.

Of the scrubbing liquids embodied by this process, perchloroethylene is the preferred liquid because of its availability and ease of recovery from a soot slurry by low-temperature distillation.

The scrubbing liquids contemplated for use in this process may, of course, be used in conjunction with any process which produces noxious emissions. These liquids can be substituted for water in any gas scrubbing application. The preferred mode of application, however, is contemplated to be in applications where fissile materials are present in a contiguous relationship to the scrubbing system. In such a situation, the avoidance of moderating material such as hydrogenous material is important for reasons previously mentioned.

The process and scrubbing liquid of this process is particularly useful in combination with furnaces for coating nuclear fuel particles with carbon.

In actual operation, gas providing the carbon coating on the fuel particles enters the coating furnace at the bottom thereof and acts as a fluidizing medium for the fuel particles. Any effluent gas then passes into the scrubber. This gas is at a temperature of about 600°C and contains, for example, decomposition products of the pyrolytic decomposition of the coating gas including soot and various hydrocarbons, including long-chain species such as tars. Within the scrubber, the coater effluent stream is contacted with several sprays of C_2Cl_4 at a temperature of about 20°C, and the combined materials drop into the reservoir which acts as a gas-liquid separator. The gaseous stream therefrom passes through the demister and a cooler to remove residual liquids and condensible materials prior to exit through absolute filters.

A portion of the C_2Cl_4 is removed from the separator by a pump, passed through the heat exchanger, and reinjected through the spray nozzles. The solids and dissolved vapors remain in the C_2Cl_4 solution and thus are reinjected into the scrubber. This continues until the solution is loaded sufficiently to warrant regeneration. Perchloroethylene accommodates about 180 grams soot per liter and remains sufficiently fluid to permit pumping and recirculation at soot concentrations of 125 to 150 grams per liter. Regeneration is accomplished by distillation or centrifugation. For a coating furnace having a five-inch diameter, the gas effluent exits at a rate of about 100 to 500 ℓ/min. A total volume of C_2Cl_4 of about 70 liters is sufficient to scrub the gas without overheating of the liquid. The volume flow rate of the liquid corresponding to that of the gas volume flow rate above is within the range of 25 to 50 liters per minute.

The effectiveness of perchloro- and perfluoroalkenes for scrubbing effluent gases is illustrated by the high efficiency of perchloroethylene for removing very fine soot particles and tars from the particle coating furnace effluent gas. When acetylene and/or propylene is decomposed within the furnace, the effluent contains numerous hydrocarbons including long-chain species such as tars as well as 30 to 50% by weight soot.

The soot is present as very fine particles, less than one micron, with the average size being about 0.5 micron as measured by scanning electron microscopy. The effluent gas contains enough of this fine soot to plug an absolute filter in a matter of seconds. Furthermore, experience has shown that the effluent gas contains sufficient tars to plug an absolute filter in about one hour. When the effluent gas stream at about 600°C was passed downwardly through a spray scrubber where it was contacted with three successive cocurrent sprays of perchloroethylene and then to an absolute filter, the filter remained clean for many hours of operation and lasted for weeks without plugging.

Qualitative estimates place the efficiency of the three-stage perchloroethylene scrubber at over 99% for this very fine soot. This is particularly significant when it is considered that one manufacturer of spray-type gas filters using conventional scrubbing liquids reports a single-stage particle removal efficiency of only about 40% for particles of 0.5 micron, which would correspond to no more than a 79% overall removal efficiency, for a three-stage scrubber. Actually the overall efficiency of a three-stage scrubber using prior art liquids would be expected to be less than 79% since the smaller, more difficultly removable particles would escape the first stage and pass to the second and third stages which would then be correspondingly less efficient. Accordingly, it is seen that perchloroethylene is substantially more effective than prior art liquids for removing soot particles from gas streams wherein a major portion (more than 50% by weight) of the soot is present as particles smaller than 1 micron in diameter (equivalent sphere diameter).

Implanting of Radioactive Gas in Base Material

M. Terasawa; K. Mawatari and O. Morimiya; U.S. Patent 4,124,802; November 7,

1978; assigned to Tokyo Shibaura Electric Co., Ltd., Japan describe a method and apparatus capable of safely detaining radioactive gas over a long period, instead of storing the radioactive gas in the original form as has been practiced in the past.

To this end, the method comprises the steps of ionizing radioactive gas, accelerating the ionized radioactive gas into a high energy form and implanting the high energy radioactive gas in a base material for permanent detainment. To attain the abovementioned object, the apparatus is provided with a radioactive gas reservoir; an ion source for ionizing the radioactive gas delivered from the reservoir; an accelerator for accelerating the ions produced in the ion surce; and an ion implantation unit connected to the accelerator through the extension thereof and designed to implant high energy ion beams conducted through the accelerator for detainment in a base material received in the ion implantation unit.

With a preferred embodiment, first and second exhaust units are connected to the extension of the accelerator and ion implantation unit respectively, enabling residual gas to be drawn off from the accelerator extension by the first exhaust unit and also radioactive gas released from the base material to be taken out by the second exhaust unit. This arrangement allows high energy ion beams sent forth from the accelerator to be smoothly implanted in the base material without being scattered by residual gas and the gas released from the base material in the ion implantation unit and a passage thereto.

The base material is formed of, for example, a band-shaped stainless steel foil. The foil is made to travel relative to the high energy ion beams by driving means, thereby enabling a large amount of radioactive gas to be detained in the base material. This base material may be prepared from not only a stainless steel foil, but also other foils of metals such as aluminum and copper.

An electromagnetic deflection unit is provided in an intermediate unit following the aforesaid accelerator extension to create an ac magnetic filed or ac electric field acting perpendicular to the running course of the ionized beam. The ion beam is oscillated in zigzags across the metal foil, causing the ion beam to be fully implanted in the foil even when the foil is made to travel slowly.

Where a radioactive gas is mixed with nonradioactive gases having a different mass number, all the gases are ionized into ion beams, and a separation magnet is provided in a passage of the mixed ion beams to separate the radioactive component from the ion beam, enabling a large amount of the radioactive ion beam alone to be securely implanted in the metal foil. In this case, a stopper made of heat-resistant material such as graphite, pyrocarbon, magnesia, calcia or zirconia is placed in an intermediate stopper chamber formed in the intermediate unit following the ion separation magnet or in the ion implantation unit. The nonradioactive ion beams are ejected on the stopper to be partly implanted therein. The remainder of the nonradioactive ion beams is again released for gasification. The gasified nonradioactive ion beam is drawn off by an exhaust unit connected to the intermediate stopper chamber and ion implantation unit. Therefore, the radioactive ion beam can be smoothly implanted in the metal foil without being scattered during passage by the abovementioned released nonradioactive gas.

This process which securely detains radioactive gas in a proper base material instead of storing it in the original form eliminates the necessity of providing a pressure vessel as has been practiced in the past.

Separation and Sequestering of Radioactive Gases

H. Hesky and A. Wunderer; U.S. Patent 4,206,073; assigned to Hoechst AG, Germany describe a process for separating volatile, radioactive substances obtained in the reprocessing of nuclear fuel, which comprises:

  (a) reducing the higher nitrogen oxides formed in the dissolution and the denitration and contained in the respective gas mixtures to give nitric oxide (NO) and absorbing the iodine contained in the gases;

  (b) separating the gas mixture obtained into a first fraction containing the volatile radioactive substances and nitric oxide, a second fraction essentially consisting of nitrogen and a third fraction consisting of xenon;

  (c) adding oxygen to the liquid, radioactive wastes and separating the oxygen enriched with the volatile radioactive substances from the wastes;

  (d) subjecting to a chemical treatment the enriched oxygen of stage (c) together with the first fraction of stage (b) and reacting the oxygen with the nitric oxide and water of the aqueous nitric acid to form nitric acid; and

  (e) conducting the reaction product of stage (d) obtained into the dissolution process and passing the volatile, radioactive substances to a common storage.

In the reduction of the higher nitrogen oxides to nitric oxide, nitric acid is formed which absorbs iodine. The nitric acid contaminated with iodine is subjected to an iodine desorption, the desorbed iodine is blown out with oxygen and passed to an iodine filtration. The oxygen leaving the filtration can be combined with the contaminated oxygen set free from the liquid radioactive waste.

It may be expedient to pass the gas mixture leaving the reduction stage through an adsorber chain to be regenerated with oxygen. In this adsorber chain aerosols, carbon dioxide and suspended matter which may be present inter alia are retained. The oxygen used for the regeneration can be combined with the contaminated oxygen set free from the liquid, radioactive waste.

It proved advantageous to concentrate by distillation with subsequent electrolysis the radioactive liquid wastes essentially consisting of tritium-containing water, formed in the recovery of nitric acid which has not been consumed. For concentration there may also be used a distillation and/or an electrolytic enrichment cascade.

Still further, it can be of advantage to pass the oxygen contaminated with radioactive substances and set free from the liquid radioactive wastes over a catalyst to convert tritium-containing hydrogen formed by radiolysis to tritium-containing water. The tritium-containing water formed can then be subjected to an electrolytic enrichment process and the water depleted of tritium can be recycled into the reprocessing, for example, into the dissolution process or it can be used to adjust the concentration of nitric acid.

The contaminated oxygen which has not been reacted in the catalytic hydrogen oxidation can be subjected to a chemical treatment together with the fraction containing the volatile, radioactive substances, essentially krypton, and nitric oxide. It proved also advantageous to recycle the separated nitrogen at least partly as scavenging gas into the comminution and dissolution of the nuclear fuel.

In order to retain traces of iodine, if any, the separated xenon can be passed through an appropriate filter and released into the atmosphere or it can be utilized for a convenient purpose. As far as the nitrogen is to be released into the atmosphere, it should be passed through an appropriate filter to remove and isolate traces of iodine possibly present and, if tritium is still contained in the nitrogen, it should be subjected to an appropriate further after-purification, for example, a tritium oxidation.

It is the advantage of the process that for the sequestration of volatile, radioactive substances obtained in the reprocessing of nuclear fuel only oxygen need be added which is reacted with the NO formed in the process to give higher nitrogen oxides. A further advantage resides in the fact that the gas mixtures remain in the process and that chiefly only xenon, the excess amount of nitrogen and the volatile, radioactive substances are withdrawn from the cycle. The common storage of krypton and tritium constitutes a further advantage since tritium forms a solid with rubidium, a decomposition product of krypton, so that the stored volume and hence the storage pressure are reduced. Still further, it is of advantage that all volatile, radioactive substances to be withdrawn from the process are jointly obtained at one point and can be stored.

Treatment of Liquid Waste Containing Volatile Substances

H. Stünkel; U.S. Patent 4,139,420; February 13, 1979; assigned to Kraftwerk Union AG, Germany describes a method for treating contaminated aqueous liquid to remove volatile radioactive substances contained therein.

The method includes maintaining a body of liquid containing contaminants in a still zone, heating the liquid to generate vapors and volatilize radioactive substances, passing the generated vapors and volatilized radioactive substances upwardly in a column containing a plurality of spaced plates in intimate contact with reflux condensate on the plates resulting from the partial condensation of the vapors in the column, withdrawing liquid from the lower plate of an intermediate group of at least two successive plates and recirculating the liquid to the upper plate of the intermediate group, maintaining the recirculating liquid at a pH different from the body of liquid in the still zone to aid in retaining volatile radioactive substances passing up through the intermediate group of plates in contact with the liquid of a different pH on the plates, and releasing purified vapor free of volatile radioactive substances from the top of the column.

In accordance with the process, reflux condensate on a plate above the intermediate group of plates is withdrawn from the column and introduced into the column at a point below the intermediate group of plates.

A portion of the withdrawn reflux condensate in an amount of less than 10% is diverted and directed as reflux to the intermediate group of plates.

As a further feature, a small amount of less than about 10% is bled from the recirculating liquid to prevent build-up of volatile radioactive substances in the recirculating liquid and in the column.

There is provided apparatus for treating contaminated aqueous liquid to remove volatile radioactive substances contained therein. This includes a still for contain-a body of contaminated liquid, means for heating the liquid and generating vapors and volatilizing the radioactive substances, a vertical column disposed above and in open communication with the top of the still, a plurality of spaced plates in the column and passageways in the plates through which vapor passes up in intimate contact with liquid on the plate, an intermediate group of at least two successive plates in the column, conduit means for withdrawing liquid from the lower plate of the intermediate group and pump and conduit means for recirculating the withdrawn liquid to the upper plate of the intermediate group.

Also included are inlet means for introducing a substance to make the pH of the recirculating liquid different from the body of liquid in the still, conduit means connected to a plate above the intermediate group of plates for the withdrawal of liquid reflux on the plate, and conduit means for passing the withdrawn liquid reflux from the plate above into the column at a point below the intermediate group of plates, an opening at the top of the column for the release of vapor, a cooler and condenser through which the vapors pass and are condensed, and a return line to the top of the column for the return of a portion of the condensed vapors.

There is provided means for determining the pH of the recirculating liquid, and control means connected with the means for regulating the introduction of the substance to maintain the pH in the recirculating liquid.

Figure 7.1 illustrates the apparatus used in the process.

THE NOBLE GASES

Storage of Radioactive Gases

The encapsulation of gases in zeolites is known and it has been taught that the encapsulation of radioactive krypton (^{85}Kr) takes place under high temperatures and pressures. Following the absorption and cooling, the pressure may be lowered without the loaded zeolite releasing the krypton. The process suggested to date for encapsulation has a distinct drawback; namely, it cannot be carried out without contamination of the autoclave reaction vessel; and/or the atmosphere.

For example, it has been proposed to place the zeolite in a mesh basket, to lower the basket into an autoclave and to pressurize the autoclave with krypton. After loading the zeolite, the excess krypton is pumped out of the autoclave and the basket is removed through the atmosphere and placed in a sealable container for storage. The possibilities for contamination of the atmosphere are many. Worse yet, the loaded zeolite can be exposed to moisture in the air. As krypton 85 decays, rubidium 85 is produced which can react with the absorbed moisture to form a strong caustic and hydrogen. The caustic can cause corrosion of the container and the hydrogen can result in gas pressure buildup in the container.

J.M. Berty; U.S. Patent 4,158,639; June 19, 1979; assigned to Autoclave Engineers, Inc. describes a process of storing gas by high temperature and pressure absorption, adsorption or reaction with a bed of capturing solids comprising a first step of placing the capturing solids in a relatively thin-walled container having an opening therein connectable to a conduit. The container need only be able to withstand small pressures across its body of 25 psi. Preferably the container has a relatively

Figure 7.1: Apparatus for Treating Waste Liquids Containing Volatile Substances

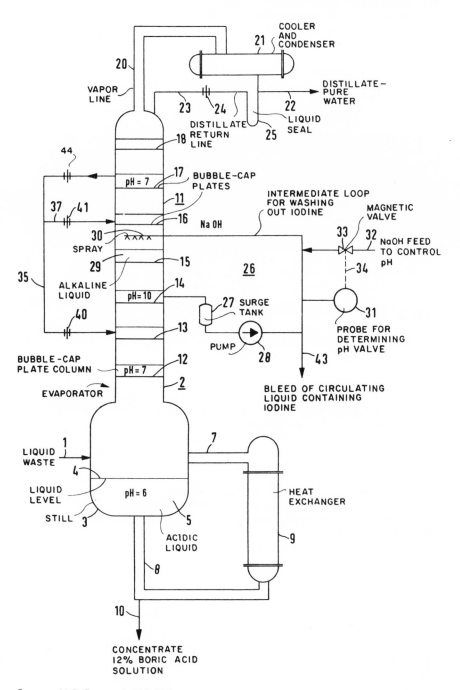

Source: U.S. Patent 4,139,420

large opening at the top for introducing the capturing solids to the container and a lid sealing the large opening. Built into the lid is a valve which, when opened, provides communication between the interior of the container or canister and a fitting connectable to a conduit. A second step of the process comprises placing the thin-walled container in a pressurizable autoclave. A third step comprises bringing the interior of the thin-walled canister into communication with a conduit extending through the walls of the autoclave and communicating with a source of the gas to be stored.

Typically, this comprises a connection between the fitting described above and a fitting in the wall of the autoclave. A fourth step comprises simultaneously pressurizing the autoclave and the interior of the thin-walled vessel by pumping gas to be stored into the thin-walled vessel and inert gas into the autoclave external the thin-walled vessel. The gas to be stored is continuously pumped into the thin-walled vessel as it is being absorbed, adsorbed or reacted with the capturing solids. When the bed of solids can no longer capture additional gas, a fifth step comprises first cooling and thereafter depressurizing the autoclave and the thin-walled vessel. In a final step, the autoclave is opened; the conduit from the thin-walled vessel is disconnected; the vessel is sealed and removed to provide a substantially nonpressurized container loaded with absorbed, adsorbed or reacted gases.

Encasement of Krypton 85 in Metal

S. Ozaki; U.S. Patent 4,250,832; February 17, 1981; assigned to Tokyo Shibaura Denki KK, Japan describes a method of storing radioactive materials, comprising introducing a metal halide compound or a metal carbonyl compound to the vicinity of a solid material into which the radioactive material is implanted while a gaseous radioactive material, ionized and accelerated, is being implanted into the solid material so as to allow the elemental metal liberated from the metal compound to be deposited on the surface of the solid material, thereby detaining the radioactive material in the solid material and, at the same time, providing a new metal layer for multilayer implantation on the surface of the solid material.

The metal halide compound or the metal carbonyl compound may be introduced singly or together with a carrier gas such as nitrogen gas. Also, the metal halide may be introduced together with hydrogen gas. In this case, the metal halide is reduced by the hydrogen so as to deposit the elemental metal.

There is also provided an apparatus for storing radioactive materials, comprising means for introducing a metal halide compound or a metal carbonyl compound to the vicinity of a solid material into which the radioactive material is implanted, means for controlling the temperature of the solid material, and means for introducing a carrier gas or hydrogen gas, as required.

The metal carbonyl compounds used in this process include, for example, nickel carbonyl, iron carbonyl and chromium carbonyl. These metal carbonyl compounds are decomposed at temperatures lower than 200° to 300°C. For example, nickel carbonyl is decomposed at 140° to 240°C to deposit nickel metal as shown below:

$$Ni(CO)_4 \rightarrow Ni + 4CO$$

This metal deposition method, which is called the carbonyl method, is employed for the production of high-purity metal. Since the surface of the solid material is heated to a considerably high temperature in general by the bombardment of accelerated ions, it is unnecessary to supply from outside the heat required for the thermal decomposition of the metal carbonyl compound. On the contrary, it is preferred to employ cooling for protecting the solid material from thermal damage. In this process, it is desirable to control the temperature of the solid material to fall within the range of from 300° to 800°C. Besides the thermal decomposition, irradiation of high-speed particles and radiations such as x-rays, ultraviolet light, etc., may be used for decomposing the metal carbonyl compound to deposit the elemental metal. Where, for example, the metal carbonyl compound is introduced during the step of implanting the accelerated ions of the radioactive material into the solid material, the accelerated ions are irradiated to the metal carbonyl compound, promoting the decomposition of the metal carbonyl compound.

The metal halide compounds used in this process include, for example, chlorides, bromides and iodides of silicon, germanium, titanium, zirconium, hafnium, vanadium, chromium, tantalum, iron, cobalt, copper, beryllium, niobium, molybdenum and tungsten. The metal halide is thermally decomposed or reduced by hydrogen under temperatures lower than 1000° to 3000°C so as to deposit the elemental metal. Also, irradiation of high-speed particles and radiations may be employed for decomposing the metal halide as is the case with the metal carbonyl compound. In general, the temperature at which the metal halide is thermally decomposed or reduced by hydrogen is higher than the temperature at which the metal carbonyl is thermally decomposed.

However, it is desirable to control the temperature of the solid material to fall within the range from 300° to 800°C as mentioned previously even for the case of using the metal halide, because irradiation of the ionized krypton particles to the metal halide also contributes to the thermal decomposition or reduction by hydrogen of the metal halide. Needless to say, hydrogen gas should be supplied together with the metal halide where reduction by hydrogen is used for the metal deposition. It should be noted that metal halides such as chromium iodide, titanium iodide, iron chloride, cobalt chloride, zirconium chloride and titanium chloride tend to be readily decomposed by irradiation of high-speed particles. Particularly, metal halides having a low melting point are readily decomposed to deposit the elemental metal by irradiation of the accelerated ions of the radioactive material regardless of the ambient temperature.

Where accelerated ions of a radioactive material are implanted into a solid material, the implantation chamber is kept in general at a vacuum of less than 10^{-3} torr. In order that the implantation of the accelerated ions may not be interrupted, it is preferred to introduce the metal halide or metal carbonyl at a partial pressure of 10^{-1} to 10^{-4} torr.

In this process, at least one compound selected from metal halides and metal carbonyls is used as the metal source, and at least one of thermal decomposition, decomposition utilizing irradiation of high-speed particles, radiolysis, and reduction by hydrogen is employed for liberating the elemental metal from the metal compound. It is important to note that this process permits depositing the elemental metal without interrupting the implantation of the accelerated ions such as ionized krypton 85.

Entrapping Within Metal by Sputtering

R.S. Nelson, S.F. Pugh and M.J.S. Smith; U.S. Patent 4,051,063; September 27, 1977; assigned to the United Kingdom Atomic Energy Authority, England describe a process whereby krypton containing the isotope krypton 85 is entrapped within a solid, for example, nickel metal or copper metal, by bombarding the solid with krypton ions so as to produce krypton bubbles within the solid.

In one embodiment of the process sputtering is utilized to build up the solid.

Bubble size is related to temperature, but typically bubble diameter will be in the region of a few hundred angstroms.

The bubbles would be stable at least up to the temperature at which they were formed. Thus, if the bombarding of the solid with krypton ions was carried out at elevated temperature (e.g., 500°C), the temperature at which release would occur would be well above ambient storage temperatures.

Bombarding the solid at elevated temperatures is therefore believed to provide a means for entrapping gas such that the risk of release is reduced if the solid is accidentally subjected to heat during storage, as for example, during a fire.

In principle, this process is applicable to the entrapping of a wide range of materials for storage. Thus, in addition to krypton, it is believed for example, that xenon, helium and tritium could be conveniently entrapped.

In fact, xenon and krypton are produced together during nuclear fuel reprocessing. However, since xenon has a short half life and commercial value, it would be separated from a krypton/xenon mixture prior to the storage of the krypton.

In carrying out this process to entrap light materials such as helium and tritium it is necessary to take into account that while light materials can be implanted readily, sputtering by light materials is small. Thus to enable sputtering to be utilized to build up the solid a further gaseous material, such as argon, which can be used to give sputtering, is included with the light material to be implanted.

It is believed that up to about 340 liters of krypton at STP could be stored in 1,000 cc (8.9 kg) of nickel in accordance with this process. Using a conventional gas cylinder storage technique only about 170 liters of krypton at STP per liter gas space could be achieved at a cylinder pressure of about 2,200 psi.

Use of a metal to entrap krypton has the advantage that radioactive decay heat during storage would be dissipated.

Once the gas has been entrapped in the solid, the solid itself could be encapsulated to reduce further the risk of release during prolonged storage.

This process also provides apparatus for use in entrapping a material to be stored within a solid by bombarding the solid with ions of the material so as to form a concentration of the material within the solid, comprising a pair of electrodes forming part of a discharge system and means for maintaining about the electrodes an atmosphere containing material to be stored, the arrangement being such that the electrodes can be so energized from an electrical supply that ions of the material to be stored can be implanted and the material thereby entrapped in the solid.

One electrode can form the solid within which the material to be stored will be entrapped.

In one embodiment the apparatus can be in combination with an electrical supply which is controllable so as to build up one of the electrodes by sputtering from the other electrode.

Example 1: Argon was implanted into solid by glow discharge using two plane nickel electrodes separated by a 16 mm gap.

An electrical supply was used to deliver 4 mA over 1.22 cm^2 at 6kV and an atmosphere of argon at a pressure of 100 microns was used to surround the electrodes.

The deposition rate of sputtered nickel was found to be 3.5×10^{-4} g/cm^2/mA-hour.

Example 2: Argon was implanted into nickel by electron-supported discharge using two plane electrodes, a grid and a filament. The electrodes were 5 x 4 cm separated by a gap of approximately 5 cm and the filament was inside a cylindrical grid of approximately 1 cm diameter.

An atmosphere of argon at 12 microns pressure was used to surround the electrodes, grid and filament. The discharge within the grid was 125 mA at 50 V and the electrodes were arranged to receive 30 mA at 500 V negative with respect to the grid.

The deposition rate of sputtered nickel was 1.42×10^{-5} g/cm^2/mA-hour.

Removal of Oxygen with Titanium or Zirconium

In a process described by *A. Tani, Y. Yuasa, A. Watanabe, B. An, M. Soya and H. Tanabe; U.S. Patent 4,125,477; November 14, 1978; assigned to Tokyo Shibaura Denki KK; Nippon Genshiryoku Jigyo KK; and KK Kobe Seikosho, all of Japan* a method is provided for treating radioactive waste gas by treating a waste gas containing oxygen and radioactive gases with one or more of the reactive metals selected from the group consisting of titanium, zirconium and alloys thereof.

In this method the oxygen contained in the radioactive waste gases is adsorbed and hence fixed by the reactive metal thereby avoiding the danger of an explosion caused by the oxygen being converted to ozone. Furthermore, since the oxygen is substantially removed, the degree of concentration of the recovered rare gases can be improved. An additional advantage of the process is that it is not necessary to install a hydrogen tank or a hydrogen-generating apparatus near the recovery apparatus, thus increasing the overall safety of the recovery system.

The gas treated by the method may be any gas as long as it contains oxygen and one or more radioactive rare gases. Thus, the gas may be waste gas from a nuclear power plant. The gas is concentrated by a primary condenser by a factor of about 5,000. The gas concentrated by the primary condenser consists of 10 parts by volume of air (that is a mixture of N_2, O_2, etc.) and 1 part by volume of radioactive gases, (for example, Xe, Kr, etc.).

The reactive metal utilized in this process may be any metal as long as it can remove oxygen from the gas mixture but not fix the rare gases by absorption. More particularly, titanium, zirconium, titanium-base alloys, zirconium-base alloys, zirconium-titanium alloys, and alloys consisting essentially of titanium or zirconium are preferred. These reactive metals remove oxygen from the rare gases by forming such oxides as ZrO_2, and TiO_2 but do not absorb and fix the rare gases such as krypton and xenon. These reactive metals can be used either singly or in combination.

Certain components of the reactive metals other than titanium and zirconium may be used to remove gas components other than oxygen, for example, nitrogen, the types of reactive metals used being determined by the composition of a particular gas to be treated.

One important factor necessary to remove oxygen from the gas mixture is the surface area of the reactive metal; that is, the area on which the reaction with oxygen is effected. The reactivity of the reactive metal increases in proportion to its surface area. As a result, when the reactive metal is used in the secondary condenser for treating the waste gas from the nuclear reactor (in this case the volumes of oxygen and radioactive rare gases contained in the waste gas are substantially equal) the required surface area is larger than 10 cm^2 per 1 ml of the oxygen. In this case, titanium and or zirconium can efficiently remove the oxygen component without appreciable absorption of the radioactive rare gases.

Although the shape of the reactive metal is not limited, in order to improve its surface area and hence the adsorption capability, spherical, granular, or powdery shapes, and wire net, fiber, foil or plate shapes are all preferred.

Where a reactive metal containing one or more of titanium, titanium- or zirconium-base alloys and titanium-zirconium alloys is used, it is possible to remove not only oxygen but also nitrogen to some extent, thus increasing further the concentration of the recovered rare gases.

When incorporated with a suitable component, zirconium-base alloys can remove not only oxygen, but also nitrogen as ZrN. Thus, when a reactive metal containing a desired amount of a zirconium-base alloy incorporated with about 16% by weight of aluminum is used at a temperature of about 400°C, it is possible to remove oxygen together with a considerable amount of nitrogen.

In this reaction system, it is possible to use a mixture of reactive metals just described and a component that can remove gas components other than oygen. For example, for the purpose of removing nitrogen, it is possible to admix the reactive metal with another metal component.

The reaction temperature for absorbing oxygen or nitrogen generally ranges from 400° to 800°C, although it differs in dependence on the type of the reactive metal used. Use of a temperature in the range is not only economical, but it also assures a satisfactory reaction of the reactive metal with oxygen and nitrogen.

Although it may differ depending upon the reaction temperature and the type of the reactive metal used the reaction pressure is generally equal to atmospheric pressure but it may be slightly higher or lower than it for economically carrying out the reaction. Thus, in this process, the reaction pressure is not a critical factor.

Generally, the reaction time ranges from 1 to 50 hours although it may differ depending upon the reaction temperature and the type of the reaction metal used. With a reaction time of less than 1 hour, it is impossible to sufficiently fix oxygen or nitrogen. Too long a time is not economical because it takes a long time to recover the radioactive rare gases.

When the activity of the reactive metal against oxygen and nitrogen decreases substantially, in other words, when the surface of the reactive metal is entirely or substantially covered by ZrO_2 or TiO_2, the reactive metal is discarded.

Sparging Gas Cycle

H. Schnez; U.S. Patent 4,123,484; October 31, 1978; assigned to Kernforschungs-anlage Jülich Gesellschaft mit beschränkter Haftung, Germany describes a process for the removal of fission-product inert gases from the exhaust gases of a plant for the reprocessing of nuclear fuel materials whereby the disadvantages of the various systems are avoided and which can be carried out with technological simplicity and high economy.

The process for the treatment of nuclear fuel materials involves the comminution of the nuclear fuel elements, the combustion or burning of the graphitic components thereof and the chemical solubilization of the nuclear fuel residue in a solubilizing unit through which a sparging gas is passed. The sparging or flushing gas, according to the process, is a part of the filtered fission-product gases recovered earlier and recycled to the solubilizer as the rinsing, sparging or entraining gas.

The fission-product inert gases released in the solubilizing unit can be cleaned in conventional ways and gathered in a compensating storage reservoir or expansion tank. From the latter the sparging gas is withdrawn as required for proper functioning of the chemical solubilizer so that a portion of the fission-product inert gases is recycled continuously. The fission-product gas quantity required for the recirculation to the chemical solubilizer may be supplied thereto via control means maintaining the supplied quantity per unit time substantially constant. The gas recovered from the chemical reactor is separated from the acid, compressed and supplied to a storage vessel from which the reactor is fed with the recycled gas and from which gas bottles may be filled for discharge of the inert gases from the system.

The advantages of the process lie primarily in its simplicity and economy since the sparging gas cycle uses only the fission-product inert gases which are produced by solubilization of the nuclear fuel material. The system is readily controlled and explosive gas mixtures which may result from the formation of ozone or hydrocarbon/oxygen mixtures cannot arise. The formation of such explosive mixtures is known in conventional reprocessing plants.

Still another and highly important advantage of the system is that with the recirculation of fission-product inert gases as the exclusive sparging gas stream, no corrosive gas is developed and the sparging gas undergoes no chemical reactions in the chemical solubilizer or reactor.

According to a feature of the process, before the filtered fission-product inert gas is recycled to the chemical reactor, krypton is removed. This can be effected most advantageously by compressing the fission-product inert gases, cooling them and subjecting them to rectification into a liquid phase consisting predominantly of xenon and a gas or vapor phase consisting predominantly of krypton mixtures. The vapor phase removed from the rectification column thus consists primarily of krypton mixtures at an elevated pressure and can be stored at this elevated pressure. The liquid phase, consisting predominantly of xenon, is at least partially recycled to the chemical solubilizer.

Precipitation of Xenon

W. Koeppe, J. Bohnenstingl and S.G.J. Mastera; U.S. Patent 4,080,429, March 21, 1978; assigned to Kernforschungsanlage Jülich Gesellschaft mit beschränkter Haftung, Germany describe a method of and device for separating krypton from waste gases which become free during the chemical dissolution of core fuel particles to be regenerated, and which method and device will permit a continuous removal of the krypton gas from the waste gas mixture.

The method is characterized primarily in that to the substances bringing about the chemical dissociation, disintegration or dissolution, there is simultaneously together with the core fuel particles added to the chemical dissolution such a quantity of chemically inactive carrier gas which mixes with the waste gases that after purification of the waste gas mixture of gas components such as oxygen, carbon dioxide and nitrous oxide as well as hydrocarbons and water steam and subsequent cooling of the waste gas mixture to the boiling temperature of the liquid nitrogen the quantity of xenon contained in the waste gas mixture is precipitated quantitatively in solid form and that subsequently thereto the waste gas mixture is by means of liquid nitrogen with simultaneous increase in pressure cooled to such an extent that krypton is separated.

By the addition of carrier gas during the chemical dissociation, disintegration or dissolution of the core fuel particles, two things are accomplished according to the process: on one hand, during the chemical dissociation, disintegration or dissolution which is carried out in the low underpressure in order to prevent an uncontrolled discharge or escape of radioactive waste gases, smaller quantities of air are supplied. The oxygen components of the air due to the gamma rays of the isotope krypton 85 will contribute to the formation of ozone so that with greater quantities of ozone, a considerable explosion danger would exist. On the other hand, due to the slight quantity of carrier gas, the total partial pressure for krypton and xenon is reduced to such an extent that during the first subsequent cooling of the gas mixture by means of liquid nitrogen, xenon is precipitated in solid form.

Prior to the cooling-off of the waste gas mixture by means of liquid nitrogen, according to the method, there is provided in an advantageous manner a purification of the waste gas mixture of such gas components as oxygen, carbon dioxide, nitrous dioxide, and hydrocarbons such as water steam which would cause a freezing of the devices in response to a cooling-off with liquid nitrogen. Due to the precipitation of xenon in solid form, not only will the quantity of waste gas contained in krypton be considerably reduced, but it is also advantageous that the precipitated xenon can be used economically as a product.

Heretofore, a natural appearance of xenon has been known only in the atmospheric air. In the air, xenon is contained in very minute quantities, namely in quantities of 8.6×10^{-6} percent by volume. This method opens up a new source of xenon which not only brings about a reduction in price of the industrially interesting xenon gas but also considerably increases the economy of the method for separating krypton.

In order during the separation of xenon in addition to the krypton dissolved in the solid xenon, to freeze out only as little krypton as possible, it is expedient to keep the partial pressures of krypton and xenon in the waste gas mixture as low as possible. It has proved advantageous to add to the waste gas such a quantity of carrier gas that the total partial pressure of krypton and xenon will be reduced to about 16.7 mb. With such a pressure, in addition to the krypton dissolved in solid xenon, additionally only 0.35% by volume of krypton is precipitated. As carrier gas, helium is used advantageously.

A further development of the method consists in that from the waste gas mixture a portion of the krypton content therein is precipitated in solid form, and that the remaining waste gas mixture is substantially conveyed into an adsorber which consists of active carbon and is cooled by liquid nitrogen. On the active carbon, the residual krypton and gas components, still contained in the waste gas such as nitrogen and argon, adsorb so that from the adsorber highly purified carrier gas flows off which expediently is again utilized with the chemical dissolution of the core fuel particles. Those skilled in the art are sufficiently aware of the chemical dissolution process of nuclear fuel particles as preconditioned for this process with the regeneration of fuel elements.

For the economic exploitation of the obtained xenon, it is advantageous to purify the precipitated xenon quantity from the krypton contained in xenon. This is expediently effected by rectification. In this connection, the quantity of krypton still contained in the precipitated xenon is reduced to such an extent that utilization of the xenon is possible without a danger due to radioisotopes.

A gas-separating plant to be connected to a regenerating plant for core fuels for carrying out the method is characterized in that there is provided a chamber with a connection to the dissolver, the chamber being in communication with a supply for the fuel particles and for the carrier gas. The plant is furthermore characterized in that in a waste gas conduit connected to the dissolver there are successively interposed a gas purifier, a nitrogen-cooled container for precipitating solid xenon, a compressor conveying the waste gas mixture to a pressure of approximately four bars, and a nitrogen-cooled separator for krypton with subsequent shut-off valve inserted in the waste gas conduit.

Concentration in Liquid Oxygen

P. Faugeras, P. Lecoq, P. Miquel, H. Rouyer and G. Simonet; U.S. Patent 4,055,625; October 25, 1977; assigned to Commissariat a l'Energie Atomique, France describe a method of treatment of effluent gases derived from the reprocessing of irradiated nuclear fuels constituted by a gaseous mixture containing at least xenon and krypton which is radioactive in air. The method essentially comprises in succession a first stage of purification of the gaseous mixture with a view to removing any impurities such as the nitrogen oxides, carbon dioxide gas, water vapor, and more particularly the hydrocarbons, a second stage of concentration of the xenon and the krypton of the gaseous mixture, the concentration being carried out in solution in liquid oxygen by distillation of the light gases and mainly nitrogen from the liquefied mixture, a third stage of separation of the xenon and the krypton by catalytic combustion of the oxygen of the mixture previously obtained with hydrogen and removal of the water which is formed, and a fourth stage of recovery of the xenon by cryogenic distillation of the mixture of xenon and krypton.

The catalytic combustion of oxygen which preferably forms part of the method can take place in the presence of an excess quantity of hydrogen. One of the advantages of this combustion lies in the fact that it facilitates the final stage of xenon recovery. In fact, the distillation of the xenon-krypton mixture takes place in the absence of oxygen. Although at least the greater part is preferably eliminated, especially by trapping of the krypton and the xenon by cold production, the hydrogen which remains in the mixture distills more readily with krypton than would be the case if oxygen were employed or even more readily than is the case with the use of nitrogen in conventional methods. Moreover, the fact that the xenon-krypton-hydrogen mixture has a reducing action is advantageous in one preferred mode of execution of the method in which the mixture is subjected to a complementary catalytic treatment for removing any traces of nitrogen oxides.

In an alternative embodiment of the process, the catalytic combustion of oxygen takes place in the presence of a slight excess quantity of hydrogen in a gas mixture which mainly contains krypton and xenon derived from recycling of the gases at the end of processing.

The method in accordance with this alternative embodiment permits dilution of the oxygen in the mixture which circulates within the reactor and this dilution limits the heat build-up of the reactor.

The problem of heat build-up in a reactor is of primary importance: if the dilution of oxygen circulating within the reactor is not carried out, there is a considerable temperature rise due to the exothermic reaction of hydrogen with oxygen to form water. In one practical example, the temperature rise due to the exothermic reaction of combustion of hydrogen in oxygen is approximately 160°C per one volume percent of oxygen in the case of combustion in hydrogen and would attain 220°C per one volume percent of oxygen in the case of combustion in a krypton-xenon mixture.

Thus the object of this alternative embodiment is to provide a loop system for the recirculation of gases discharged from the reactor with a view on the one hand to operating with a slight excess quantity of hydrogen only at the reactor outlet and on the other hand with a view to strongly diluting the oxygen which passes through the reactor with inert gases, namely krypton and xenon; the nonrecycled fraction of gases obtained at the reactor outlet is continuously returned after removal of water into a cryogenic distillation column in order to ensure separation and recovery of the xenon in the column.

As already mentioned, the stage of concentration by distillation of the light gases and mainly nitrogen can take place in liquid oxygen, in spite of the radioactive medium. It is even a remarkable fact that the presence of a large quantity of oxygen, as is likely to be the case if the preliminary operations provided in known processes for the removal of oxygen are completely suppressed, offers the advantage of promoting efficient operation at the level of the concentration by ensuring enhanced solubilization of the xenon and the krypton and facilitating complete removal of nitrogen, which also proves beneficial in the final stage of xenon recovery.

It can prove an advantage, however, to prevent accumulation of the formed ozone by means of a preferred mode of execution of the method whereby a fraction of the mixture which has become liquefied during concentration is withdrawn preferably in continuous operation and passed into a trap for dissociation of the ozone which may be present in the fraction before returning this latter into the liquefied mixture.

Moreover, it is preferable to subject the mixture of air and rare gases to purification prior to the concentration stage with a view in particular to removing any hydrocarbons which may be present. This purification process can also be more complete and make it possible in particular to remove the nitrogen oxides, the carbon dioxide gas and the water vapor as in the methods of the prior art for the treatment of gaseous effluents derived from reprocessing of nuclear fuels.

Concentration in Liquid Argon

In a related process, *J. Duhayon, J.-P. Goumondy, A. Leudet and J.-C. Rousseau; U.S. Patent 4,277,363; July 7, 1981; assigned to Commissariat a l'Energie Atomique, France* describe a method of processing of effluent gases arising from the reprocessing of irradiated nuclear fuels constituted by a mixture containing at least radioactive krypton and xenon in air.

The method essentially and successively comprises a first step involving removal of any impurities such as hydrocarbons, nitrogen oxides, carbon dioxide gas, water vapor, a second step involving concentration of xenon and krypton in solution in liquid argon by distillation of the light gases and mainly nitrogen from the liquefied mixture, a third step involving removal of argon by cryogenic distillation of the separated mixture of argon, xenon and krypton which has previously been liquefied, and a fourth step involving separation of xenon and krypton by cryogenic distillation of the separated mixture of xenon and krypton which has previously been liquefied.

In the method under consideration, the argon required for the second concentration step can be introduced into the gas mixture to be processed either before the first step involving removal of various impurities or before the second concentration step.

According to an advantageous feature of the method, part of the krypton which has been separated from the xenon and obtained during the fourth step is introduced into the mixture obtained at the end of the first step before proceeding to the second step of concentration of the xenon and krypton in solution in liquid argon. This reintroduction of a certain quantity of the krypton obtained at the end of the process makes it possible to reduce the pressure at which the second step of concentration of xenon and krypton in liquid argon is carried out.

It is readily apparent that the process also extends to suitable installations for the practical application of the method. Installations of this type comprise in particular catalytic reactors and cryogenic distillation columns with all their ancillary equipment units which are known per se, arranged and connected together by means of pipes for the circulation of different products, thus making it possible to perform the successive operations of the method defined in the foregoing.

Recombiner System

A. Lofredo; U.S. Patent 4,012,490; March 15, 1977; assigned to Airco, Inc. describes a process concerned with providing an improved and simplified system for safely isolating radioactive noble gases from the environment and is applicable to systems for removing krypton and xenon from air, argon, nitrogen, CH_4, CO and mixtures thereof as applied to nuclear power plants of the boiling water reactor, pressurized water reactor, fast flux and breeder reactor types or any other reactors that generate the above gas mixtures. In addition, the process is applicable to the treatment of off-gases such as the foregoing mixture developed by nuclear fuel processing, reprocessing and manufacturing operations.

In accordance with the process, the off-gas from nuclear systems such as a boiling water reactor which is mainly air, is passed first through a so-called recombiner stage wherein the oxygen and ozone and nitrogen oxides that may be formed in the radiation field react with added hydrogen to form water, etc., that is easily removed from the gas stream such as by condensation. The resulting oxygen-free stream from the recombiner composed in general of N_2, H_2O, H_2, CO_2, Kr and Xe including radioactive isotopes, and trace impurities, is dried and then passed through adsorbers for removing both CO_2 and Xe, these two gases being subsequently removed from the adsorbers, separated and collected. The gas stream now free of CO_2 and Xe, is cooled to low temperature and passed for cryogenic distillation to a primary separation column, the Kr with some nitrogen being concentrated as condensate at the bottom of the column. The effluent gas from the column, essentially N_2 and H_2, is advantageously used for system cooling and to elute and remove the CO_2 and Xe from the adsorbers above. The condensate is fed to one of two monitor tanks for temporary storage, the second tank from a previous filling meanwhile feeding a batch still (secondary separation column) which further concentrates the Kr and returns the now "clean" gas to the first column.

By this method, the long-lived krypton is concentrated to a small volume and held in the system where it presents no hazard to the environment. The eluted Xe and CO_2 from the off-stream adsorber are deposited in a freeze-out heat exchanger and later restored to vapor phase by sublimation. The combined gases are then fed to a batch still where they are finally decontaminated of residual traces of Kr, then compressed and bottled for storage and eventual venting after decay of radioactive Xe, or for further commercial use.

Figure 7.2a is a generalized schematic illustration of the off-gas receiving and processing sections of the system.

Figure 7.2: Process to Remove Noble Gases from Off-Gas Stream

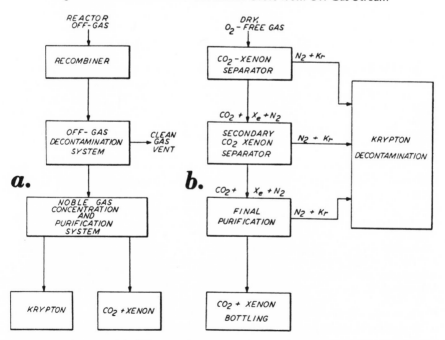

(a) Off-gas receiving and processing sections.
(b) CO_2-Xe separation and N_2-Kr separation.

Source: U.S. Patent 4,012,490

Referring briefly to the generalized schematic of Figure 7.2a, the process essentially comprises as labeled, three major subsystems, namely:

> (1) A "recombiner" which receives the reactor off-gas and wherein hydrogn and oxygen and the like, including radiolytic oxygen and hydrogen are removed by catalytic reaction. Hydrogen is added to react with the elemental oxygen component of air in the feed stream;
>
> (2) An "off-gas decontamination system" wherein carbon dioxide, water and higher hydrocarbons are removed by condensation for precluding degradation of system performance, and wherein xenon and krypton are separated from the process stream by adsorption and cryogenic distillation techniques, respectively, and
>
> (3) a "noble gas concentration and purification system" for concentrating the krypton and xenon, and for producing highly decontaminated vent gas. As indicated, the krypton is separately concentrated and stored, whereas the xenon and carbon dioxide, which together are removed from the process stream by adsorption in the decontaminated system, are subsequently recovered, concentrated and stored together.

Figure 7.2b is a generalized schematic illustration of carbon dioxide-xenon separation and nitrogen-krypton separation respectively, after passage of the process gas stream from the recombiner section.

IODINE

Distillate Recovery System

The processing of radioactive liquid wastes for disposal requires concentration of such wastes to the smallest volume practical. When aqueous radioactive wastes are being processed, the volatility of some isotopes, such as radioactive iodine gas, precludes them from remaining in the concentrate solution resulting in the release of these gases along with the boiled-off water or vapor stream. During the subsequent condensing of these water vapors, a small portion of these volatile gases are often redissolved in the liquid condensate that is otherwise safe for disposal or reuse. This can lower the decontamination factor of such liquid condensate to an undesirable level.

A.N. Chirico; U.S. Patent 4,084,944; April 18, 1978; assigned to Ecodyne Corporation describes a process for the segregation of a gas that prevents such gas from being redissolved in a liquid effluent.

The method for disposing of the radioactive waste gas of a gas-water vapor mixture comprises:

> (a) cooling the mixture so as to condense the water vapor;
>
> (b) passing any uncondensed water vapor and the radioactive gas to a chamber where the mixture is cooled to a lower temperature;

(c)   flowing condensed water vapors downwardly to and
through a sealed boiler; heating the condensed water
vapors in the sealed boiler so as to drive off radioactive gas dissolved in the condensed water vapors and
also to produce additional water vapors which flow
upwardly countercurrent to the condensed water
vapors and thereby strip radioactive gas from the condensed water vapors; passing the radioactive gas and
additional water vapors directly from the sealed boiler
to the chamber without recombining with the mixture;

(d)   filtering liquid droplets from the radioactive gas and
any uncondensed water vapors exiting from the chamber;

(e)   chilling the filtered radioactive gas; and

(f)   recovering the condensed water vapors and the chilled
radioactive gas at different locations.

Adsorption on Silver-Exchanged Zeolite

*T.R. Thomas, B.A. Staples and L.P. Murphy; U.S. Patent 4,088,737; May 9, 1978;
assigned to the U.S. Department of Energy* describe a method for removing radioactive iodines from a waste gas stream and providing long-term permanent storage
for the iodines.

This is accomplished by passing the waste gas stream containing radioactive iodine
through a bed of silver-exchanged zeolite until the zeolite is loaded with iodine,
passing hydrogen gas through the iodine-loaded silver-exchanged zeolite at a temperature of 400° to 550°C to desorb the iodine as hydrogen iodide and regenerate
the zeolite, passing the hydrogen gas stream containing the hydrogen iodide at a
temperature of 80° to 250°C through a lead-exchanged zeolite bed whereby the
iodine is adsorbed on the lead-exchanged zeolite, and permanently storing the thus-
formed zeolite loaded with radioactive iodine.

Synthetic zeolites of the sodalite and mordenite groups may be used to practice
this process. The pore size of the zeolite must be of adequate size to permit the
passage of iodine therethrough. One such zeolite having an adequate pore size has
the following general formula expressed in terms of mol fractions of oxides:

$$0.9 \pm 0.2 M_{2/n}O \cdot Al_2O_3 \cdot 2.5 + 0.5 SiO_2 \cdot 0\text{-}8H_2O$$

where M represents a cation, for example, hydrogen or a metal, and "n" its valence.
This compound is known as zeolite X and is described in U.S. Patent 2,882,244.
Zeolite X is available commercially under the trade name of Linde Molecular Sieve
13X. Zeolite Z has also been found satisfactory (commercially available under the
trade name of Norton Zealon-900) and is more thermally stable and thus with-
stands regeneration better than Zeolite X. Both the granular and spherical forms
of the zeolite are satisfactory.

The composition of a waste gas stream from the process off-gas of a nuclear reactor
fuel-reprocessing plant will depend upon the type of reactor fuel which is under-
going reprocessing. For example, the reprocessing of a high-temperature gas reactor
fuel will produce an off-gas which is approximately 85% CO_2 with the remainder
CO, O_2, N_2, NO and the radioactive contaminants of iodine, tritium, radon and

krypton. The reprocessing of a light water reactor fuel will result in an off-gas which is air, containing water vapor, NO_x, and radioactive contaminants of iodine, tritium and krypton.

The waste gas stream containing the radioactive iodine and other contaminants is first passed through a bed of silver-exchange zeolite which adsorbs and removes the iodine contaminants. The gas stream is then passed through other beds and undergoes additional processing for removal of the other contaminants before it is exhausted to the atmosphere. The waste stream is passed through the silver-ex-changed zeolite until the zeolite is loaded with iodine, at which time the flow of the stream may be diverted to a second bed or stopped. The bed is generally maintained at the same temperature as the gas stream which is generally about 150°C as it comes from the dissolver, to prevent condensation of any moisture which may be present.

However, bed temperatures from ambient to about 300°C have been shown to have no adverse effect upon the ability of the silver-exchanged zeolite to adsorb elemental iodine or methyl iodide except for a slight reduction in bed capacity at the higher temperatures. It might be noted that removal of the iodine from the waste gas stream is by chemisorption rather than physical adsorption, and that the AgI formation is not restricted in CO_2 or air atmospheres but is affected by the increased presence of moisture and NO_x.

The iodine is desorbed from the iodine-loaded silver-exchanged zeolite and the zeolite regenerated by passing hydrogen gas through the zeolite bed at a temperature from 400° to 550°C, the iodine being desorbed as hydrogen iodide which passes into the H_2 gas stream. Temperatures below 400°C produce a very slow desorption rate, while temperatures about 550°C cause a phase change in the silver iodide to a liquid which will result in structural damage to the zeolite and reduced subsequent capacity of the silver zeolite for iodine. Preferably the temperatures of the bed and the gas are about the same.

The hydrogen stream may be pure hydrogen or it may be diluted with an inert gas such as nitrogen, argon, helium, etc., which will slow the hydrogen iodide desorption process. The presence of moisture in the gas will adversely affect the process; therefore as a general practice the system is dried before starting the desorption process. The presence of oxidizing gases such as oxygen are to be avoided as it presents an explosion hazard in the presence of hydrogen. The face velocity of the hydrogen through the zeolite should preferably be at least 100 cm/min in order to attain a reasonable rate of iodine desorption. There is no apparent upper limit as to face velocity of the gas.

The hydrogen gas stream containing the desorbed iodine as hydrogen iodide is then passed through a lead-exchanged zeolite at a temperature of 80° to 250°C, preferably 80° to 200°C, where the iodine is adsorbed on the zeolite and thus removed from the hydrogen gas stream which may then be recycled.

The iodine-loaded lead-exchanged zeolite is then stored in any available appropriate permanent storage facility, which may be underground, in open air in an appropriate container or in some other approved facility for long-term storage.

Ceramic Impregnated with Metallic Salt-Amine Mixture

M. Pasha, D.K. O'Hara and J.A. French; U.S. Patent 4,204,980; May 27, 1980;

assigned to American Air Filter Company, Inc. have found that certain ceramic materials impregnated with mixtures of metallic salts and water-soluble secondary amines when coming in contact with radioactive iodine and radioactive organic iodides trap or adsorb these radioactive materials.

It has been found that many ceramic materials of preselected surface area ranges when impregnated with the salt-amine mixture are susceptible to radioactive iodine adsorption when contacted by a gas stream containing radioactive iodine and organic iodide compounds, the most effective sorbent materials being those wherein the surface area of the ceramic material is between from about 5 to about 250 m^2/g. Furthermore, it has been found that the nitrate salts of silver, copper, cadmium, lead, zinc, nickel, cobalt, cerium, and mixtures thereof when mixed with water-soluble secondary amines and particularly morpholine, piperidine, piperazine, and triethylenediamine and mixtures thereof have strong affinity for chemisorption and binding of the radioactive iodines with the nitrate salts of silver in admixture with triethylenediamine being the most effective.

Even though the exact reaction mechanism is not known, it is believed that the products of reaction between the amine-containing compound upon going into solution with a metallic salt forms a metal-ammonium complex which is positively charged and upon contact with the radioactive iodine and organic iodides, particularly methyl iodide, these radioactive materials have an affinity to react whereby the I^{131} replaces the ammonium in the complex according to the following equation:

$$M(NH_3)_2^+ + (I^{131})^- \rightarrow MI^{131} + 2NH_3\uparrow$$

Even further, it is noted that the percent by weight of the mixture on the ceramic material also has an effect on the affinity of the ceramic material for adsorbing radioactive iodine and iodide. For example, it has been found that ceramic materials impregnated with the salt-amine mixture wherein the salt is in the range of from about 1 to 4% by weight of the ceramic material and the amine is in the range of from about 2 to 5% by weight of the ceramic material have proved to be the most efficient. Also, it has been found that the most effective adsorbing materials are those containing silica or alumina and mixtures thereof and the salt-amine mixture is prepared by dissolving the salt and amine compounds with an organic acid, the salt and amine compounds being present in an aqueous solution.

One preferred radioactive iodine adsorbing material found useful in this process is a ceramic material having a surface area of from about 5 to about 250 m^2/g wherein the ceramic material is selected from the group consisting of silica and alumina and mixtures thereof, the ceramic material being impregnated with a mixture of silver nitrate and triethylenediamine wherein the silver nitrate-triethylenediamine is in the range of from about 3 to 9% by weight of the ceramic material and the weight ratio of silver nitrate to triethylenediamine is from about 0.3 to 1.0.

One preferred method for making a sorbent material for removal of iodine and organic iodide from an iodine-iodide-containing off-gas stream comprises the steps of:

(a) adding a preselected amount of metallic salt to a container containing a preselected volume of water with stirring until all of the metallic salt has gone into solution;

(b)   mixing the two prepared solutions and adding a predetermined amount of an organic acid, preferably acetic
or propionic, to form a homogeneous solution of the
salt and amine;

(c)   impregnating a ceramic adsorbent material with the salt-
amine solution by dipping a predetermined amount of
the ceramic adsorbent material into the solution for a
period from about 10 to 20 minutes; and

(d)   drying the impregnated material for about 6 to 8 hours
at a temperature of from about 60° to 80°C.

Example: A first aqueous solution was prepared by adding with stirring, 3.0 g of
silver nitrate to a first beaker containing 50 ml of water. Stirring continued until
all of the silver nitrate had gone into solution. A second aqueous solution was then
prepared by adding 4.0 g of triethylenediamine to a second beaker containing 50
ml of water, this solution also being stirred until all of the triethylenediamine had
gone into solution. In a third beaker, the first aqueous solution and the second
aqueous solution were admixed. Upon mixing the two solutions, a brown precipitate was formed and the pH of the resulting solution was found to be 8.5. About
0.4 ml of 0.1 normal acetic acid was then added to the solution with the pH rising
to 13.0 during the acetic acid addition. In about 10 minutes the solution was found
to be clear and the pH was 11.0.

To the resulting solution, 100 g of $\frac{1}{16}$" extrusions of silica-alumina having a surface
area of 230 m^2/g was added. This mixture was stirred for about 10 minutes, then
dried in an air-circulating oven for 6 hours at about 65°C.

The product was formed into a bed two inches in depth and placed into a radioactive gas stream containing 1.10 mg/m^3 of radioactive methyl iodide for 2.0 hours,
the temperature of the gas stream being about 24.4°C at a relative humidity of
95.5%. The exposed product was then purged with clean moist air at the aforementioned conditions for 2 hours.

It was found that the radioactive material removed from the gas stream was about
99.77%.

Trapping of Iodine Using Nitric Acid

*M. Anav, J. Duhayon, J.-P. Goumondy, A. Leseur and E. Zellner; U.S. Patent
4,275,045; June 23, 1981; assigned to Commissariat a l'Energie Atomique, France*
describe a method of extraction, trapping and storage of radioactive iodine contained in irradiated nuclear fuels distinguished by the fact that, after dissolving the
fuels in a nitric acid medium, the vapors resulting from this dissolution and consisting of water, nitrogen oxides and iodine are passed into a condenser, then into a
column for the absorption of the nitrous vapors in which is formed recombined
nitric acid containing iodine and nitrous ions, the iodine contained in the recombined acid being then separated out under the favorable influence of the nitrous
ions present in the recombined acid.

In a first embodiment of the method, separation of the iodine contained in the recombined acid is carried out by passing the recombined acid solution into a desorption column in counterflow to a carrier gas which is thereby loaded with iodine
and the iodine thus trapped in the gas is then recovered.

The iodine can be extracted from the carrier gas by circulating this latter in counterflow to an alkaline solution to which is added a reducing agent in an absorption column.

The alkaline solution which leaves the absorption column receives an addition of a lead salt which precipitates the iodine in the form of lead iodide.

As an alternative to the above two steps, the iodine-loaded carrier gas may be passed into a scrubbing tower in counterflow to an alkaline and reducing solution containing a lead salt so as to form lead iodide in a single step.

As an alternative, the iodine-loaded carrier gas may be passed through a column containing a solid adsorbent exchanged with silver and chosen from the zeolites, the molecular sieves, the catalytic supports.

As still another alternative, iodine-loaded carrier gas is passed through a column in counterflow to a wash liquor which contains Pb^{2+} ions and to which is added a make-up quantity of lead nitrate and hydrazine nitrate so as to recover the iodine in the form of lead iodide crystals.

In a second embodiment of the method, in order to separate the iodine contained in the recombined acid, the solution which leaves the recombination column is passed into a distillation unit in which the iodine distills quantitatively in the presence of nitrous ions contained in the solution and is then separated out by cooling.

The iodine crystals formed at the time of cooling are separated from the distillate for the purpose of long-term storage.

The iodine crystals are redissolved in an alkaline solution to which is added a reducing agent in order to transfer this solution to the effluent treatment station.

The alkaline solution containing the dissolved iodine receives an addition of a lead salt so as to precipitate the lead iodide which is conditioned for long-term storage in accordance with a third embodiment of the method.

The recombined acid is introduced into an organic solvent extractor which is then treated by being passed in counterflow to a wash liquor which contains Pb^{2+} ions and to which is added a make-up quantity of lead nitrate and of hydrazine nitrate so as to recover the iodine in the form of lead iodide crystals.

A wash liquor containing a copper salt can also be employed, in which case the iodine is precipitated in the form of copper iodide.

Volatilization of Iodine from Nitric Acid Using Peroxide

G.I. Cathers and C.J. Shipman; U.S. Patent 3,914,388; October 21, 1975; assigned to the U.S. Energy Research and Development Administration have found that residual traces of radioactive iodine may be removed from nitric acid solution by adding hydrogen peroxide to the solution while concurrently holding the solution at the boiling point to give off hydrogen iodide vapors and collecting the iodine thus given off.

In general, iodine may be removed from nitric acid solution by conventional distillation. However, when the iodine level is reduced to a residual level of about

10^{-5} to 10^{-6} M or less, the remaining iodine cannot be removed by such conventional distillation techniques. It is at this point where this process begins. Experimental evidence has indicated that the small amount of iodine which remains in the nitric acid solution is in an oxidized state.

It has been unexpectedly found that hydrogen peroxide appears to act as a reducing agent when added to a nitric acid solution containing iodine. This enables the formation of hydrogen iodide which can be distilled from the solution by boiling. The particular reactions which produce hydrogen iodide are thermodynamically unfavorable; however, it has been found that by keeping the solution at boiling and allowing hydrogen iodide vapors to escape, the reaction will proceed. It is critical to the process that the solution be kept at boiling or otherwise the reaction which produces hydrogen iodide will not proceed. It is further critical that the hydrogen peroxide be added only at boiling or otherwise the hydrogen peroxide will be lost before the reaction can begin.

The formation of elemental iodine is visually apparent in the process. Hydrogen iodide is given off as a colorless vapor. As the vapor nears the collection point, it contacts air and immediately oxidizes to release elemental iodine vapor which is distinctively reddish-violet in color.

The process is generally carried out with a four to six molar nitric acid solution having an iodine molarity of less than 5×10^{-4} M. Hydrogen peroxide is generally added to the solution as a 6 wt % water solution (about 2 M). However, any concentration of hydrogen peroxide may be used so as to maintain superficially an H_2O_2 concentration after dilution and before reaction or decomposition of at least 4×10^{-3} M (0.012 wt %). Preferably the hydrogen peroxide is added continuously during the process to replace that which is lost due to reaction. The solution is maintained at boiling during the entire process. Boiling generally occurs in the system at a temperature between about 105° to 115°C.

The reaction which occurs to create the hydrogen iodide is thought to be as follows:

$$I_2 + H_2O_2 \rightarrow 2HI\uparrow + O_2$$

A similar reaction is thought to occur with iodine in the higher oxidation states. As pointed out before, this reaction is thermodynamically unfavorable (i.e., $\Delta G > 0$), and the solution must be kept at boiling in order for it to proceed.

The process is preferably carried out in a stepwise fashion of successive distillations followed by replenishing the solution to its initial volume and initial iodine concentration. Nonradioactive iodine is added to replace the lost radioactive iodine and to act as a carrier for the remaining radioactive iodine. Prior to distillation and hydrogen peroxide addition the solution may be pretreated by bubbling with ozone to minimize the effect of organic impurities which have the effect of tying up the radioactive iodine as organic iodide species. The hydrogen peroxide solution may also be treated with ozone. The hydrogen peroxide also appears to promote isotopic exchange between the nonradioactive iodine diluent and the radioactive iodine which is tied up by the organic species. By this exchange the radioactive iodine may be distilled leaving the nonradioactive iodine with the organic species.

Metal Iodine Adsorbent on Carrier

K. Kamiya, H. Yusa, M. Kitamura, M. Takeshima and T. Ishidate; U.S. Patent

4,234,456; November 18, 1980; assigned to Hitachi, Ltd., Nippon Engelhard Ltd., Japan describe an adsorbent whose iodine removal efficiency is not greatly lowered in a highly humid atmosphere.

The process is characterized by supporting an iodine-adsorbing metal on a carrier having a large number of pores having a mean pore size of 200 to 2000 Å, and preferably by a pore volume of a carrier being at least 0.1 cc/g.

Procedure for preparing a silver-alumina adsorbent will be described in detail below.

Alumina carrier is prepared in the following manner: 4 N aqua ammonia is slowly added to an aqueous 10 wt % aluminum sulfate solution, and pH is adjusted to 5 to 10. The resulting mixture is left standing for 24 hours, and then filtered, and the resulting cakes are washed with clean water. The resulting alumina gel is dried at about 120°C, whereby alumina powders are obtained. 1 kg of the alumina powders and 30 g of bentonite are added to 15 cc of nitric acid/ℓ H_2O, and thoroughly kneaded by a kneader mixer. Then, the kneaded mixture is extruded from dies having an opening diameter of 1 mm, shredded, and then granulated into spheres by a granulator. The resulting spherical particles are calcined at a temperature of 800° to 1400°C, whereby alumina carrier particles having diameters of 1 to 2 mm are obtained, where mean pore sizes of the carriers can be adjusted to a range of 200 to 2000 Å by setting the calcining temperature in the stated range, and the pore volume of the carrier at pore sizes of 200 to 2000 Å can be adjusted by adjusting the calcining temperature and pH in the step of alumina gel formation.

Silver is supported on the alumina carrier particles by the following treatment: 50 g of alumina carrier particles is placed in 50 cc of 0.1 N nitric acid, retained at room temperature for 15 minutes, then filtered, admixed with 8.2 g of silver nitrate/20 cc of H_2O, and dried at about 100°C.

In the foregoing description, alumina and silver are used as the carrier and the metal supported, respectively, but the adsorbent is not restricted to the alumina and the silver. That is, the percent iodine removal can be enhanced to the level of the alumina by using other carriers such as silica gel, etc., so long as their mean pore sizes are adjusted to 200 to 2000 Å, and their pore volumes are adjusted to at least 0.1 cc/g.

500 cc of water glass comprising silicon oxide and sodium oxide is dissolved in 1 ℓ of water, and a solution of 245 cc of 4 N hhydrochloric acid/300 cc of H_2O is prepared separately at the same time. These two solutions are joined together, and the resulting mixture is left standing for about 1 hour, while gels are formed in the mixture. Then, the mixture is treated by an aqueous 1 N ammonium nitrate solution, washed with water, dried at 150°C for 4 hours, then granulated, and calcined at 500° to 1000°C in air, whereby silica gel carrier particles having mean pore sizes of 200 to 2000 Å are obtained. The pore size of the carrier depends mainly upon a calcining temperature.

Silver is supported on the silica gel carrier particles in the following manner: 100 g of the silica gel carrier particles are placed in 100 cc of 0.1 N nitric acid, maintained at room temperature for 15 minutes, then filtered, admixed with 16.4 g of silver nitrate/40 cc of H_2O, and then dried at about 100°C, whereby a silver-silica gel adsorbent is obtained.

It is more difficult in the case of the silica gel carrier to adjust the mean pore size to 200 to 2000 Å than in the case of alumina carrier, and the silica gel is more fragile when it absorbs water.

As an iodine-adsorbing metal to be supported on the carrier, copper, lead, etc., are also effective besides silver. Amounts of CH_3I adsorbed by the respective metals are shown in the following table:

Metal Supported in Nitrate Form	Amount of CH_3I Adsorbed (mmol/mol metal)
Ag	0.26
Cu	0.16
Pb	0.15
Zn	0.13
Cd	0.12
Ni	0.10
Co	0.08

These values are measured according to a static adsorption test at a gas temperature of 56°C for an adsorption time of 90 minutes.

Charcoal Impregnated with Salts of Iodine Oxyacids

V.R. Deitz and C.H. Blachly; U.S. Patent 4,016,242; April 5, 1977; assigned to the U.S. Energy Research and Development Administration describe a method of removing methyliodide[131] gas from the effluent of a reactor comprising passing the effluent gas through a charcoal sorbent formed by first contacting charcoal with an aqueous mixture of a first component comprising a salt of the iodine oxyacids selected from the group consisting of periodate, iodate and hypoiodite and a second component selected from the groups consisting of iodine and iodide salt, the aqueous mixture being adjusted to a pH of about 10, and then contacting the resulting impregnated charcoal with a tertiary amine.

Preferably, the alkali salts of the iodine oxyacids, for example, KIO (potassium hypoiodite), KIO_3 (potassium iodate), KIO_4 (potassium periodate) are used together with an alkali iodide, such as KI (potassium iodide) and/or elementary iodine (I_2). Although hypoiodites are not particularly stable compounds, the hypoiodite group was found to be stable when adsorbed on charcoals, especially when adsorbed from solutions of pH 10 and above.

In accordance with the process, this stable hypoiodite group is capable of entering into exchange with radioactive methyliodide. The hypoiodite group forms from a redistribution among the iodine components of the mixture. That is, the hypoiodite species is formed and even resupplied on standing or after exchange as a result of the iodine in this solution being distributed among the following charges: I^-, I^0, I^{+1}, I^{+5}, I^{+7}. This occurs in the mixture as one of the reactant species is chemically oxidized and the second species is chemically reduced, forming an equilibrium mixture with different charges on the iodine.

An advantage of this process is in the choice of one of several specified chemical operations that yields hypoiodite reactive species in the impregnating solution and also in the use of a two-step process of impregnation constituting contact of charcoal with the reactive mixture followed by contact with a tertiary amine.

The two-step process of impregnation results in the minimum penetration of radio-active methyliodide through the charcoal bed. An example is the use of KIO_3 + KI (or I_2) + KOH (to form the hypoiodite species) as the first step of impregnation and hexamethylenetetramine (HMTA) as a second step of impregnation. In this example, the total iodine was maintained at between 0.5 to 4 g/100 g of charcoal; KOH was added to form a clear solution at pH ~10. The amine was maintained at 5 g/100 g of charcoal. The solutions in droplet form in pressured air were directed into a slowly rotating bed of charcoal in a cylinder equipped with inclined lifts.

The results shown in the table below are for independent impregnations using a coal-base charcoal. The table shows figures which show a penetration well within the regulatory requirements set forth in Regulatory Guide 1.52 published by the Nuclear Regulatory Commission. The requirements call for passage of less than 5% of radioactive methyliodide through a 2-inch test bed. The test conditions used were as follows: bed diameter 2-inch; height 2-inch; contact time 0.26 second; prehumidification 18 hours at 96 relative humidity.

Sample	Impregnant for 100 g charcoal		% Penetration
	Step 1	Step 2	
4250	KIO_3, KI, KOH	HMTA, 1g	0.26
4251	KIO_3, I_2, KOH	HMTA, 2g	0.47
4254	KIO_3, KI, KOH	HMTA, 1g	0.54
4260	KIO_3, KI, KOH	HMTA, 2g	0.30
4264	KIO_3, KI, KOH	HMTA, 2g	0.36
4266	KIO_3, KI, KOH	HMTA, 2g	0.50
4267	KIO_3, KI, KOH	HMTA, 2g	0.25
4268	KIO_3, I_2, KOH	HMTA, 2g	0.38
4269	KIO_3, I_2, KOH	HMTA, 2g	0.16
4271	KIO_3, I_2, KOH	HMTA, 2g	0.43
4272	KIO_3, I_2, KOH	HMTA, 2g	0.22
4273	KIO_3, I_2, KOH	HMTA, 2g	0.18

A series of impregnations were made with the source of the iodine in the following binary mixtures: KIO_3 and KI, KIO_3 and I_2, KIO_4 and KI, KIO_4 and I_2. A two-step impregnation was made on the coal-base charcoal; the total iodine was again maintained at between 0.5 and 4 weight percent of the charcoal. The first step of the impregnation was with one of the above mixtures and the second step was with 1% HMTA by weight of charcoal. To show the contribution of the tertiary amine in this process, the impregnation was repeated without the HMTA. The results are given in the table below. The tertiary amine participates by way of an addition reaction in the removal of radioactive methyliodide as opposed to exchange of radio-active methyliodide with the hypoiodite species. The test conditions used were as follows: 2-inch diameter; 2-inch height; contact time 0.25 second; prehumidification 18 hours at 96 relative humidity.

Sample	Step 1	Step 2	% Penetration
4250	KIO_3 and KI	1% HMTA	0.26
4251	KIO_3 and I_2	1% HMTA	0.47
4252	KIO_4 and KI	1% HMTA	0.18
4253	KIO_4 and I_2	1% HMTA	0.54
4254	KIO_3 and KI	none	0.54
4255	KIO_3 and I_2	none	0.56
4256	KIO_4 and KI	none	0.44
4257	KIO_4 and I_2	none	0.51

The iodine in all of the iodine-containing compounds has been varied from a total of 0.5 to a total of 4 wt % of the charcoal; about 2% appears to be optimum. The tertiary amine was varied from 1 to 5 wt % of the charcoal, the source of the charcoal being an important factor in establishing the optimum amount to use.

The high-flash-point amines such as HMTA do make significant contributions to a successful formulation with respect to high-temperature desorption, to an elevation in the ignition temperature, and to the prevention of iodine leakage in the event of decomposition at some elevated temperature before ignition is reached. Examples of the high-flash-point amines which can be used are hexamethylenetetramine, triethanolamine, N-methylmorpholine, N,N,N',N'-tetramethylethylenediamine, 1,-dimethylamino-2-propanol, N-methylpiperazine, etc.

Effective trapping of radioactive methyliodide can be achieved by using for step 1 a water mixture of $KI + I_2 + KOH$ inasmuch as the hypoiodite group can also be formed in this mixture, but a greater amount of this reactive species will be generated by adding the oxyacid initially to the solution made up for step 1.

Example: One hundred grams of coal-base charcoal, dried at 100°C was impregnated in sequence with the following two solutions: Solution 1 was made up of 1.29 g of KIO_3 (potassium iodate), 1.00 g of elementary I_2, 0.06 g of KOH and 25 ml of water. Solution 2 contained 5 g of HMTA in 25 ml of water. Solution 1 was injected as a fine spray into the rotating bed of charcoal and followed immediately by injection of the second solution as a fine spray. This operation was followed by a fine spray of 20 ml of water. The bed of charcoal was continued to be rotated for 20 minutes. The product was free-flowing with no apparent film of liquid.

The product was then dried in an open air oven at 95°C overnight. This dried product was placed in a filter 2 inches in diameter and 2 inches in height and prehumidified with air at 96% relative humidity at a flow of 25 liters per minute. (This step was performed to conform with stringent regulatory test procedures to have the charcoal perform under extreme adverse humidity conditions.) After humidification 5.49 mg of radioactive methyliodide was injected into the air stream at intervals of 2 hours, and the flow of air continued for 2 hours. (The passage of this amount of radioactive methyliodide, 5.49 mg, constitutes a very large amount of radioactive contaminant from a nuclear reactor, and is an amount recommended by the Nuclear Regulatory Commission for testing the suitability of filters for recovery of this type of contaminant.) The test charcoal, and the additional charcoal in the back-up beds were then counted for radioactivity, and the penetration calculated. The penetration in the test charcoal was found to be 0.30%. While the example is limited to HMTA, other high-flash-point tertiary amines can be used.

Immobilization of Iodine in Sodalite

H. Babad and D.M. Strachan; U.S. Patent 4,229,317; October 21, 1980; assigned to the U.S. Department of Energy describe a process whereby radioactive iodine is incorporated into an inert solid material for long-term storage. The radioactive iodine which is present as alkali metal iodide and/or iodates in an aqueous solution is incorporated into an inert solid material by adding to the solution a stoichiometric amount, with respect to sodalite ($3M_2O \cdot 3Al_2O_3 \cdot 6SiO_2 \cdot 2MX$, where M = alkali metal; $X = I^-$ or IO_3^-) of an alkali metal, alumina and silica, stirring the solution to form a homogeneous mixture, drying the mixture to a powder, and compacting and sintering the powder or hot pressing the powder at 1073° to 1373°K (800° to 1100°C) and 6.9 to 17.2 MPa (1,000 to 2,500 psi) for a time sufficient to form a sodalite. The alkali metal is preferably sodium, potassium or a mixture thereof and may be added as a hydroxide while the alumina and silica may be added as a kaolinic clay. Alternately, the alkali metal and alumina may be added together as an alkali metal aluminate while the silica is added as a sol.

The aqueous solution of sodium or potassium iodide and/or iodate may be readily prepared from any of the processes for recovering iodine from off-gas streams with the exception of the mercury process. For example, the HI recovered from silver-exchanged zeolite may be bubbled through a solution of sodium or potassium hydroxide. The HI_3O_8 recovered from the Iodex process could be dissolved in sodium or potassium hydroxide. Iodine which has been recovered using nitric acid will have to be neutralized and any nitrates present removed to prevent oxidation of the I^- to the more volatile I_2 during heating. Any radioiodine in a solid form can be readily dissolved in an alkali metal hydroxide. The concentration of iodine in the aqueous solution is not critical nor is the concentration of alkali metal since these can be adjusted for when adding the other constituents for preparing sodalite.

To this solution is added a stoichiometric amount or a slight excess of alkali metal, alumina and silica to form iodide or iodate sodalite. These may be added as several different compounds in order to obtain the stoichiometric quantities necessary to prepare the sodalite. For example, the alkali metal may be added to the solution as the hydroxide while the alumina and silica can be added together as a hydrated kaolinic clay. Alternately, the alkali metal and alumina can be added together as an alkali metal aluminate while the silica is added as a sol.

It is important that the aqueous solution containing the compounds to make sodalite be stirred well to form a homogeneous mixture of the compounds in order to ensure the formation of sodalite and because it helps to reduce iodine volatility during the heating step. After the homogeneous mixture is formed, the solution is dried to form a powder, preferably with the application of heat. The dried homogeneous powder can then either be hot pressed or compacted and sintered for a time sufficient to form sodalite. Hot pressing at 6.9 to 17.2 MPa (1,000 to 2,500 psi) with a die body temperature of 1073° to 1373°K (800° to 1100°C) was adequate to form a sodalite having greater than 90% of the theoretical density while cold pressing and sintering at the same temperature will provide similar results. Care should be taken to prevent a temperature which is too high to prevent volatilization of the iodine.

Example 1: To a solution containing 26.02 g NaI and 20.8 g NaOH (27.4 ml, 19 M NaOH) was added 67.17 g KCS Kaolin Clay (Georgia Kaolin Co.) and the resulting mixture thoroughly mixed. The solution was allowed to stand for 16 hours and evaporated to dryness in an oven at 100°C. The resulting product was hot pressed at 1080°C (die surface temperature) at 2,000 psi ram pressure to form iodide sodalite. Upon testing, the sample was found to contain 12.7% iodine and the product density was found to be 2.45 g/cm^3 which is greater than 90% of the theoretical density.

The 96-hour leach rate for this material at room tempeature was found to be about 2.62×10^{-6} kg/m^2·s [2.9×10^{-5} g/(cm^2·day)] based on analysis for iodine. This leach rate was comparable to that of the average leach rate for glass. A Soxhlet leach test at 100°C was also made and was determined to be 6.4×10^{-4} kg/m^2·s) [7.0×10^{-3} g/(cm^2·day)] based on the analysis for iodine and on sample mass loss.

Example 2: To 30 ml H_2O was added 4.27 g $NaAlO_2$ followed by 2.6 g NaI. The resulting solution has a pH of 11.7. To this was added 7.8 g Ludox As-40 which is a 30% solution of colloidal silica stabilized with ammonia and the solution heated and stirred. A gel formed which was then dried at about 343°K (70°C). The resulting powder was hot pressed at 1173°K (900°C) and 13.8 MPa (2,000 psi) to form sodalite.

Simultaneous Recovery of Iodine with Cesium, Rubidium and Tritium

M. Anav, A. Chesne, A. Leseur, P. Miquel and R. Pascard; U.S. Patent 4,180,476; December 25, 1979; assigned to Commissariat a l'Energie Atomique, France describe a process for the extraction of fission products which have been subjected to a temperature of at least 1200°C during their irradiation, prior to dissolving the fuel by the wet process, characterized in that after mechanically treating the elements in order to decan and/or cut them the treated elements are brought into contact with water so that the fission products pass into an aqueous solution, the treated elements are then separated from the thus-obtained aqueous solution and at least one of the fission products is recovered from this aqueous solution.

According to the process, the fission products are iodine, cesium, rubidium and tritium and at least the iodine is recovered from this aqueous solution.

The treated elements are brought into contact with water at a temperature between 20° and 100°C and preferably at a temperature close to 100°C.

The above-defined process has the particular advantage of making it possible to ensure by treatment with water a simultaneous extraction of all the iodine and most of the cesium, rubidium and tritium contained in irradiated nuclear fuel elements prior to dissolving the latter, thus making it impossible for the radioactive iodine to be given off during the subsequent processing stages of the irradiated fuel elements.

The elimination of most of the radioactive cesium, rubidium and tritium also decreases the radioactivity of the dissolving solution.

Moreover, the obtaining by this process of an aqueous solution containing at least 95% of the radioactive iodine present in the irradiated fuels makes it possible to ensure in simple manner the recovery of the radioactive iodine, followed by its treatment in a stable and concentrated form with a view to its long-term storage.

According to another feature of the process, the aqueous solution obtained is clarified by bringing the elements into contact with water prior to recovering at least one of the fission products from the aqueous solution in order to avoid the entrainment in the latter of fine fuel particles.

According to another feature of the process, the treated elements undergo rinsing with water after separating them from the aqueous solution and at least one of the fission products is recovered from the rinsing water.

In this case the aqueous solution obtained during the contacting of the treated elements with water is advantageously concentrated by distillation and the condensed vapor obtained during this distillation is used for the rinsing of the treated elements.

The aqueous solution obtained can be treated in different ways in order to simultaneously or separately recover the various fission products contained therein.

According to a first embodiment of the process, the radioactive iodine is recovered from the aqueous solution, preferably by precipitation and the aqueous effluents obtained after separating the radioactive iodine are added to the solution for dissolving the fuel.

According to a second embodiment of the process, the iodine and cesium are successively recovered from the aqueous solution and the aqueous effluents obtained after separating the iodine and cesium are added to the dissolving solution for the fuel and/or to the rinsing water for the treated elements.

According to a third embodiment of the process certain of the fission products are recovered from the aqueous solution by concentrating it by distillation in order to obtain a concentrate containing the fission products and the concentrate is treated in order to ensure a long-term storage of the fission products, for example, by subjecting the concentrate to vitrification or covering with bitumen.

This latter embodiment of the process takes advantage of the fact that radioactive iodine is present in the irradiated fuel in iodide form, for example, cesium or rubidium iodide, and that these alkaline iodides are very soluble in water and very stable thermally. Moreover rubidium and cesium, which may be present in the irradiated fuel without being combined with the iodine, form in aqueous solution hydroxides which are also very soluble and thermally very stable, making it possible to obtain a concentrate containing most of the fission products present in the aqueous solution and in particular iodine, cesium and rubidium, whereby they can be directly treated by vitrification or covering with bitumen in order to bring about the long-term storage of iodine. Furthermore due to the thermal stability of alkaline iodides vitrification or covering with bitumen of the concentrates obtained can be effected without there being any liberation of the corresponding halogen.

Moreover, this latter embodiment of the process is advantageous because it does not require the introduction of extraneous elements such as lead or copper for bringing about the recovery of the radioactive iodine and also makes it possible to obviate certain technological difficulties linked with the separation of the precipitates formed, as well as problems caused by the recycling and treatment of precipitation mother liquors.

In the latter embodiment of the process the treated elements separated from the aqueous solution can also be rinsed with water. In this case it is advantageous to use the water vapor obtained during concentration by distillation of the aqueous solution for rinsing the elements.

Purification Bypass Flow System

R.P. Colburn; U.S. Patent 4,075,060; February 21, 1978; assigned to Westinghouse Electric Corporation describes a primary system of a nuclear reactor such as a liquid metal-cooled fast breeder nuclear reactor which includes a purification bypass flow system having a means which effectively removes radioactive fission products from the reactor coolant. The purification bypass flow system comprises a separate flow system coupled in parallel with the primary system of the nuclear reactor. It allows a small portion of the contaminated reactor coolant to be bypassed from the primary system, purified of the contaminants and then returned to the primary system. Continued operation of the bypass flow system reduces the concentration of the fission products so that the overall radioactivity of the reactor coolant is reduced to a safe level. Consequently, the deposition of the fission products is reduced and the resulting radioactivity of the primary system components is also reduced to a safe level.

The bypass flow system includes apparatus to introduce hydrogen, at a controlled rate, into the reactor coolant flowing therein. A suitable example of such apparatus comprises thin-walled stainless steel tubing which becomes permeable to hydrogen at the operating temperature of the reactor coolant. The reactor coolant flows past the thin-walled tubing and absorbs a predetermined quantity of hydrogen. The hydrogen, tritium, cesium and iodine comprising the reactor coolant contaminants remain as ions in the liquid sodium. Upon lowering the temperature of the liquid sodium and the contaminants contained therein, sodium hydride precipitates out of solution. It has been found that this precipitate contains cesium 137, iodine 131 and tritium ions. The precipitation occurs in a solid form in an apparatus provided therefor and commonly referred to as a cold trap.

Thus, by intentionally adding hydrogen, which itself is a contaminant, above the normal concentration present in the sodium, and then precipitating the hydrogen out of solution the coprecipitation of the iodine, cesium and tritium radioactive isotopes is enhanced. In this manner, the reactor coolant is effectively purified of radioactive contaminants, thereby preventing the primary and the intermediate systems from becoming radioactive and eliminates the possible health and safety problem of the prior art.

M.H. Cooper; U.S. Patent 4,204,911; May 27, 1980; assigned to Westinghouse Electric Corporation describes a related process in which the bypass flow system includes apparatus to introduce nonradioactive isotopes of the radioactive element into the reactor coolant.

For example, natural iodine is added where the radioactive nuclide is a radioactive isotope of iodine such as iodine 125 or iodine 131. A wire mesh basket containing the natural iodine and housed in a conventional sodium sample tube may be used for this purpose. The reactor coolant flowing past the basket dissolves the iodine until the solubility limit of iodine in the sodium at the temperature of the sodium is reached. Then the temperature of the liquid sodium is lowered as it flows through a cold trap. This causes the sodium iodide to precipitate out of solution onto a wire mesh surface provided in the cold trap. It has been found that the precipitated sodium iodide contains the fission product nuclides, iodine 131 and iodine 125. In this manner, the reactor coolant is effectively and simply purified of iodine.

TRITIUM

Combining of Coolant with Oxygen

J.N. Ridgely; U.S. Patent 3,937,649; February 10, 1976 describes a safe process and system for the removal of radioactive tritium from high-temperature gas-cooled atomic reactors without polluting the environment and for removal of tritium from any system containing an inert-to-oxygen circulating fluid which becomes tritiated.

These and other objects are achieved by a process and system in which part of the reactor coolant which becomes permeated with tritium is continually removed and processed to remove the tritium. The process involves combining the removed reactor coolant under conditions of elevated temperature and pressure with gaseous oxygen, so as to result in a tritiated water vapor formation reaction from the tritium in the reactor coolant and the gaseous oxygen. The tritiated water vapor and the remaining gaseous oxygen are then successively removed by fractional liquefaction steps. The liquefied tritiated water vapor is then removed from the processing system and safely disposed of; the liquefied gaseous oxygen is used as cooling means in the water vapor liquefaction step and then used as the gaseous oxygen combined to form the water vapor; and the now untritiated reactor coolant is returned to the reactor for recirculation.

The process system is designed against accidents through the inclusion of radiation monitors at points immediately after removal of the reactor coolant from circulation and immediately prior to its return to recirculation, and through pressure and temperature sensors connected through electronic controls to fast-acting pneumatic valves which immediately shut the processing system down in case of any malfunction. An additional provision is the heating of the reactor coolant prior to its return to be recirculated by means of a heat exchange with ordinary circulating water, thus resulting in a supply of chilled water for use elsewhere in the reactor system and supporting environment.

Addition of Oxygen

In the process of *H. Hesky and A. Wunderer; U.S. Patent 4,206,185; June 3, 1980; assigned to Hoechst AG, Germany* a method is detailed for retaining and separating tritium and other volatile radioactive fission products, if any, obtained in the processing of spent nuclear fuel, in which method no contaminated gases are formed that have to be purified and/or stored.

To solve the aforesaid problem a process is provided which comprises adding oxygen to the liquid radioactive waste products, reducing in a subsequent stage the oxygen enriched with tritium and separating the reduction product from the tritium.

To separate the tritium from the oxygen the latter can be reduced and the higher nitrogen oxides formed can be condensed or reacted to give nitric acid. Condensed tritium water possibly formed is recycled and used to dilute the liquid, radioactive waste products. The tritium water dissolved in the nitric acid formed can be passed into the working up process together with the nitric acid. The gaseous tritium and the liquid, radioactive fission products contained therein are stored.

Rubidium, a decay product of krypton, which is likewise formed in the nuclear fission of uranium and contained in the gases formed in the dissolution of spent nuclear fuel, forms hydrides with hydrogen and tritides with tritium. It is, therefore, advisable to store the tritium together with krypton. To this end the oxygen charged with tritium is contacted with a mixture of krypton and NO as obtained in the working up of the gases formed in the dissolution of the nuclear fuel. The tritium and krypton are then separated from oxygen or NO in the manner described above.

Precious Metal Catalyst

M.F. Collins and R. Michalek; U.S. Patent 4,178,350; December 11, 1979; assigned to Engelhard Minerals & Chemicals Corporation describe the removal from a process of other gas stream of the radioactive tritium, and/or compounds of tritium, by reacting the tritium or tritium compounds such as tritiated water, and/or tritiated hydrocarbons, hereinafter known as "tritiated species," over a precious metal catalyst with sufficient air or oxygen to convert all of the tritiated species to tritiated water and, in the case of hydrocarbons, to carbon dioxide. At this point in time, all of the tritium has been converted to tritiated water. The tritiated water fraction is removed by serially adsorbing it on a pair of desiccant dryers wherein the effluent from the first desiccant dryer is diluted with nonradioactive water at a point upstream of the inlet of the second desiccant dryer, in which second dryer the mixture of tritiated and nonradioactive waters is adsorbed. Preferably, the effluent is swamped or flooded with ordinary nonradioactive water (protium oxide, or even deuterium oxide for that matter) to dilute the tritiated water by a factor of at least about 1,000:1. Since the isotopes of hydrogen have transport properties similar to protium, the tritiated water behaves exactly like the nonradioactive water and the total water mixture can then be adsorbed on a second desiccant dryer to a concentration of less than one-half part per billion tritiated water. By diluting to a value more than about 1,000:1, or by multiple-stage dilution, with nonradioactive water, concentration levels in the order of parts per trillion are readily achieved.

Extraction of Tritium from Heavy Water

M. Shimizu; U.S. Patent 4,173,620; November 6, 1979; assigned to Doryokuro Kakunenryo Kaihatsu Jigyodan, Japan describes a method of producing a high purity tritium gas by extracting tritium from tritium-containing heavy water. According to this process, a part of the tritium-containing heavy water from a heavy water source is first led to an exchange reaction column. In the exchange reaction column, the heavy water is brought into countercurrent contact with a tritium-containing heavy hydrogen to thereby transfer tritium in the heavy hydrogen into the heavy water by way of the exchange reaction.

From the bottom of the exchange reaction column is withdrawn the heavy water having an enriched tritium content, and from the top of the exchange reaction column is withdrawn the heavy hydrogen having a reduced tritium content. The thus-resulting tritium-enriched heavy water is led to a heavy hydrogen gas generator, such as an electrolytic cell, to generate a tritium-enriched heavy hydrogen. A part of the tritium-enriched heavy hydrogen is recycled to the exchange reaction column to carry out the countercurrent contact between the heavy water and the heavy hydrogen. The remaining part of the tritium-enriched heavy hydrogen is led into a hot-wire type thermal diffusion column, e.g., cascade to enrich tritium, and the thus-enriched tritium gas is withdrawn from the thermal diffusion column cascade as a product. The tritium-depleted heavy hydrogen withdrawn from the top of the exchange reaction column is burnt to produce a tritium-depleted heavy water which is then recycled to the heavy water source.

Reextraction of Tritiated Water

C. Bernard; U.S. Patent 3,954,654; May 4, 1976; assigned to Saint-Gobain Techniques Nouvelles, France describes a process for separating tritium in a plant for the treatment of spent fuel elements in which:

(1) After shearing into fragments, the fuel elements are dissolved in nitric acid.

(2) The solution thus obtained is extracted by a suitable solvent which is thereafter washed by an aqueous solution of nitric acid. There is thus obtained an aqueous phase containing tritiated water and the major portion of the fission products and, in addition, an organic phase containing uranium, plutonium and a small proportion of fission products.

(3) Simultaneously or separately there are extracted the uranium and the plutonium in aqueous phases which are treated in subsequent extraction cycles for separation of the uranium, plutonium and residual fission products.

(4) The organic phase is treated to purify it prior to recycling, the first step in which treatment may consist of a washing with a sodium carbonate solution and subsequent steps being optional alkaline and acidic washings.

(5) The various fractions containing fission products are then treated in order to concentrate at least the most active ones and optionally to solidify them, the tritiated water and the nitric acid being thereby separated for recycling in the plant.

The process is characterized by the fact that after a preliminary extraction and dissolution of the irradiated fuels and an initial washing of the organic phase obtained in this extraction, the washing being effected with aqueous nitric acid solutions containing tritiated water, the organic phase is subjected to a second washing with an aqeuous solution of nitric acid which is free of tritium and the amount of which solution involved is substantially lower than that of the organic phase, all for the purpose of reextracting the tritiated water dissolved in the contaminated solvent so as to avoid contamination with tritium in the subsequent stages of the process.

According to one feature of the process, the quantity of nitric acid solution free of tritium is desirably of the order of 1/100 the volume of the organic phase. This ratio may, however, vary from an average value therefore by a factor between 1 and 3.

According to another feature of this process, the quantity of water which must be introduced into the part of the plant containing tritiated products is reduced by use of anhydrous and/or concentrated reactants.

According to still a further feature of the process, the excess water to be eliminated from the cycle under tritiated conditions is adjusted as a function of the quantity of tritium present. Advantageously, this water is removed by a process of rectification from the nitric acid solutions.

Removal of Tritium Using Basic Materials

C.J. O'Brien; U.S. Patent 4,085,061; April 18, 1978; assigned to Aerojet-General Corporation describes a method and apparatus for removing and concentrating tritium from the cooling water of a nuclear reactor without using excessive material or energy. Tritium is first transferred from the reactor cooling water to a more concentrated distribution, and then removed from that more concentrated distribution.

The system illustrated herein provides a concentrated tritium distribution by a two-step transfer process in which each step comprises contacting a more strongly basic, labile, hydrogen-containing material. But, any number of transfer steps using one or more materials with different affinities for tritium could be used. Tritium will transfer to any material that contains labile hydrogen atoms. Materials that have high pH values have a greater affinity for tritium and will thus support higher tritium concentrations than those that have lower pH values, or in other words are less strongly basic. The two-step transfer process described herein, transfers tritium from the reactor water to an intermediate solid, and then to a more strongly basic liquid. This provides good tritium concentration. The concentrated tritium distribution can be efficiently and conveniently removed from a liquid medium by processes such as distillation without requiring excessive energy.

The system illustrated herein includes two exchange cells or containers holding material somewhat more basic than the reactor water. Those cells are interconnected with the reactor cooling system such that either one cell can be used to remove tritium from the reactor cooling water, while tritium buildup is being removed from the other by washing with a more strongly basic liquid. The exchange cells are also interconnected to distillation apparatus for removing tritium from the wash liquid. This interconnection provides convenient recycling and reuse of the wash liquid.

Figure 7.3 is a schematic diagram showing one embodiment of the system for removing tritium from nuclear reactor cooling water.

Figure 7.3: System for Removing Tritium from Nuclear Reactor Cooling Water

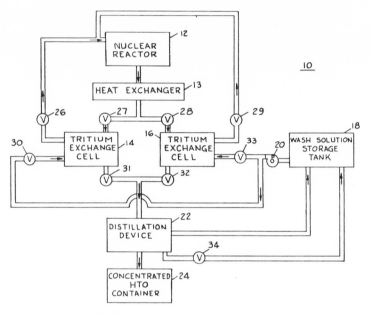

Source: U.S. Patent 4,085,061

Any number of different materials can be used in the exchange cells and for the wash liquid. Materials containing substituent alcohol (—OH), acid (—COOH), ketone (>C=O), aldehyde (—HC=O), ether (≡COC≡) and amine (—NH$_2$) groups were used to remove the tritium from the reactor cooling water with positive results.

Cellulosic materials, such as cotton and wood which contain both the alcohol and ether groups, and proteinlike materials, such as muscle tissue which contains both ketone and amine groups, provided the best results. These materials are all more basic than the reactor cooling water, and therefore extract tritium from the water. The tritium extracted by these materials is then best removed therefrom by a very strongly basic wash liquid. Sodium hydroxide is an example of a good wash liquid.

Fixation of Tritiated Acetylene

M. Steinberg, P. Colombo and J. Pruzansky; U.S. Patent 4,020,003; April 26, 1977; assigned to the U.S. Energy Research and Development Administration describe a process whereby tritiated water is reacted with calcium carbide to produce calcium hydroxide and acetylene, separating the final products, and polymerizing the acetylene. To improve the yield of the process, the calcium hydroxide may be calcinated to remove the tritiated water and the water-calcium carbide reaction is repeated.

Tritiated water is concentrated in respect to tritium by isotopic separation or by any other method as is known in the art. This allows the dispersion of a large quantity of very low activity water to the environment and the required fixation of a small quantity of relatively high activity tritiated water. It also allows the recovery of deuterium for tritiated heavy water wastes from certain types of fission reactors. The concentrated tritium waste is converted to tritiated acetylene by reaction with calcium carbide, in accordance with the following reaction:

$$CaC_2 + 2H_2O \rightarrow Ca(OH)_2 + C_2H_2 \text{ (gas)}$$

The tritiated acetylene is then polymerized to form polyacetylene which is then incorporated into a water-cement mix to provide a massive monolithic encapsulation.

The reaction of tritiated water with calcium carbide partitions only one-half of the initial tritium as tritiated acetylene; the remainder is contained in tritiated calcium hydroxide. Tritiated water can be removed from this calcium hydroxide by calcination at 350° to 400°C and recycled to the tritiated water-calcium carbide reaction. If desired, on the other hand, the tritiated calcium hydroxide can be directly incorporated into cement with the tritiated polyacetylene produced.

Polymerization of the acetylene gas can be accomplished either by exposure to ^{60}Co gamma radiation or the use of a catalyst.

Polymerization by gamma radiation may be accomplished by exposing a vessel containing acetylene maintained under pressure to gamma rays. As the acetylene is polymerized and collects as a powder, more gas enters the vessel due to a reduction in pressure. An example of this follows.

Example 1: Tritiated acetylene gas was produced by the reaction of 6.0 g of tritiated water (1.0 Ci/ml) with 12.0 g of calcium carbide (calcium carbide in excess) as shown in the following equation:

$$CaC_2 + \text{tritiated } 2H_2O \rightarrow \text{tritiated } C_2H_2 + \text{tritiated } Ca(OH)_2$$

Tritiated acetylene was introduced into an evacuated 1,250 cm^3 glass reaction vessel to a total acetylene pressure of 1,374 mm Hg absolute.

The reaction vessel was placed in a ^{60}Co gamma source with an intensity of 1.8 x 10^6 rads/hr.

$G(-C_2H_2)$ initial = 147 molecules/100eV. With water present in the reaction vessel, $G(-C_2H_2)$ initial = 168 molecules/100eV.

The polyacetylene produced was in the form of a yellow powder which on SEM examination was found to have a spherical morphology with a particle diameter ranging from approximately 1.0 to 2.0 μ. This powder was determined to be inert, nonvolatile, insoluble and thermally stable to approximately 325°C. The bulk leach rate for tritium release was 1.8×10^{-8} g/cm^2-day.

Polymerization by catalyst may be effected by passing the acetylene gas over a suitable catalyst such as copper oxide. Typically, the reaction vessel could consist of a glass tube into which the catalyst is placed and a Chromel-Alumel thermocouple is inserted. The thermocouple would act to control the current flowing through the elements of a clamshell furnace in which the reaction vessel is placed and thus the reaction temperature desired.

The gas flow would be determined by a rotometer placed in-line between the gas sources and the reaction vessel. Since the gas flow is determined, the amount of acetylene flowing into the reaction vessel is known. Gas flowing through the reaction vessel without reacting flows through a bubbler submersed in acetone, which is used to remove acetylene and also prevents the back flow of air into the system. Prior to the initiation of the reaction, the system is purged with nitrogen to remove residual oxygen and air. The production of tritiated acetylene gas from tritiated water can be accomplished by use of an acetylene gas generator.

The following is an example of the polymerization by catalyst.

Example 2: Tritiated acetylene gas was produced in a commercial acetylene generator by the addition of tritiated water (1.0 μ Ci/ml) to calcium carbide.

A glass reaction vessel was used which supported the catalyst bed on a copper mesh screen. A thermocouple was inserted into the catalyst bed allowing thermostatic temperature control by a clamshell furnace placed around the reaction vessel.

Tritiated acetylene gas was allowed to flow for five hours at 50 cm^3/min over a 30.0 g cupric oxide catalyst bed heated to 260°C.

The polyacetylene produced was in the form of a brown powder which had a contorted cylindrical morphology. The powder was determined to be inert, insoluble, nonvolatile and thermally stable to approximately 325°C. An entrained copper content was evident.

After the acetylene is polymerized by any of the methods described above, the powdered C_2H_2 can then be mixed with water and cement or any other solidifying matrix and then cured into hardened form, thereby immobilizing the fixed tritiated polyacetylene.

Reduction of Tritium Released Through Natural Gas Production

R.L. Gotchy; U.S. Patent 3,932,300; January 13, 1976; assigned to the U.S. Energy Research and Development Administration describes a process to reduce the amount of radioactivity released to the biosphere by projects involving stimulation of natural gas recovery by use of nuclear explosives. It is a further object of this process to reduce the amount of tritium released to the biosphere through the production of gases from a nuclearly stimulated natural gas reservoir. It is an additional object to reduce the amount of tritium released through flaring gas produced from a nuclearly stimulated gas reservoir.

Briefly summarized, the above and additional objects and advantages are accomplished by separating liquid water and water vapor from the gaseous mixture produced from a nuclear chimney and pumping this water, which will contain most of the tritium, back into a chimney through a string of tubing positioned within the well casing that connects the chimney with ground surface. Water formed upon combustion of the gas during flaring opertions is condensed out of the combustion products and pumped into the chimney along with the separated water.

Pumping the water down a string of tubing which is positioned within the well casing while the chimney gases flow upwardly in countercurrent relationship within the annulus between the tubing and the well casing removes heat from the gases thereby enhancing the effectiveness of the separation of the water from the gas. That arrangement also permits the end of the tubing to be positioned some distance below the opening of the well casing at the top of the chimney formation in order to minimize the entrainment of the reinjected water into the upwardly flowing, high velocity gases entering the annulus in the well casing.

CARBON DIOXIDE

Immobilization of CO_2

There is growing concern that carbon 14 may constitute a significant radiological hazard in the gaseous effluent, or off-gas, from a nuclear facility, such as a nuclear power plant or a nuclear fuel reprocessing plant. For example, in a proposed facility for reprocessing fuel from a light-water reactor, the off-gas would be a stream consisting of air containing CO_2 in a concentration in the range of from about 0.03 to 1.0%. (The concentration of CO_2 in ambient air is about 0.033%, or 330 ppm.) Part of the CO_2 in the off-gas would comprise radioactive carbon 14, owing to neutron reactions with isotopes of elements present in the light-water reactor cooling water, fuel, and structural materials.

Thus, in this instance and in the case of certain other nuclear facilities as well, there is a pressing need for an efficient, convenient, and relatively inexpensive process capable of (1) removing as much as possible of the radioactive carbon dioxide from the off-gas and (2) immobilizing the removed carbon dioxide in a compact form well suited for long-term storage.

D.W. Holladay and G.L. Haag; U.S. Patent 4,162,298; July 24, 1979; assigned to the U.S. Department of Energy describe a process for disposing of fission-product radioactive carbon dioxide contained in the off-gas from a nuclear facility, which comprises removing the carbon dioxide and then immobilizing it in dry, stable, and substantially water-insoluble form by directing the off-gas through a bed of particulate hydrated barium hydroxide while continuously introducing water vapor to the bed in an amount in the range of from 1 to 100% of the saturation value for the off-gas to convert at least 75% of the hydroxide to barium carbonate.

The process is based on the finding that beds of selected hydrated barium hydroxides are highly effective for the continuous removal of carbon dioxide from gas streams in the presence of water vapor. It was found that with proper selection of process variables, the reaction with carbon dioxide is very rapid, even at room temperature, and that the extent of bed conversion to barium carbonate can be very high. In some of the runs essentially 99% conversion of the bed to $BaCO_3$ was achieved—a highly desirable feature for applications where radioactive carbon dioxide is to be immobilized, since ideally 100% of the used bed material should be in the highly insoluble stable carbonate form to ensure long-term immobilization.

It was found that under selected reaction conditions the hydrated barium hydroxides have a very high capacity for carbon dioxide. For instance, a packed bed of 190 g of the monohydrate [$Ba(OH)_2 \cdot H_2O$] will absorb about 22 liters of CO_2 from a water-saturated stream of air at room temperature. The kinetics for carbon dioxide and selected hydrated barium hydroxides are highly favorable for industrial applications such as the immobilization of carbon dioxide from the off-gas streams from nuclear or fossil-fuel facilities.

The following are the major advantages of the process:

(1)   The process avoids difficulties associated with liquid pumping and handling; such difficulties are especially objectionable where radionuclides are present in the feed gas.

(2)   The product is a solid which has very favorable properties for maintenance of stability in long-term storage.

(3)   An especially important advantage is that the removal reaction can be conducted at relatively low temperatures in the range of from ambient to about 105°C.

(4)   The process provides an extremely high removal efficiency (>0.999) for CO_2 in air streams containing 0.03 to 3% CO_2 at ambient conditions.

(5)   It was found that if a suitable amount of water vapor is provided continuously to the reaction zone, the conversion of $Ba(OH)_2$ to $BaCO_3$ can be maximized, significantly extending the useful life of the bed.

(6)   Preliminary studies indicated that the process is much more economical than competitive processes.

Dissolution of Nuclear Fuel in Nitric Acid Under Carbon Dioxide Atmosphere

In one known process for the reprocessing of irradiated nuclear fuel the irradiated nuclear fuel material is first dissolved in nitric acid and the acid solution which contains uranium, plutonium and fission products is contacted with an organic solvent in a solvent extraction process which effects separation of the uranium and plutonium values from the fission products and from each other. During the dissolution of the irradiated fuel material gaseous fission products such as iodine 129, krypton and xenon are released. These gaseous fission products may present a hazard if released into the environment and they are preferably retained in the plant.

A.L. Mills and J.A. Williams; U.S. Patent 4,201,690; May 6, 1980; assigned to the United Kingdom Atomic Energy Authority, England describe a process for the dissolution of irradiated nuclear fuel in nitric acid which is operated under an atmosphere consisting essentially of carbon dioxide.

The carbon dioxide may be circulated inside a plant in which the carbon dioxide is contained by alternatively condensing and evaporating at least a part of the atmosphere circulating in the plant in each of a pair of heat exchangers, one of the pair of heat exchangers being used to condense the part of the atmosphere while the other of the pair of heat exchangers is used to evaporate condensed carbon dioxide to provide a continuous driving force to effect circulation of the atmosphere around the plant. The carbon dioxide may be dried to remove moisture and cooled to remove condensible contaminants such as iodine before passing to the heat exchangers which may be purged to remove gases such as Kr and Xe which do not condense during the condensation of the CO_2.

OTHER RADIOACTIVE GASES

Formation of Ruthenates, Molybdates and Technetates

J.M. Longo and E.T. Maas, Jr.; U.S. Patent 4,092,265; May 30, 1978; assigned to Exxon Research & Engineering Company describe a process for the removal of volatile radioactive oxides of ruthenium, molybdenum and technetium from the process gases and effluent gases produced during nuclear fuel reprocessing procedures which comprises passing the gaseous stream containing the volatile radioactive ruthenium, molybdenum and/or technetium oxides over a trapping agent selected from the group consisting of alkaline earth compounds, lanthanide compounds and lead oxides at a temperature of from 400° to 1000°C, preferably 500° to 700°C which results in the formation of nonvolatile ruthenates, molybdates and/or technetates. The nonvolatile compounds thus formed can be easily handled and kept isolated from the environment during the period of maximum radioactive decay. Alternatively, the trapping agent may be supported.

In a typical nonlimiting embodiment, the process can be described as follows. An alkaline earth compound represented by $BaCO_3$ deposited on a ceramic honeycomb such as cordierite ($Mg_2Al_4Si_5O_{18}$) is heated to about 600°C and a gas stream containing the volatile radioactive Ru oxide passes through it. The volatile ruthenium oxide reacts with the $BaCO_3$ to form the stable nonvolatile $BaRuO_3$ which can be disposed of.

GAS-FILTERING APPARATUS

Filter Assembly

J.R. Catlin, R.E. Strong and W.R. Guest; U.S. Patent 4,198,221; April 15, 1980; assigned to British Nuclear Fuels Limited, England describe a process which relates to particle-in-gas filtration and particularly relates to changing ventilation filters in hazardous, toxic or radioactive environments. In such environments it is necessary to remove particulate matter from the atmosphere and to minimize redispersion of such particulate matter during the replacement of the filters. It is particularly important in nuclear installations that particulate radioactive material is removed from the atmosphere of the installation and is not redispersed so as to protect the operators of the plant from exposure to radioactivity.

According to the process, a filter assembly for particle-in-gas filtration comprises an enclosure divided into inlet and outlet chambers by a horizontal partition, the horizontal partition having one or more apertures communicating between the inlet and outlet chambers, one or more filter units which are adapted to be located in a filtering position on the partition in such a way that gas passing from the inlet chamber to the outlet chamber passes through the filter units, means for moving the one or more filter units vertically from the filtering position to a second position in which the one or more filter units are displaced from the partition and means for moving the one or more filter units horizontally when they are in the second position.

Conveniently the one or more filter units have outwardly extending flanges along opposite sides thereof and the means for moving the one or more filter units vertically from the filtering position to the second position comprise supporting members which are movable between a lower position in which the supporting member rests below the flanges of the filter units when the filter units are in the filtering position and an upper position in which the one or more filter units are suspended from the supporting members by the flanges the one or more filter units being in their second position when the supporting members are in the upper position. With the filter units in the second position replacement filter units may be slid horizontally along the supporting members from one end of the supporting members to displace used filters from the other end of supporting members. The filter units may be moved by a ram. Means may be provided to permit replacement filters to be introduced into the enclosure and used filters to be removed from the enclosure in such a way that there is no egress of the atmosphere from the enclosure to the environment.

A second aspect of this process provides a method of changing filter units in a filter assembly as described above comprising the steps of moving the one or more filter units vertically from the filtering position to the second position in which the one or more filter units are displaced from the partition, introducing a further filter unit into one end of the enclosure, moving the further filter unit horizontally towards the other end of the enclosure to cause any other filter units present to be pushed horizontally by the further filter unit towards the other end of the enclosure, removing a filter unit from the other end of the enclosure, and lowering the filter units remaining within the enclosure into contact with the partition so that gas passing from the inlet chamber to the outlet chamber passes through the filter units.

Compressible Filter Elements

H.-H. Stiehl; U.S. Patent 4,225,328; September 30, 1980; assigned to Delbag-Luftfilter GmbH, Germany describes a process concerned with exchangeable filter elements for use in nuclear installations for the purification of airstreams or gas streams which contain toxic or radioactive dust.

The system comprises a rigid outer frame and a compressible filter paper element which is received in the outer frame and removed from the outer frame in such a way that the filter element can be compacted. This may be accomplished by bending the individual filter bag into a small size so that it will exactly fit in the inner circumference of the waste container. It can be put into the waste container and be reduced in volume by means of a radial pressing motion directly at the place where the filter is changed and without endangering the environment. The rigid constructable frame, which receives the filter element, remains in the housing without having to be stored and decontaminated. In this way the structural frame can again be used, the necessary exchange time and exposure time at the filter for the operating personnel is reduced to a minimum and the costly preparation, which was necessary to date for the disposal of the filter, is substantially reduced.

The problem is solved by using a filter element consisting of a plurality of V-shaped filter pockets which are individually located side by side. The flanges of these filter pockets, which surround their open side, are connected with each other in a removable or break-away fashion. The flanges of all V-shaped pockets are tightly enclosed by a common elastic material which will break under pressure, such as a mounting frame made from a pliable plastic. The mounting frame is inserted with its circumferential edge and by means of an embedded soft gasket tightly between the top of the receiving frame which is loosely inserted into the housing and the upper inner contact area of the housing. Due to the fact that the filter elements consist of separable V-shaped filter pockets, the filter elements can be pressed together sideways without a substantial effort after they have been received in a protective bag such that the pointed ends of the V-shaped filter layer come in contact with each other. In this way a circle-shaped object of variable size is formed, depending on the number of the V-shaped filter pockets, which circle-shaped object can be inserted in the drum-shaped receiving container. The circle-shaped object adapts itself to the inner circle-shaped circumference of the drum-shaped receiving container. If the number of the V-shaped filter pockets of the filter element is so large that a circular object is formed that does not correspond to the inner circumference of the waste container, one or more filter pockets can be broken off within the protective bag and can be located in the middle of the circle-shaped filter object.

In order to facilitate a tightly fitting insertion into the waste container, the filter element that has been put into the waste container in this way and which has the dimensions of, for instance, 610 ± 222 millimeters is reduced in its volume by approximately one-third by means of pressure from a press in the direction of the opening of the container from top to bottom. In this way at least four filter elements, one on top of the other, can be put into a waste container of, for instance, 200-ℓ capacity and one after the other can be reduced in volume by means of pressure. In this way the filter elements will completely fill the container and, after the filling of the waste container is completed, it can be closed gas-tight in the usual manner. The upper mounting frame of the flexible filter element is equipped with a gasket and is positioned at the top of the loose, rigid structural receiving frame whereupon the tightening element of the filter housing, which is located below the structural receiving frame, can push and tighten the receiving frame against the tightening area of the housing.

The tightening mounting frame of the V-shaped filter pockets of the filter element is positioned between the receiving and the tightening area of the housing. Since the toxic or radioactive matter is received in the suspended V-shaped filter pockets which are closed at the bottom, they cannot come in contact with the receiving frame. Thus, a decontamination of the receiving frame is not necessary for each exchange of the filter element but only for a general cleaning.

This system is furthermore characterized by the fact that the filter pockets show two layers of filter paper which are folded in a zigzag fashion with a thickness of 10 to 40 millimeters. The folds are sealed and embedded with a sealant with their edges which form the open side of the pickets and which are closed at their opposite ends and which touch each other by a common edge, which is formed with an elastic sealant; and that the side areas of the filter pockets are closed with a sealant by means of a V-shaped support, which is filled with a plastic material and which V-shaped support consists of a perforated sheet of metal, expanded metal or cardboard or an elastic sheet of plastic or a similar material. The material is tightly embedded in the flange of the filter pocket which closes the edges of those filter pockets. Due to these provisions, the edges of the paper filter layers which form the sides of the filter pockets are embedded in plastic in their entire circumference in such a way that closed elastic and easily shapeable filter pockets result, the handling of which is not difficult even when it is done in a transparent bag.

It is furthermore characterized that both sides of the filter layers of the filter pockets are equipped with a perforated protective cover made from a perforated sheet of metal or similar material, the edges of which protective cover are embedded by bending in the sealant of the flanges and the V-shaped support. The perforated protective covers on both sides of the filter layers of the filter pockets prevent damage to the sensitive paper filter layers due to touching during the manufacturing and the later handling of the filter elements during the exchange and, in addition, they serve to achieve a sufficient stiffness of the filter element without influencing the elastic shapeability and removability from each other in a disadvantageous way.

By means of the sealant strips which are located on the touching edges of the flanges of the filter pockets and which are, for instance, equipped with a known tear tape, the individual filter pockets can be separated from each other by operating the tear lines in such a way that, even in the case of a larger number of filter pockets of a connected filter element, an approximately circle-shaped object can be achieved by shaping inside the protective bag, which circle-shaped object is approximately in its circumference the inner circumference of the drum-shaped receiving container. The same is achieved by the V-shaped separating joints or grooves which are filled with an elastic sealant and which are located between the individual filter pockets.

The system is furthermore characterized by the fact that the structural receiving frame of the filter element is constructed in the shape of an open box, the frontal part of which open box is constructed in such a way that it can be flipped down like a hinge from top to bottom for the insertion of the filter element.

The system is furthermore characterized by the flipping of a side brace of the receiving frame which is made of structural metal. In this way the filter element can be inserted into the receiving frame without any effort and whereby its tightening edge contacts the upper circumventing edge of the closed structural receiving frame. After the insertion of the filter element, the side brace is again flipped back and is locked in position. The structural frame for the reception of the filter element remains always in the filter receiving housing during this operation.

The method for the exchange and for the waste disposal of the filter element is carried out by employing the protective bag technique, which was described in the main conception; and it is characterized by the fact that, before the protective bag which contains the used filter element is put into the waste container, the filter element in the protective bag can be adapted to the inner circumference of the drum-shaped container by bending or breaking or tearing off the tear strings of all filter pockets or off individual filter pockets in such a way that it almost fills the container; and that subsequently the reduction of the volume of the filter element is achieved by vertical pressure exerted by a press directly in the waste container.

However, the used filter element can directly, after putting it into the enclosing protective bag and before putting it into waste container, be shaped to form a circular object in such a way that it will fit to the inner circumference of the drum-shaped waste container, whereupon the reduction of the volume by means of a press can also be carried out in the waste container.

Another characteristic of the method consists in the fact that the filter element with the receiving frame is pulled out of the housing into a protective bag. The filter element in the protective bag is pulled out of the receiving frame in a known fashion and is separated and sealed off from it and is disposed of; and then a new protective bag which contains a new filter element is attached to the housing. The remainder of the bag which contains the receiving frame is pulled off the housing, the filter element is inserted into the receiving frame and the receiving frame with the filter element together is again put into the housing.

A part of the method which is modified and substantially simplified is characterized by the fact that the structural receiving frame, which is connected with the tightening device of the filter receiving housing, remains during the exchange of the filter element in the frame which can be flipped down and all the filter elements are exchanged employing the protective bag technique.

Filter Bed Assembly Without Z-Bar Stiffeners

In the manufacture of adsorbent filtration beds, such as those including charcoal or silver zeolite, which are used in radioactive-gas filtration systems, one requirement is that a close tolerance of spacing between screens, or perforate sheets, which comprise the beds be maintaned. The perforate sheets are often as wide as 31 inches so that reinforcing is required to give them structural rigidity and to maintain desired spacing tolerances.

Generally, this is done by putting Z-bar stiffeners in the beds as described below. The perforate sheets are spot-welded to the end sections of the Z-bar stiffeners. However, a problem with using Z-bar stiffeners is that the end sections thereof create blank areas on the screens which are undesirable because they necessitate making the filter beds larger than is necessary to have the required surface areas.

L.J. Rose; U.S. Patent 4,048,073; September 13, 1977; assigned to Helix Technology Corporation describes a process to provide stiffeners for giving rigidity to perforated sheets or screens of filtration beds and for maintaining spacings therebetween which do not create blank areas.

According to this process, folded upstanding "flutes" or pleats in perforate sheets or screens of filtration beds substitute for Z-bars. The flutes extend across unsupported spans of otherwise flat screens which are placed in an opening of a filter frame. There are two types of pleats: straight pleats with form spacers between beds, and creased pleats which extend into the beds and have zigzag cross sections. The straight pleats or flutes are used for spacing the beds from one another. The primary function of the creased pleats is to impart rigidity to the screens, but they have a secondary function of causing fluid to flow along a tortuous path through adsorbent material.

Figure 7.4 is an isometric, sectional partially cutaway view of the filtration bed assembly.

Figure 7.4: Filter Bed Assembly

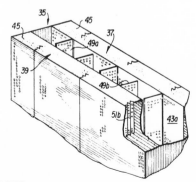

Source: U.S. Patent 4,048,073

Self-Contained Ceramic Filter Element Exchange Apparatus

I. Ito and Y. Karita; U.S. Patent 4,050,638; September 27, 1977; assigned to NGK Insulators, Ltd., Japan describe a ceramic filter element exchange apparatus for a radioactive-matter-containing waste gas treating installation, in which the apparatus can prevent radioactive contamination caused by scattering radioactive dust and prevent an operator from being subjected to radiation internal exposure and at the same time rapidly exchange a spent ceramic filter element for a new one by remote control.

The apparatus can mechanically finely pulverize a spent ceramic filter element in a short time without relying on human power so as to considerably reduce in volume and change the spent ceramic filter element into a form which can easily be solidified.

The ceramic fllter element exchange apparatus comprises a filter element exchange airtight box means having a filter element supplying box and a sight glass, a traveling crane means driven by driving shaft means arranged in the upper part of the box means so as to be moved in forward and backward directions as well as left and right directions, and a lift-up means carried by the traveling crane means and for suspending a chuck for holding a filter element and raising and lowering it, whereby the filter element is exchanged for a new one while observing through the sight glass by remote control.

The ceramic filter treating apparatus comprises a crusher means having an upper dimension which permits the bottom plate of the treating installation to be opened and closed in an airtight manner for pulverizing the filter element, and a container means arranged closely adjacent to the bottom of the crusher means for receiving crushed pieces therein.

The treating apparatus is arranged closely adjacent to the bottom of the treating installation.

The upper part of the crusher means, i.e., the part having a dimension which permits the bottom plate of the treating installation, may preferably be separated from the crusher means and constitutes a glove box means.

Provision may be made of a device having a sealed joint structure which includes a heat-resistant packing sandwiched between the support plate and the flange formed at the top of the ceramic filter element. The packing may be a ring-shaped sheet made of a heat-resistant ceramic filter. The heat-resistant packing may be provided on its upper part with a heat-resistant mortar layer secured thereto in a sealed manner. The heat-resistant mortar layer may eventually be omitted and use may be made of the heat-resistant packing only for sealing the gap between the ceramic filter element and the support plate. In this case, the ceramic filter does not lower its collection efficiency for collecting submicron aerosol. This face renders it possible to omit the heat-resistant mortar layer.

As stated hereinbefore, the system has a number of advantages. In the first place, the ceramic filter element exchange apparatus causes the filter element to be maintained, inspected and exchanged mechanically by remote control from outside in a simple, rapid and reliable manner. Secondly, it is possible to prevent the radioactive dust adhering to the inside wall of the filter chamber from being scattered toward outside during exchange of the filter element, and prevent radioactive contamination of the operating house and operator's clothes. Third, there is no risk of an operator being subjected to radiation internal exposure even when he works without wearing a dust mask. The ceramic filter element treating apparatus causes the ceramic filter element contaminated with the radioactive matter to be mechanically pulverized, and as a result, there is no risk of radiation external exposure occurring. The finely divided particles render it possible to effect aftertreatment of solidifying these particles in an extremely easy manner, thereby densely collecting these particles into a container. The crushing power for pulverizing the ceramic filter element can be made significantly higher than that of the prior art method.

In addition, the radioactive-matter-containing waste gas treating installation according to the process is capable of sealing the gap formed between the ceramic filter element and the support plate with the aid of the heat-resistant packing which also serves to secure the ceramic filter element to the support plate, mounting the ceramic filter element on the support plate in a simple manner while inspecting the mounting operation and hence mechanically performing the mounting operation by remote control, and completing the mounting operation in a short time even when it is effected in the glove box.

Quasi-Hermetically Sealed Compartments

M.J. Szulinski; U.S. Patent 4,277,361; July 7, 1981; assigned to Atlantic Richfield Company describes a ventilating system for a nuclear facility such as a plant for reprocessing nuclear fuel rods in which the central air cleaner for the zones for remotely controlled units is designed for relatively small flow rates. A manifold duct system supplies ventilating air to such central air cleaning unit from each of a plurality of remotely controlled compartments. Particular attention is directed to the fact that the air in each remotely controlled compartment is locally processed with a high ratio of gas recirculation.

The fresh gas injection rate and the stale gas withdrawal rate are low so that the turnover rate of the air in such compartment is longer than a day but less than a year. Such slow turnover rate is designated as quasi-hermetic sealing of compartments. Each compartment is maintained in such quasi-hermetically sealed condition with a signficant amount of recirculation and recycling of the inventory of air within such compartment. The air pressure within each compartment is controlled to be slightly lower than the air pressure in the area adjacent (usually above) such compartment. Automatic controls provide for the flow of fresh air or air from such adjacent area into the compartment when necessary to maintain such acceptable range of air pressure within the compartment. Similarly automatic controls provide for the withdrawal of ventilating gas from the compartment into the manifold duct system leading to the central air cleaner for the purpose of preserving such gas pressure in the compartment at the desired range below the air pressure adjacent the compartment.

Whatever random variations in the atmospheric pressure in the access zones adjacent the remote control compartment may occur, may lead to withdrawal of the air at a rate corresponding to a replacement rate which for the moment may exceed the average replacement interval of more than one day. However, such random pressure variations are not frequent enough to bring about the turnover of the air inventory more than 365 times per year because the air pressure in such adjacent areas are regulated to maintain a substantially constant air pressure differential.

In modifications of the process, the intake volume and exhaust volume for a compartment are controlled so that the marginal pressure difference between the air adjacent the remote control compartment and the air pressure within the remote control compartment can be broadened or narrowed to preserve an acceptable replacement rate. Supplemental gas injection and supplemental ventilating discharge control desirably are included with such volume monitoring system.

In the localized air processing unit, the air from the remote control compartment can be processed in any of the ways deemed appropriate for the processing and purification of air in a radioactive processing zone. For example, heat exchangers can be employed to cool the air inventory in the unit. Filters can be provided for the removal of mists and/or particulate material and for the control of the humidity of the air prior to its recirculation to the principal zone of the compartment. Of particular importance, the gas processing is custom-matched to respond to the problems attributable to the specialized processing conducted in that particular compartment. Thus iodine removal can be very complete close to the source of radioactive iodine evolution. This avoids dealing with iodine contamination of long ducts and decreases the problems of removing extremely diluted iodine from a large gas volume.

It is sometimes desirable to employ two duplicate facilities for the recirculation, pumping, cooling and filtering of the air for a compartment so that the compartment can continue to be operated during the time when attention is being given to the change of filters, caring for the maintenance or correcting any kind of malfunction in one of the two swing purifiers for the air recirculated within a remote control compartment.

The pumps employed for circulating the air within the remote control compartment and through the purifier and back into the remote control compartment provide a turnover rate for such air which is within a range from a few minutes to a few hours, but is significantly less than a day, whereby the assured circulation of air throughout all portions of the remote control compartment is achieved. The pressure within the remote control compartment is regulated to be within an appropriate narrow range slightly below the reference pressure, that is the pressure in the access zone immediately adjacent the remote control compartment. In some embodiments a stabilized regulated pressure isolated from the fluctuating pressure of the weather-influenced atmosphere is maintained in such access area.

Ordinarily it is cheaper to maintain pressure differentials among the selected zones and to permit a series of zones to have pressures fluctuating according to a pattern resembling the weather-induced fluctuations of atmospheric pressure. Such gas zone or air zone immediately adjacent the remote control compartment is the inside of a factory inasmuch as all portions of a reprocessing plant, including even the piping between processing units, are desirably protected from the atmosphere by appropriate walls and roofs for the fuel rod reprocessing facility.

It is conventional to control the pressure of the air in every portion of such reprocessing facilities to be slightly less than the pressure of the exterior atmosphere so that if any air leakage occurs, such leakage will be from the atmosphere into the reprocessing facility rather than routinely permitting any air within the reprocessing facility to flow to the atmosphere except through the air cleaning systems. Thus the general effort is that of controlling the overall combination of ventilating systems so that air discharged to the atmosphere from the reprocessing facilties will be subjected to appropriate cleaners prior to such discharge.

An important feature of the use of the localized purifiers is that each local purifier can be adapted to cope with the particular problems of the processing unit which it serves. In some portions of a reprocessing plant, radioactive krypton may be in-involved. By isolating substantially all of the radioactive krypton in filters at the localized purifier, the degree of removal of krypton can be greater. The problems of dispersion of the radioactivity through the ventilating system can thus be de-creased. The cost of operating the central air cleaner system for the remote control compartments of the reprocessing plant thus can be significantly reduced.

A crystalline zeolite having silver in the ion exchange position can be employed in filters adapted to capture radioactive iodine vapor in the air circulated through a local purifier. By filtering out a major portion of the radioactive iodine and/or other iodine vapor in the air of a remotely controlled compartment, the concentra-tion and amount of iodine directed to the central air cleaner can be significantly decreased.

In most portions of the reprocessing plant in which trace amounts of gas contain-ing tritium might be found, a water synthesis catalyst can be provided for convert-ing all hydrogen, deuterium, and tritium to water and the water can be directed to a purification system for the recovery of tritium and/or deuterium. Other types of filters adapted to capture radioactive hydrogen, i.e., tritium, can be utilized in the local purification zone.

In most portions of the reprocessing plant in which mists of aqueous solutions of radioactive components might exist, the purification zone can utilize appropriate demisting filters adapted for the separation of such mists from the recirculating air.

In preferred embodiments of the process, each remotely controlled compartment is located in a caisson maintained beneath a standard level at the reprocessing plant. Overhead cranes, rubber-tired cranes and/or other appropriate vehicles are adapted to move within a vehicle access zone so that a cover can be removed from a caisson during times of construction, during times of monitoring of operations, or during times of maintenance or replacement of units.

All pipe connections, valve manipulations, etc., in the reprocessing unit are remotely controlled, such as by crane operators. Not only the reaction vessels and processing units, but also substantially all valves and/or other components which might mal-function or leak are positioned within the caisson-type compartment. A sump pit at the bottom of a compartment is provided with sensors sending emergency warn-ings in the event of leakage or malfunctioning. Certain compartments can also be provided with annular storage tanks beneath the sump pit.

Drainage systems directing liquid from the sump pit toward appropriate storage in a remote area can be provided when desirable. The caisson cover for each compart-ment is separate from the cover for each of the purification units and/or pair of swing purification units. Thus, there can be selective access to whatever processing unit may require attention while retaining the covers on the other compartments. Near the top of the compartments and just below the vehicle access zone a system of a series of pipes and related communication lines serves to connect the various compartments. The vehicle access zone includes free space so that the cranes can manipulate the tools, vessels and/or other equipment which might need to be moved after start-up of the reprocessing facility. The general operation of the reprocessing plant is not interrupted by reason of making repairs to one or two swing units of an air purifier for a compartment.

The ventilating system for the access zone of the reprocessing plant, including the area about which the vehicles (e.g., overhead cranes) move to the covers giving access to the various air purification units and various processing compartments, can be maintained at a controlled and stabilized pressure differential. Inasmuch as some contaminants and mists may enter such zone at the time of opening a compartment and contamination can occur during periods of malfunctioning, any flow of air from the access zone to the atmosphere would be through the central air cleaner.

VENTILATING SYSTEMS

Compensation of Rapid Gas Volume Fluctuations

H. Queiser; U.S. Patent 4,157,248; June 5, 1979; assigned to Kraftwerk Union AG, Germany describes a system for venting tanks containing radioactive liquid with the emission of additional large storage volumes, and in which rapid fluctuations are compensated without the discharge of gas of high radioactivity.

With the foregoing and other objects in view, there is provided a method for treating and venting gaseous nuclides in the gas space above the liquid level in tanks, which includes connecting the gas spaces of the tanks to provide a chain of tanks with the gas space in the tank at one end of the chain in communication with the gas space in the tank at the other end of the chain through the intermediate gas spaces, in the other tanks in the chain, normally discharging limited amounts of gas from the gas spaces at the first end of the chain into a waste gas system, discharging excess gas in the gas spaces resulting from filling a tank with liquid, from the gas space at the other end of the chain, and when liquid is drained from a tank thereby creating a void, introducing inert gas into the gas space at the other end of the chain to fill the void.

A primary objective is to keep the emission of radioactivity into the environment low. Pretreatment of the gas to remove nuclides in a waste gas system may be employed. Such systems will satisfactorily handle the normal emission of gas, but are unsatisfactory for handling emissions resulting from sudden fluctuations, unless extra large systems at great cost are employed. Large storage tanks have been suggested for holding back the gas for a time long enough to permit the activity of the gas to decay to a large degree, but here again large increased costs are involved and in addition such large tanks present an additional hazard. In this process, the waste gas system may be operated throughout at substantially normal capacity despite sudden or rapid fluctuations in gas volume. With the arrangement of gas spaces of the tanks in accordance with the process, large storage tanks are not required.

In this method for treating gaseous nuclides from tanks with different liquid levels, a limited quantity of exhaust air is continuously drawn from the air spaces of the tanks. Further quantities of excess air which occur for brief periods of time when the tanks are filled up with liquid are led to a stack. The tanks which are connected to form a chain act as a delay system. Thus, when the liquid content is emptied out of the tanks, the ventilation takes place from that end of the chain of tanks which is opposite the connection to the waste gas system. The method is preferably carried out so that a filter is used as a reversibly operating storage device for removing the nuclides discharged from the tanks with the exhaust air in the direction toward the stack. The nuclides are held back from the air leaving the chain of tanks and are later released from the storage device into the air flowing back into the chain of tanks.

The movements of the quantities of liquid are largely predetermined by the operating program of the nuclear power station. However, it has been found that practically always arrangements can be made so that the quantity of air drawn into the chain of tanks, through the continuous discharge of air into the waste gas system, is kept larger, as integrated over the time, than the quantities of motive air given off in the direction toward the stack.

The gas system particularly well suited for treating gaseous nuclides has the gas spaces of the tank connected with the tanks lined to form a chain arranged in accordance with the degree of radioactivity of the gas in each tank. The high-activity end of this chain or series of tanks is connected to a waste gas system and the low-activity chain leads to the stack.

The gas volumes which are present in the tanks above the liquid level, are used as storage chambers which cause a delay of the gases before they leave for the stack, and thereby make possible a decay of the radioactivity. By virtue of the order of the tanks in accordance with the degree of radioactivity of the respective gas present, it is ensured that highly active gases take the longest time before they are discharged into the stack. At the same time, the connection of the high-activity end of the chain to a waste gas system which is equipped with filters for radioactive gases and, in particular, rare gases, ensures substantially continuous purging. The discharge of gases into the waste gas system influences the balance of the amount of air entering the gas spaces at the tank of low activity in the chain. Because the discharge of gases from the high activity end of the chain does not exceed limited amounts even during fluctuations, there is no overloading of the waste gas system with contamination of the environment.

It is advantageous to arrange a filter which can absorb long-life activity carriers, particularly rare gases, at the stack-side end of the chain. The filter, which provides for practically "zero emission" from the area of the ventilated tanks into the stack, may be designed as an absorption filter, for instance, in the form of finely granulated activated carbon. Such a filter may be located in a storage tank or may be combined with one. This device, called a filter, is suitable for holding back long-life activity carriers, i.e., storage of rare gas.

A purge gas line connected at the stack-side end of the chain makes it possible to provide a slow flow, with, for instance, a quantity of 10 m^3 gas per hour. This purge gas flows through the gas spaces of the tanks connected to form a chain, in the direction from the stack toward the waste gas system. Not only is the overall activity level kept low through such purging, but also because of the direction of flow of the purge gas, any gas leaving the stack side has the lowest activity. Purified and dried waste gas that comes from the waste gas system can be used, for instance as purge gas. However, fresh air, which likewsie may be dried, can also be used for this purpose.

A particularly advantageous embodiment is obtained by connecting the purge gas line to the chain between the abovementioned filter and the stack. In this manner, the purge gas cleans the filter as a part of the normal operation, of radioactive components which had been stored in previous operating phases when gas leaves the chain for the stack. These radioactive products are then returned by the purge gas through the gas spaces of the tanks to the waste gas system. Air, which is drawn into the chain of tanks when one or several tanks are emptied, can act in the same sense as the purge gas.

The purge gas may be fed to the chain via drying apparatus, in order to prevent moisture from being carried into the tanks of the chain. In this regard, it is advantageous to design the drying apparatus as a cooling trap, which is structurally combined with a cooling device for the filter. Particularly high absorption rates can be achieved for a given volume with filters cooled in this manner.

The connections of the tanks lined up to form a chain are located advantageously on opposite sides of the gas space above the liquid, or are spaced a distance of at least the radius of the tank. Thereby substantially the entire volume of the gas space in the tanks acts as a delay tank for the flow of the gases with decay of radioactivity. The connections to the tanks are preferably mounted on the top side of the tanks or at a point close to the top, so that the highest conceivable liquid level will not block off the flow of gas between gas spaces in connecting tanks.

The pressure in the chain of tanks may be below atmospheric pressure to avoid an escape of radioactive gases at leakage points. The chain of tanks are frequently connected to a part of the waste gas system which is also below atmospheric pressure. A pressure reduction device may further be provided here in order to bridge the pressure difference between the waste gas system and the chain of tanks. It is intended thereby to ensure that the pressure in the chain is higher than that in the waste gas system, so that the generally higher radioactivity of the waste gas system cannot be dragged into the chain of tanks.

WASTE TREATMENT
APPARATUS AND EQUIPMENT

WASTE DISPOSAL APPARATUS

Use of Purified Wastewater to Seal Gland Seal Sections of Turbine

In a nuclear power plant, wastewater which is commonly referred to as floor drain is produced. For example, the wastewater is produced when the drain dropping onto the floor from various machines and instruments of the nuclear power plant is washed away. Such wastewater carries radioactively contaminated impurities. The wastewater is also produced when steam seeping through a valve and condensed back into water drops onto the floor and the floor is cleaned by washing away the leak, when machines and instruments making up the nuclear power plant are cleaned before they are repaired, and when objects, as working clothes polluted with radio-activity or radioactive impurities are cleaned or laundered. It is not permissible to allow such wastewater to flow out of the nuclear power plant without giving any treatment thereto. Thus the wastewater is usually treated by wastewater ion-exchange treating means so as to reduce the radioactive impurities carried thereby. The wastewater is supplied to a condensation storage tank after the concentration of the radioactivity carried thereby is lowered to an average of 1×10^{-6} μCi/cc by passing it through the wastewater ion-exchange treating means.

Meanwhile a portion of the water flowing through the main circulation system is introduced into the condensation storage tank immediately after the water has passed through a condensation desalinator. The main circulation system constitutes a steam-feed water cycle which connects a nuclear reactor, a turbine, a condenser, a feed water pump, and a feed water heater in the indicated order, the feed water heater being connected to the nuclear reactor.

Besides being used to cope with an accident in case of emergency, the water contained in the condensation storage tank is also used for operating the control rod drive apparatus during normal operation. The water used for operating the control rod drive apparatus is returned to the nuclear reactor. The water in the condensation storage tank is also used for providing sealing steam to the gland seal sections of the turbine. More specifically, the water in the condensation storage tank is supplied to a steam generator where it is converted into steam which is supplied to the gland seal sections of the turbine. The steam passing through the gland seal sections is condensed back into water in gland steam condenser, and the water produced by condensation is returned to the condenser through a condensation recovery tank. The volume of water flowing from the main circulation system into the condensation storage tank is substantially equal to the sum of the water used for operating the control rod drive apparatus and the volume of water supplied to the steam generator. Thus the water from the main circulation system accounts for the major portion of water introduced into the condensation storage tank. Because of the fact that the water passing through the main circulation system flows into the condensation storage tank as aforesaid, the concentration of the radioactivity carried by the water in the condensation storage tank is at a level of 1×10^{-4} μCi/cc on an average.

Water which is substantially equal in volume to the wastewater introduced into the condensation storage tank is withdrawn from the condensation storage tank and mixed in the seawater used in the condenser for cooling purposes. Thus the cleaning wastewater is diluted and released into the sea. The water thus released into the sea carries radioactivity which has a concentration of 1×10^{-9} μCi/cc. It has been proposed at the National Academy of Sciences in the United States that the allowable concentration of the radioactivity carried by the seawater returned from a nuclear power plant to the sea be less than 4×10^{-9} μCi/cc. In recent years, 10 CFR 50 Appendix 1 which is a law of the U.S. has set the goal of reducing a dose of radiation from the radioactive gas to less than 1/100 of the prevailing value at the boundary of the site. As to the radioactive liquid, it is stipulated that a dose of radiation therefrom be less than 5 Ci (except for tritium) per year.

N. Kita; U.S. Patent 4,138,329; February 6, 1979; assigned to Hitachi, Ltd., Japan describes a method of treating wastewater of a nuclear power plant including a steam turbine system constituting a closed loop, comprising the steps of:

(a) subjecting the wastewater polluted with radioactivity to an ion-exchange treatment to lower the concentration of radioactive substances contained therein;

(b) heating the wastewater to generate steam after the wastewater is treated;

(c) supplying the steam obtained in step (b) to the gland seal sections of the turbine to provide a seal thereto;

(d) condensing the steam back into water; and

(e) recirculating independently of the closed loop at least a portion of the water obtained in step (d) as feed water for generating the steam in step (b).

The outstanding characteristic of the process is that radioactively contaminated wastewater that has passed through wastewater ion-exchange treating means is vaporized to produce steam which is supplied to the gland seal sections of a turbine provided in the steam-feed water cycle. The steam provides a seal to the gland seal sections, and is thereafter condensed into water and mixed with the cooling water for a condenser mounted in the steam-feed water cycle, so that the wastewater is released to the outside together with the condenser cooling water.

The process offers the advantage of markedly lowering the concentration of the radioactivity carried by the wastewater produced in a nuclear power plant before being released to the outside. Figure 8.1 is a flow sheet showing one embodiment of the process.

Figure 8.1: Production of Steam to Seal Gland Seal Sections of Steam Turbine

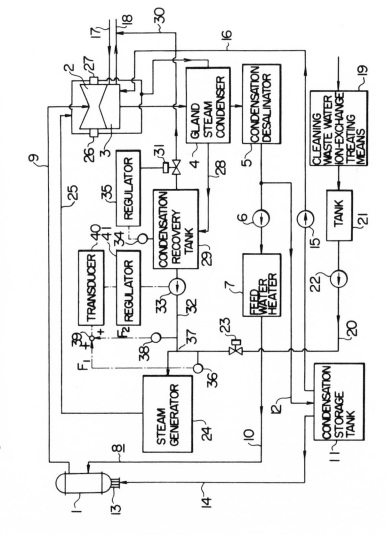

Source: U.S. Patent 4,138,329

Low Maintenance Apparatus for Fixing of Waste

G. Meier; U.S. Patent 3,971,732; July 27, 1976; assigned to Gesellschaft fur Kernforschung mbH and Firma Werner & Pfleiderer, both of Germany describes an apparatus for fixing radioactive and/or toxic wastes creating conditions for a disturbance-free operation for long periods without the necessity of external interference involving long inoperative periods and further, which insures superior operational safety.

The apparatus for fixing radioactive or toxic waste has an extruder including a mixing mechanism for intermingling and advancing the waste and a carrier material introduced into the extruder. The extruder has a heating zone with which there communicates a vapor outlet device having an observation window. Within the vapor outlet device there is disposed an arrangement for cleaning the window and an arrangement for removing deposits from locations of the vapor outlet device that are adjacent the mixing mechanism. The condenser is coupled to a distillate accumulator with the interposition of two alternatingly operating filters for removing particles from the condensate obtained from the condenser. To the outlet of the extruder there is coupled a loading device in which containers are successively filled with the material discharged by the extruder. The loading device includes an interrupter bowl which receives the material discharged by the extruder during an exchange of an empty container for a filled container below the extruder outlet.

The advantages accomplished by the system reside particularly in that the utilization factor of an apparatus for fixing radioactive and/or toxic waste is significantly increased with the aid of simple means which effect a substantial reduction in the soiling of components with radioactive materials. Thus, the supplemental work necessary in conventional apparatus for removing such soiling material and the operational pauses involved, are substantially or entirely eliminated. By practicing the process, the radioactive waste fixing apparatus associated, for example, with a nuclear installation of known waste yield may be of reduced capacity compared to conventional apparatus. In this manner, better efficiency is ensured, involving economy of space and capital investment. A further important advantage resides in the substantially reduced radiation exposure of personnel, due to significantly reduced periods of maintenance.

Fluidized Bed Type Waste Disposal System

L.E. White and E. Schmauderer, Jr.; U.S. Patent 3,994,824; November 30, 1976; assigned to Aerojet-General Corporation describe a nonmechanical, dynamic air flow system for controllably removing bed material from an operating fluidized bed reaction chamber, and mixing and agglomerating bed solids with fine particles to provide a mass of material having a uniform bulk density. The term "air flow" is used herein to designate any gaseous flow. The bed removal system comprises a gravity flow exit pathway for bed material reaching a preselected elevation in the reaction chamber, that intersects an exit pathway for air and fine particles above that preselected level. Air flow through the two pathways causes fine particles to mix with larger bed-sized particles, sometimes called solids herein.

A centrifugal solids-air separator is connected to receive flow from the exit conduits, centrifugally agglomerate mixed particles, and deposit the agglomerated particles into a storage container. And, control valve apparatus is disposed to inject air into the gravity flow exit conduit.

Air injected into the gravity flow pipeline provides both an air flow through that pipeline and a pressure seal that prevents uncontrolled release of fluidized gas from the reaction chamber, and also controls the quantity of the bed material in the reaction chamber. The pressure seal thus prevents unstable operation of the reaction chamber. In one embodiment, air is injected along a direction substantially toward the reaction chamber in order to minimize the air injection required to provide a desired limitation to flow from the chamber. The quantity of injected air required to control different systems is generally between about 2-10% of the fluidization air in the reaction chamber, depending upon the diameter of the exit pipeline and the position of the entrance to that pipeline relative to the intended height of the fluidized bed.

Air flow through the two conduits produces fluidic mixing at the intersection of the two exit passageways. This mixing prior to entry into the cyclone provides a relatively uniform mixture of bed solids and fine particles which, when agglomerated and separated from the air stream by centrifugal action results in an agglomerated bed of material having a relatively uniform bulk density that approaches the bulk density of the bed material. The high uniform density of the deposited material facilitates further handling and processing. Another advantage of the system is that the mixed, agglomerated particles have significantly less tendency to clog the cyclone separator than is the case for cyclone separators handling only fine particles in prior art systems. In addition, the system eliminates mechanical parts that wear and break down. And, it is not subject to leakage.

A particularly advantageous fluidized-bed-type waste disposal system for treating radioactively contaminated salt solutions is provided by utilizing the bed removal structure in combination with a high-temperature fluidized bed reaction chamber. The system maintains a predetermined quantity of bed material and a predetermined fluidization level in the reaction chamber, even though the rate at which salt solution is fed to the chamber, and the concentration of dissolved salts might be subject to variation. These variations, which change the rate at which material builds in the chamber, merely cause the rate at which material flows into the gravity flow exit passage to also change in a corresponding manner. It is thus not necessary to adjust operation of the particle removal apparatus in response to each small change in the rate at which material is fed into the reaction chamber.

The system described herein, for controlling bed product removal from a fluid bed reactor, mixing the fines particulates with bed product, and centrifugally agglomerating fines and bed product to produce a highly uniform bulk density end product solid is a physically simple and nonmechanical system that provides overall cost reduction benefits to users of fluid bed systems and devices, provides full contol capability to maintain bed height at a desired level prescribed for a process, provides the inherent high reliability associated with nonmechanical devices or systems, and is easily maintained since the number of components involved are limited and common to gas handling industry applications.

The operation of the system in a fluid bed application provides for bed withdrawal rate and bed depth control in a large self-controlling fashion when operated in a continuous mode of operation. However, bed withdrawal and bed height control can equally be achieved by providing modulated or intermittent removal from the reaction chamber. And, product transfer can be stopped by simply increasing or decreasing the air control flow rate. Further, in a start-up mode, no bed product material is consumed or required to form the pressure seal, the absence of which would preclude correct operation of the fluid bed reactor, since this function is controlled solely by the control air flow. And, the inlet to the bed removal pipeline can thus be located at any height between the floor of the reaction chamber and the highest intended level of the fluidized bed during operation of the reactor.

Passing of Vapor Through Electric Field

A. Schiffers, W. Oschmann, J. Brandt and D. Leith; U.S. Patent 4,308,105; Dec. 29, 1981; assigned to Apparatebau Rothemühle Brandt & Kritzler, Germany describe a process which comprises subjecting radioactive wastewater containing radionuclides adapted to form a dispersed liquid or solid aerosol phase in the water vapor upon evaporation, to vaporization, thereby producing a stream of water vapor in which the radionuclides are dispersed as an aerosol of droplets and/or solids, and removing the particles of the aerosol from the vapor by subjecting this aerosol to an electric field.

In the treatment of radioactive nuclear waste one should distinguish between the so-called system-decontamination factor (DF_s) and the vapor decontamination factor (DF_v).

The system decontamination factor DF_s is the ratio of the specific activity of the wastewater to that of the distillate and is of a value or quantity which is significant in dealings with respective control authorities such as governmental units. The evaporator decontamination factor DF_v establishes a ratio between the specific activities of the concentrate and the distillate and can correspond approximately to 100 times (10^2) the system decontamination factor DF_s. The evaporator decontamination factor DF_v is established by the parameters of the evaporator unit and hence by the manufacturer. Practical experience with known decontamination systems has shown that the assumed DF_s values of 10^8 to 10^{12} are not attained in operation but that, at best, DF_s values of 10^4 to 10^6 can be obtained. The actual DF_s values, moreover, decrease with duration of operation of the system.

Considerable investigation has shown that a primary reason for the low DF_s values actually obtained is the continuous contamination of the fittings of the separator column because the particles to be separated from the vapor tend to desosit on the fittings or to promote deposition on surfaces thereof.

For high DF_s values, therefore, efforts have been made to effect cyclical cleaning of the fittings, i.e., cleaning of the fittings after the lapse of a predetermined operating period, or replacement of the fittings. Both of these operations are expensive and time consuming, necessitating shutdown of the evaporation installation for long periods and possible exposure of operating personnel to high radiation dosages.

It was found, quite surprisingly, that a process of this type allows the removal or deposition of the radionuclides from the vapor of the wastewater in an optimum manner without reduction of the DF_s value over relatively long periods of operation. In fact, tests have shown that by using a system in which an aerosol dispersion is first formed from the wastewater and the radionuclides are selectively deposited from the dispersion by electrostatic field, the vaporizer decontamination factor DF_v is improved by a factor of at least 100 (10^2) by comparison with vaporization processes used heretofore and employing separator columns. The vaporizer decontamination factors (DF_v) obtained with this system can correspond approximately to the system decontamination factors (DF_s) previously achieved.

According to a feature of the process, the stream generated by vaporization of the radioactive wastewater and containing the radionuclides in a dispersed aerosol form, is passed through at least two spaced-apart electric fields in succession, each field being controlled, established or designed so that it will result in an optimum removal of the particles present in the initial dispersion. This arrangement ensures that a failure of one of the fields will not give rise to less than optimum removal of the particles.

In accordance with another feature of the process, the system provides an evaporator or vaporizer below a column formed with at least two electrostatic precipitation zones through which the vapors and dispersion rise.

It has been found to be advantageous, in this system, to initially introduce the vapor or dispersion in a tangential manner into the column below the electrostatic deposition zone, to deflect the stream of the dispersion downwardly and only then direct the dispersion upwardly through the electrostatic precipitation zones.

This technique varies the velocity of the dispersion (in meters per second) repeatedly to a high degree during its flow to the electric field zones, an action which markedly improves the separation effect in the region of these fields.

According to another aspect of the process, the separation is carried out in an apparatus which comprises a circulation-type evaporator, i.e., an evaporator which produces the vapor and thereby forms a dispersion of radionuclides in the vapor phase by heating the radioactive wastewater to a temperature above its boiling point in one leg of the apparatus while allowing recirculation of condensate and excess liquid by convection through the other legs to the heating elements or unit.

A separator column is connected with the circulation evaporator at one side, through the wastewater or evaporator sump, in addition, through the vaporization chamber while a condenser and a water collector is connected at the outlet side of the separator column.

Between the evaporator and the condenser, one or more electrical filter zones are provided in the column in the direction of flow of the dispersion. Preferably the zones are disposed above the evaporator chamber.

According to a specific feature of the process, at least two electrical filters, structurally identical but capable of independent operation and each set for optimum separation, are disposed one above the other in the separator column in spaced-apart relation.

For the tangential introduction of the dispersion into the electrofilter zones, the outlet of the circulation evaporator is constructed so that it opens tangentially into the downstream open annular space beneath the first electrical filter. This chamber of space can have a passage coaxial with the annular chamber and the electrofilter.

Beneath the first electrofilter and above the passage there can be provided means, including a switching circuit, which activates a foam-breaking device, such as a heater, when foam or froth is formed by a short-circuit operation of the first electrical filter. The heating unit may be a grid and the foam-breaking system may also be a spray ring in which liquid is directed into the vapor to preclude the formation of foam or to destroy any quantity of foam which may have occurred.

The formation of foam has been found to be a problem especially with radioactive wastewater operating under alkali conditions, i.e., with a pH above 7 as is the case when iodine-131 must be reduced so that it can form part of the dispersed phase in the aerosol for separation.

According to yet another feature of the process, the electrodes of the electrofilters are provided with washing and liquid-collection means so that the collecting electrodes of the electrofilter can be washed down during periods in which the apparatus is idle or periodically. The washing means can be in part accommodated in the spaces between successive filter sections.

Waste Metering Device

J.D. Murphy, J. Pirro, Jr., L. Rutland and S.F. Wisla; U.S. Patent 3,883,441; May 13, 1975; assigned to Atcor Inc. describe a process relating to the disposal of any radioactive wastes which can be slurried in an aqueous medium and has for its general purpose fixing of such wastes on a production line basis avoiding the difficulties of atmospheric contamination and nonuniform mixture characterizing prior fixing methods.

In accordance with this process radioactive wastes are collected from time to time and slurried in an aqueous medium in a metering tank which is contained in a shielded enclosure. The metering tank is adapted to meter the collected slurry, when required, through a mixer which is also connected simultaneously to receive cement or other fixing materials in metered quantities proportionate to the metering rate of the slurry. The mixer is shielded and arranged continuously to discharge metered and mixed fixing material and slurry into a shipping container for shipment to a burial ground when the mixture has set.

Ordinarily the fixing process is carried out only when a sufficient quantity of radioactive waste has been collected to constitute a suitable quantity for shipment to a burial ground. This quantity may involve one or several shipping casks or other containers and the fixing process is carried out continuously until the desired number of casks have been charged. Thus expensive shipping casks are tied up at the particular plant only for that time necessary for the fixing operation and need not be kept on hand as storage vessels until an adequate quantity for shipment has been made up.

Another advantage of this process is that it is flexible in terms of achieving the proper ratio of water to fixing material in the mixture. Thus if the radioactive wastes are in slurry form containing excess water they can readily be dewatered by decanting or filtering from the metering tank prior to fixing. If they are solid wastes additional water can be added as required. The proportions of fixing material, water and waste materials are a function of density, shielding requirements, moisture content and ultimate disposition of the fixed waste.

Control of Number of Disposal Units in Operation

H. Yajima, N. Komoda and S. Horiuchi; U.S. Patent 4,278,104; July 14, 1981; assigned to Hitachi, Ltd., Japan describe a waste liquid disposal system for disposing of waste liquid in a plurality of waste liquid disposal units operated in parallel, or more in particular to a system for controlling the required minimum number of waste liquid disposal units without stand-by, no-load operation of the remaining units.

In order to achieve the abovementioned object, the amount of waste liquid to be produced is forecast over, N hours, and the forecast amount is used to calculate the maximum value m satisfying the following inequalities (1) and (2), thus newly starting additional m disposal units.

(1) $$U_1 + U_2(I) \geqslant V(I) + mV'(I)$$

(2) $$U_1 + U_2(N) \geqslant V(N) + mV'$$

where U_1: amount of waste liquid stored in the storage tank at present; $U_2(I)$: amount of waste liquid forecast to be produced before I hours later. I is a given time satisfying the relation $0 < I \leqslant N$; $V(I)$: amount of waste liquid expected to be disposed by P waste liquid disposal units in operation before I hours later (In the case where the waste in a particular waste liquid disposal unit is bumped before I hours later, however, that disposal unit is not assumed to be used after the bumping.); $V'(I)$: amount of waste liquid disposed by each disposal unit for I hours; m: a given integer; $V(N)$: amount of waste liquid capable of being disposed of for N hours by the P waste liquid disposal units now in operation. (In the case where the waste liquid in a particular waste liquid disposal unit is bumped before N hours later, however, the particular disposal unit is not used after the bumping.); V': amount of waste liquid capable of being disposed of by each disposal unit before bumping; N: a predetermined value representing a maximum time point to be forecast from the present time.

The satisfaction of inequality (1) above means the following: In view of the fact that at any time I satisfying the relation $0 < I \leqslant N$, the amount of waste liquid produced is always greater than that disposed of by the waste liquid disposal units, waste liquid is not depleted during operation of the thickeners. In other words, the stand-by, no-load operation of the thickeners is prevented for the period of time from this time to N hours.

The fact that inequality (2) is met, on the other hand, means that at the time N hours later, the amount of waste liquid to be processed before bumping by the thickeners operated at that time is secured in the storage tank, thus eliminating any stand-by, no-load operation.

Further, selection of the maximum value m satisfying both inequalities (1) and (2) means that the stand-by, no-load operation of the disposal units is prevented on the one hand and the amount of waste liquid in the storage is kept minimized to provide a margin for storing an additional great amount of waste liquid which may be produced.

The abovementioned system facilitates the operation of the waste liquid disposal units, reduces the operating cost of the thickeners, and further prevents the corrosion or scaling of the thickeners.

Generally, it requires 6 to 12 hours for a "bumped" thickener to restart for operation. If too many thickeners are operated, the bumping may occur at a number of the thickeners, making effective use of the thickeners impossible, in the event that a great amount of waste liquid is produced at the same time and is required to be disposed of quickly. According to the process, such an inconvenience is easily overcome by the fact that the required minimum number of thickeners are operated. Also, the frequency of bumping may be reduced in the process of disposal.

Dissolver Apparatus

H. Goldacker, G. Koch, H. Schmieder, E. Warnecke and W. Comper; U.S. Patent 4,246,238; January 20, 1981; assigned to Kernforschungszentrum Karlsruhe GmbH, Germany describe a process relating to a dissolver for removing nuclear fuel materials from fuel element segments during reprocessing of irradiated nuclear fuels.

In such apparatus, for reasons of criticality, the structural parts are made of a material which is a neutron absorber such as, for example, hafnium, and the dissolver is composed of a dissolving vessel into which a dissolving basket containing the fuel element sections which are to be subjected to the dissolving process can be placed so that fluid can flow therethrough. The fuel elements of nuclear reactors in the majority of cases are composed of the actual nuclear fuel and a metallic protective sleeve or shell.

At the beginning of such a reprocessing procedure, the irradiated reactor fuel elements are initially cut up mechanically into short fuel rod segments. These fuel rod segments, together with the structural components of the fuel elements, which include spacers and head and foot pieces, drop into the dissolving basket which is disposed in the dissolving vessel. In the dissolver, the nuclear fuel material is dissolved out of the fuel rod segments by means of boiling nitric acid. Upon completion of the dissolving process, the nitric acid-containing fuel material solution is extracted, the empty sleeves are washed with fresh acid and then the basket together with the empty sleeves and the other structural material is removed from the dissolving vessel, is washed with fresh water and then the basket is emptied.

The fuel material solution is subjected to further chemical processing for separating the reusable nuclear fuel material, preferably uranium and plutonium as well as possibly thorium, from the fission products. The leached sleeves and the structural material constitute solid radioactive waste and are treated and stored accordingly.

During the dissolving of the fuel materials, the presence of fissionable material, preferably uranium and/or plutonium, poses a particular problem. This involves the danger of establishment of a "critical state" in the system in which a self-sustaining nuclear chain reaction could take place. For this reason measures must be taken which prevent such critical excursion under any circumstances.

This problem has heretofore been solved by constructing the dissolver to have "geometrically critically safe" external dimensions, i.e., its dimensions were limited to values which, under consideration of the amount of fissionable material contained in the fuels to be processed, lie below the minimum critical dimensions. However, this greatly limits the volume, and thus the capacity, or fuel throughput, of the dissolver. Such dissolvers can thus be used economically only for small reprocessing systems or for nuclear fuel materials having a low content of fissionable material, such as, for example, those from heavy water natural uranium reactors.

In order to have available larger capacity dissolvers, for example for system throughputs of several tons of uranium per day, for nuclear fuel materials from modern power reactors containing greater amounts of fissionable material, e.g., light water rectors with enriched uranium oxide or with plutonium oxide/uranium oxide as the fuel material, it is known in the art to add a "soluble neutron poison," i.e., a dissolved substance having a high capture cross section for neutrons, particularly gadolinium nitrate, to the dissolving acid. However, this process has the drawback that the dissolved neutron poison is lost in the subsequent chemical reprocessing procedure together with the highly active waste and cannot be recovered. Due to the high cost and limited availability of soluble neutron poisons as well as the highly active waste connected with its use and the problems of further processing such waste, this process is very uneconomical.

Economical operation is made even more difficult by the fact that the constant presence of a soluble neutron poison in the dissolving acid must be continuously verified, involving high monitoring expenses and entailing extensive use of automated control instruments so as to exclude the possibility of a criticality accident.

The prior art dissolvers for large systems, in which head and foot portions and spacers of the fuel rod bundles are also introduced into the basket, are designed to have a diameter of 70 cm and a height of 5 m. They pose great handling problems to their users and must additionally be loaded with the abovementioned soluble and nonextractable neutron poisons, which must be continuously added and which adversely influence the highly active waste as salt formers, i.e., they limit the possible concentration by salts which are inherently inactive.

In summary, it can be stated that large reprocessing systems, those with capacities of the order of magnitude of 1500 tons per year, require either too many geometrically safe dissolvers, that is many mechanical devices subject to malfunctions and requiring many operations, or the use of soluble neutron poisons which present the abovementioned drawbacks.

This process provides a dissolver apparatus in which the interior of the dissolving basket which receives the cut up fuel element segments is divided into a plurality of individual sections or chambers, respectively, partitions being provided in the basket to create individual chambers, the partitions being in the form of screens and being made of the same material as the other parts of the apparatus.

The dimensions of the individual chambers of the dissolving basket correspond, in their cross-sectional planes, to the outer dimensions of the fuel element sections and the partitions are made of sheet metal, plates, grids or bolts with constant or variable cross sections. The cross sections may also be variable over the height of the basket or the partitions may be arranged changeably in the interior of the dissolving basket.

It is of particular advantage for the dissolving basket to have a circular cross section and the partitions which divide its interior to be arranged as radii or chords in the interior of the basket.

The dissolving basket may have a block-shaped cross section, in which case the partitions are arranged transversely to or at an angle to the side walls.

According to a further, highly favorable embodiment, the dissolving basket is formed to have an annular cross section and the partitions are oriented radially within the annular basket. The individual chambers may also be formed of juxtaposed pieces of pipe and the partitions may be perforated in an advantageous manner. Finally, a particularly advantageous feature of this embodiment resides in the provision of a distributor cone disposed in the cylindrical region formed within the annular dissolving basket at the bottom of the dissolving vessel, the distributor cone being made of the same structural material as the remainder of the dissolver and its conical tip protruding above the top of the dissolving vessel.

A dissolver presenting these features offers the advantage that periodic checking for the presence of an absorber is practically eliminated or can be performed under significantly easier conditions. In addition, it is possible, to construct a dissolver which has a significantly greater throughput than the prior art dissolvers but has the criticality safety of known conventional and smaller vessels.

Rotary Leaching Apparatus

O.O. Yarbro; U.S. Patent 4,230,675; October 28, 1980; assigned to U.S. Department of Energy describes an improved multicompartment rotary leacher of the elongated-drum type wherein liquids and solids are contacted countercurrently with relatively little back-mixing of the liquid phase.

The process may be summarized as follows: In rotary apparatus for countercurrently contacting liquids and solids, the apparatus including an elongated and generally cylindrical drum assembly which is rotatable in either direction about its longitudinal axis and which is divided by circumferentially sealed, transversely extending partitions into a solids-inlet/liquid-outlet compartment at one end of the assembly, a solids-outlet/liquid-inlet compartment at the other end thereof, and a plurality of leaching compartments therebetween, the partitions being provided with perforations for conveying liquid flow between adjacent compartments; each of the leaching compartments containing a solids-transfer assembly for advancing solids into the next compartment in the direction of solids flow when the drum assembly is rotated in a selected direction, each chute assembly including a solids-transfer chute and a perforated baffle for directing solids into the chute when the drum assembly is rotated in the selected direction; the improvement comprising:

> The partitions being formed with corresponding outer annular imperforate regions, each region extending inwardly from the rim of its respective partition to an annular array of perforations concentric with the rim, and

> The drum assembly being disposed with its solids-outlet/liquid-inlet end at a higher elevation than its other end, the longitudinal axis of the drum assembly forming an angle in the range of from about 3° to 14° with the horizontal.

EQUIPMENT DECONTAMINATION

Vacuum Cleaner for Contaminated Particles

The radioactivity of an object may be due to "fixed" contamination or "loose" contamination. While the former cannot be easily removed, the latter is often due to radioactively contaminated dust or particulate matter and decontamination may be accomplished simply by the removal of such particles from the surface of the object. Such is accomplished by using conventional vacuum cleaners.

There are several significant problems inherent in such a practice. The vacuum cleaner itself becomes contaminated, and must be taken apart and cleaned after every usage. Such cleaning generates radioactive waste that requires expensive disposal by burial. Furthermore, even after being cleaned, the vacuum cleaner must still be regarded as radioactively contaminated and therefore requires storage in a "control" area with other radioactively contaminated items. Such items must be checked in and out, and constantly accounted for. Personnel using contaminated vacuum cleaners must wear protective gear, and even then are exposed to radioactivity.

The procedure for disposing of the radioactive particles contained in the vacuum cleaner's waste bag is very time consuming because the bag must be tightly sealed prior to its removal from the vacuum cleaner to prevent the escape of radioactive particles into the ambient atmosphere. Typically, this procedure requires two men 3 hours to complete, i.e. 6 man-hours, and these men, although wearing protective clothing, are nonetheless subjected to radioactivity throughout this interval.

As conventional vacuum cleaners sometimes leak vacuumed particles into the ambient air during their operation, the hazard of radioactive exposure for all personnel in the working area is increased.

Should the vacuum cleaner become obsolete or mechanically inoperative, it would require disposal by burial.

B.L. Frye, M.G. Pittman, D.A. Runge, L.C. Souza and R.V. LaVoie; U.S. Patent 4,061,480; December 6, 1977; assigned to U.S. Secretary of the Navy describe an apparatus which effectively vacuums up and retains radioactive particles, does not leak such particles into the ambient atmosphere during its operation, and provides for quick and safe removal of filtered, contained particles (the disposal procedure requires approximately one-third of a man-hour). Furthermore, as no part of the apparatus other than the bladder-filter unit which contains the filtered radioactive particles comes into contact with such particles, the remainder of the apparatus need not be cleaned after every usage nor stored in a "control" area, and will not require disposal by burial in the event it becomes mechanically inoperative or obsolete.

Briefly, the apparatus uses a suction means to draw radioactive particles through a high efficiency filter and retains filtered particles in a disposable filter-bladder unit for subsequent disposal.

The bladder-filter unit prevents filtered particles from escaping into the ambient atmosphere during the operation of the suction means and also during its removal from the apparatus, in addition to preventing such particles from contaminating the remainder of the apparatus.

Decontaminating of Nuclear Steam Generator Channel Head

R.D. Burack and R. Shaffer; U.S. Patent 4,219,976; September 2, 1980; assigned to Westinghouse Electric Corp. describe the decontamination of the interior of a channel head of a nuclear steam generator for the purpose of providing a reasonably safe environment for other work of retubing the steam generator to proceed.

Problems of dents and potential leaks of heat transfer tubes must have been experienced in certain nuclear steam generator installations. Since this poses the possibility of contamination of the secondary heat transfer fluid, replacement of the damaged tubes is required. To accomplish this, devices which can be installed in the channel head of the generator have been and are being developed. Since the primary fluid heated by circulation through the nuclear reactor core contains radioactive particles, the channel head through which the primary fluid flows to and from the heat transfer tubes becomes relatively highly radioactively contaminated.

In accordance with the process the decontamination machine includes a first assembly comprising a vertical post and a carriage track carrying a movable carriage with an adjustable decontamination blaster thereon, the carriage track being pivotally secured to the upper end portion of the post to permit the assembly to be collapsed to a size which permits its entry through the manway of the channel head, with the assembly being sufficiently light in weight to permit its manipulation and installation thereof in the channel head by no more than two individuals in a location in which the vertical post is closely adjacent the vertical center line of the divider wall, the machine further including both a separate horizontal support beam and a separate curved track which are assembled together and to the vertical post to form a generally quadrantal shaped frame, the machine also including means for moving the carriage along the carriage track and for swinging the carriage track both horizontally and vertically so that the decontamination blaster means can be swept past substantially all of the interiorly facing walls of the channel head.

The method of effecting the decontamination contemplates the traversing of the lower face of the tube sheet, the face of the divider wall and the interior facing wall of the channel head by traversing the areas of these walls by energizing selective drive means for successive sweeps along the areas at rates according to the positioning of the blaster means and then repositioning the blaster means by other drive means and traversing at other rates in accordance with the positioning of the blaster means.

Decontamination of Metal Surfaces

H. Loewenschuss; U.S. Patent 4,162,229; July 24, 1979; assigned to Gesellschaft zur Förderung der Forschung an der Eidgenösslschen Technischen Hochschule, Switzerland describes a decontamination process of treating contaminated metal surfaces with a 0.001-1 mol cerium salt solution containing at least one cerium(IV) salt and a water-containing (aqueous) solvent.

The water-containing solvent is preferably an aqueous solution of an acid of, generally, relatively low concentration. The concentration of acid in the cerium salt solution may correspond to, e.g., the acid concentration in a 0.1-1 mol solution of this acid in water. Expediently, the concentration of acid in the cerium salt solution is at most the acid concentration in a 5 mol solution of this acid in water, in any case. The acid may advantageously be a mineral acid, preferably sulfuric acid or nitric acid.

However, this process also leads to relatively good results when instead of using an aqueous solution of an acid solvent for the cerium(IV) salt, water alone is used, or when the concentration of acid in the water-containing solvent is zero. This fact is of particularly great significance for decontaminating of reactor coolant circuits.

In addition to the advantage that this process can achieve surprisingly good decontamination results even with very dilute solutions, the process has a whole series of further advantages over known decontamination processes, particularly that—in contrast to the process which is essentially restricted to permanganate for stainless steel—it is very versatile in use and is, e.g., successfully usable also for nickel-chromium alloys envisaged for future nuclear reactors.

Also, in contrast to the permanganate process wherein the reaction of the permanganate with the chromium(III) oxide gives not only chromium(VI) oxide but also manganese oxide that is insoluble in water or aqueous solutions, in this process no insoluble reaction products result and therefore by using this process, the after-treatment with a reducing or complex-forming solution inevitably necessary in the well known permanganate process to dissolve the manganese oxide can be obviated.

This last-mentioned advantage is again also of considerable significance for decontaminating reactor coolant circuits, because it opens up the possibility of carrying out the decontamination solely with a relatively low concentration cerium salt solution. Also the after-treatment with reducing solutions which could, according to concentration, also involve problems of corrosion, is obviated. The solubility of the reaction products arising from the treatment of contaminated metal surfaces with the cerium salt solution is also advantageous because thereby it becomes possible to carry out in a simple manner the separation of the cerium and radioactive substances in the used-up solution by suitable treatment, e.g., by electrolytic separation. In this way the cerium can be recycled and used for the preparation of a fresh treatment solution, and the used-up solution remaining after the separation of the cerium and the radioactive substances can without difficulty be removed in the same way as other industrial effluents.

Expediently, the treatment according to this process takes place at a temperature lying between the freezing point and boiling point of the cerium salt solution, preferably in the range of 20° to 90°C.

If the contamination layer is of uneven thickness and/or the layer is removed unevenly during treatment, the cerium salt solution may with advantage be provided additionally with an inhibitor to prevent the metal from being dissolved at locations where it has already been freed from the contamination layer.

To achieve uniform removal of the contamination layer, it is further of advantage to produce a flow of the cerium salt solution relative to the contaminated metal surface and to maintain this flow preferably uninterruptedly during treatment. To maintain such a flow the cerium salt solution may expediently be circulated in a circuit. In this way, inter alia, the possibility arises of purifying the partially expended solution flowing off from the treated metal surfaces before it is passed to the treated metal surfaces again. Instead of maintaining a flow of cerium salt solution relative to the contaminated metal surfaces the removal of the contamination layer may also be achieved by setting the cerium salt solution into vibration; more particularly, the subjection of the cerium salt solution to sonic vibrations, preferably ultrasonic vibrations, has proved very advantageous in this connection.

After treatment with the cerium salt solution the metal surface is advantageously washed, preferably with water, to remove residue from the solution.

Decontamination of Radioactive Garments

The conventional method of cleaning radioactive particulate material from industrial worker's protective clothing is a conventional wet laundry wash. This wash entails a standard 30 to 45 minute water washing using commercial detergents followed by a separate drying cycle (usually 60 minutes) in a conventional hot air or other type textile clothes dryer. This system normally is so inefficient that from 20 to 35% of the protective clothing must be rewashed because insufficient radioactivity has been removed to permit reuse of the protective article.

Moreover, approximately 3 gallons of contaminated wash water is generated per 16 pounds of clothing washed. This water must be diluted to a safe concentration before it is released or evaporated to a concentrate, and then drummed and buried at an approved radiation waste burial facility. This makes the process very costly and time consuming. Further, the conventional wet laundry involves the wash cycle followed by a separate drying cycle in a hot air dryer.

In the event that insufficient radioactive particulate is removed, the heat fixes the contaminated dirt to the cloth fibers which makes successive cleanings much less efficient and results in an early discard of the protective garments.

J.A. Capella and D.R. Morrison; U.S. Patent 4,235,600; November 25, 1980; assigned to Health Physics Systems, Inc. describe a method of decontaminating radioactive garments which comprises the steps of depositing the garments in a cleaning drum or cage and agitating the drum during a wash cycle. Further, a dry cleaning solvent is continuously added to the drum during the wash cycle and continuously removed therefrom to flush radioactive particulate material separated from the garments into a sump. The solvent is then pumped from the sump to the drum for use to continuously flush the radioactive particulate matter therefrom. During such pumping, the solvent is filtered to remove substantially all of the radioactive particulate material suspended in the solvent.

COMPANY INDEX

The company names listed below are given exactly as they appear in the patents, despite name changes, mergers and acquisitions which have, at times, resulted in the revision of a company name.

INVENTOR INDEX

U.S. PATENT NUMBER INDEX

285

4,134,941 - 158	4,204,980 - 230	4,268,409 - 102
4,134,960 - 160	4,206,073 - 213	4,269,706 - 105
4,138,329 - 264	4,206,185 - 243	4,269,728 - 79
4,139,420 - 214	4,208,298 - 84	4,271,034 - 127
4,139,488 - 27	4,208,377 - 182	4,274,962 - 103
4,145,396 - 140	4,209,399 - 201	4,274,976 - 53
4,146,568 - 109	4,209,421 - 8	4,275,045 - 232
4,148,745 - 42	4,219,976 - 275	4,276,063 - 209
4,156,646 - 166	4,222,889 - 80	4,276,834 - 143
4,156,658 - 39	4,222,892 - 122	4,277,361 - 257
4,157,248 - 260	4,225,328 - 252	4,277,362 - 105
4,158,639 - 215	4,225,455 - 151	4,277,363 - 226
4,162,229 - 276	4,228,141 - 147	4,278,104 - 270
4,162,230 - 180	4,229,317 - 238	4,278,559 - 169
4,162,231 - 191	4,230,597 - 36	4,280,921 - 50
4,162,298 - 249	4,230,672 - 173	4,280,922 - 25
4,164,479 - 138	4,230,675 - 274	4,281,691 - 63
4,167,491 - 32	4,234,447 - 72	4,282,112 - 187
4,168,243 - 33	4,234,448 - 99	4,299,271 - 77
4,172,807 - 73	4,234,449 - 10	4,299,721 - 22
4,173,546 - 20	4,234,456 - 235	4,299,722 - 61
4,173,620 - 244	4,234,555 - 154	4,300,056 - 81
4,174,293 - 15	4,235,600 - 278	4,303,511 - 89
4,177,241 - 173	4,235,737 - 37	4,305,780 - 87
4,178,270 - 199	4,235,738 - 148	4,308,105 - 268
4,178,350 - 243	4,235,739 - 58	4,312,774 - 68
4,180,476 - 240	4,242,220 - 46	4,313,845 - 133
4,182,652 - 172	4,246,065 - 92	4,314,877 - 96
4,196,169 - 33	4,246,233 - 97	4,314,909 - 9
4,198,221 - 251	4,246,238 - 272	4,315,831 - 45
4,201,690 - 250	4,250,832 - 217	4,316,814 - 78
4,202,792 - 6	4,252,667 - 64	4,320,028 - 13
4,203,863 - 101	4,253,985 - 45	4,321,158 - 56
4,204,911 - 242	4,261,952 - 178	4,326,918 - 67
4,204,974 - 21	4,263,163 - 139	4,329,248 - 55
4,204,975 - 51	4,265,861 - 152	

NOTICE

Nothing contained in this Review shall be construed to constitute a permission or recommendation to practice any invention covered by any patent without a license from the patent owners. Further, neither the author nor the publisher assumes any liability with respect to the use of, or for damages resulting from the use of, any information, apparatus, method or process described in this Review.